T0155889

Lecture Notes
in Business Information Processing **406**

More information about this series at http://www.springer.com/series/7911

Sander Leemans · Henrik Leopold (Eds.)

Process Mining Workshops

ICPM 2020 International Workshops
Padua, Italy, October 5–8, 2020
Revised Selected Papers

 Springer

Editors
Sander Leemans (iD)
Queensland University of Technology
Brisbane, QLD, Australia

Henrik Leopold (iD)
Kühne Logistics University
Hamburg, Germany

ISSN 1865-1348   ISSN 1865-1356 (electronic)
Lecture Notes in Business Information Processing
ISBN 978-3-030-72692-8   ISBN 978-3-030-72693-5   (eBook)
https://doi.org/10.1007/978-3-030-72693-5

Preface

The International Conference on Process Mining (ICPM) was established in 2019 as the conference where people from academia and industry meet and discuss the latest developments in the area of process mining research and practice, including theory, algorithmic challenges, and applications. Although the ICPM conference series is very young, it is able to attract innovative research of high quality from scholars and industrial researchers.

While the conference was planned to take place in Padua, Italy this year, things turned out differently. Just like many other academic conferences, also ICPM had to be held online due to the COVID-19 pandemic. While this was unfortunate from many perspectives, it did not negatively affect the program that was offered by ICPM. In fact, ICPM featured co-located workshops providing a forum for novel research ideas for the first time. In total, ICPM was complemented by six workshops, each focusing on particular aspects of process mining. These proceedings present and summarize the work that was discussed in the context of the workshops. ICPM 2020 featured the following workshops:

- 1st International Workshop on Event Data and Behavioral Analytics (EDBA)
- 1st International Workshop on Leveraging Machine Learning in Process Mining (ML4PM)
- 1st International Workshop on Trust and Privacy in Process Analytics
- (TPPA)
- 3rd International Workshop on Process-Oriented Data Science for Healthcare (PODS4H)
- 1st International Workshop on Streaming Analytics for Process Mining
- (SA4PM'20)
- 5th International Workshop on Process Querying, Manipulation, and Intelligence (PQMI)

In total, the ICPM 2020 workshops attracted 59 submissions. 29 papers were accepted for publication, leading to a total acceptance rate of about 50%. Especially against the background of the COVID-19 pandemic, we consider these numbers a great success. We are very happy about the interesting contributions and discussions that took place in the context of the workshops and believe that the workshops further highlighted the importance of process mining in both research and practice.

We would like to thank all the people from the ICPM community that helped to make the ICPM 2020 workshops a success. We particularly thank the general chairs, Massimiliano de Leoni and Alessandro Sperduti, for organizing such an outstanding conference despite the COVID-19 pandemic and the associated challenges. We also thank the workshop organizers, the numerous reviewers, and, of course, the authors for making the ICPM 2020 workshops such a success.

February 2021
Sander Leemans
Henrik Leopold

Organization

Program Committee

Rafael Accorsi	PwC Digital Services
Jan Aerts	Hasselt University
Davide Aloini	Università di Pisa
Robert Andrews	Queensland University of Technology
Ahmed Awad	University of Tartu
Hyerim Bae	Pusan National University
Bruno Bogaz Zarpelão	State University of Londrina
Andrea Burattin	Technical University of Denmark
Emerson Cabrera Paraiso	Pontifícia Universidade Católica do Paraná
Daniel Capurro	University of Melbourne
Josep Carmona	Universitat Politècnica de Catalunya
Michelangelo Ceci	University of Bari Aldo Moro
Olivia Choudhury	Amazon Inc.
Marco Comuzzi	Ulsan National Institute of Science and Technology
Benjamin Dalmas	École des Mines de Saint-Étienne
Ernesto Damiani	Khalifa University
Renè de la Fuente	Pontificia Universidad Católica de Chile
Jochen De Weerdt	KU Leuven
Claudio di Ciccio	Sapienza University of Rome
Chiara Di Francescomarino	Fondazione Bruno Kessler
Matthias Ehrendorfer	University of Vienna
Stephan Fahrenkrog-Petersen	Humboldt University of Berlin
Bettina Fazinga	ICAR - Italian National Research Council
Carlos Fernández Llatas	Universitat Politècnica de Valencia
Francesco Folino	ICAR - Italian National Research Council
Luciano García-Bañuelos	Tecnológico de Monterrey
Roberto Gatta	Università degli Studi di Brescia
Kanika Goel	Queensland University of Technology
Antonella Guzzo	Università della Calabria
Marwan Hassani	Eindhoven University of Technology
Emmanuel Helm	University of Applied Sciences Upper Austria
Zhengxing Huang	Zhejiang University
Mieke Jans	Hasselt University
Owen Johnson	University of Leeds
Anna Kalenkova	The University of Melbourne
David Knuplesch	alphaQuest
Agnes Koschmider	Kiel University

Mariangela Lazoi	University of Salento
Francesco Leotta	Sapienza University of Rome
Rong Liu	Stevens Institute of Technology
Xixi Lu	Utrecht University
Fabrizio Maggi	Free University of Bozen-Bolzano
Felix Mannhardt	Eindhoven University of Technology
Ronny Mans	Philips
Mar Marcos	Universitat Jaume I
Gabriel Marques Tavares	Università degli Studi di Milano
Niels Martin	Hasselt University
Renata Medeiros de Carvalho	Eindhoven University of Technology
Mohamed Medhat Gaber	Birmingham City University
Paola Mello	Università di Bologna
Jan Mendling	Vienna University of Economics and Business
Judith Michael	RWTH Aachen
Marco Montali	Free University of Bozen-Bolzano
Catarina Moreira	Queensland University of Technology
Jorge Munoz-Gama	Pontificia Universidad Católica de Chile
Hye-Young Paik	The University of New South Wales
Simon Poon	University of Sydney
Domenico Potena	Università Politecnica delle Marche
Maurizio Proietti	CNR-IASI
Luise Pufahl	Technische Universität Berlin
Ricardo Quintano	Philips Research
Hajo Reijers	Utrecht University
David Riaño	Universitat Rovira i Virgili
Florian Richter	Ludwig-Maximilians-Universität München
Stefanie Rinderle-Ma	Technical University of Munich
Eric Rojas	Pontificia Universidad Católica de Chile
Lucia Sacchi	University of Pavia
Shazia Sadiq	The University of Queensland
Sherif Sakr	University of Tartu
Fernando Seoane	Karolinska Institutet
Marcos Sepúlveda	Pontificia Universidad Católica de Chile
Natalia Sidorova	Eindhoven University of Technology
Pnina Soffer	University of Haifa
Minseok Song	Pohang University of Science and Technology
Frederic Stahl	German Research Center for Artificial Intelligence (DFKI)
Alessandro Stefanini	Università di Pisa
Emilio Sulis	Università di Torino
Niek Tax	Booking.com
Irene Teinemaa	Booking.com
María Teresa Gómez-López	Universidad de Sevilla
Pieter Toussaint	Norwegian University of Science and Technology

Vicente Traver	Universitat Politècnica de València
Florian Tschorsch	Technical University of Berlin
Han van der Aa	University of Mannheim
Wil van der Aalst	RWTH Aachen University
Boudewijn van Dongen	Eindhoven University of Technology
Sebastiaan van Zelst	RWTH Aachen/FIT
Seppe vanden Broucke	Katholieke Universiteit Leuven
Rob Vanwersch	Maastricht University Medical Center
Eric Verbeek	Eindhoven University of Technology
Melanie Volkamer	Karlsruhe Institute of Technology
Hagen Völzer	IBM Research – Zurich
Matthias Weidlich	Humboldt University of Berlin
Moe Wynn	Queensland University of Technology

Additional Reviewers

Mathilde Boltenhagen	Université Paris-Saclay
Graziella De Martino	University of Bari
Angelo Impedovo	University of Bari
Vincenzo Pasquadibisceglie	University of Bari

Contents

1st International Workshop on Streaming Analytics for Process Mining (SA4PM'20)

5th International Workshop on Process Querying, Manipulation, and Intelligence (PQMI 2020)

**3rd International Workshop on Process-Oriented Data Science
for Healthcare (PODS4H)**

1st International Workshop on Trust and Privacy in Process Analytics (TPPA)

1st International Workshop on Event Data and Behavioral Analytics (EDBA)

First International Workshop on Event Data and Behavioral Analytics (EdbA'20)

Over the past decades, capturing, storing, and analyzing event data has gained attention in various domains such as process mining, clickstream analytics, IoT analytics, e-commerce and retail analytics, online gaming analytics, security analytics, website traffic analytics, and preventive maintenance, to name a few. The interest in event data lies in its analytical potential as it captures the dynamic behavior of people, objects, and systems at a fine-grained level.

Behavior often involves multiple entities, objects, and actors to which events can be correlated in various ways. In these situations, a unique, straightforward process notion does not exist or is unclear, or different processes or dynamics may be recorded in the same data set.

The Event Data Behavioral Analytics (EdbA) workshop's objective is to provide a forum for practitioners and researchers to study a quintessential, minimal notion of events as the common denominator for records of discrete behavior in all its forms. The workshop aims to stimulate the development of new techniques, algorithms, and data structures for recording, storing, managing, processing, analyzing, and visualizing event data in various forms. To this end, different types of submissions are welcome such as original research papers, case study reports, position papers, idea papers, challenge papers, and work in progress papers on event data and behavioral analytics. For more information, visit http://edba.science.

This first edition of the EdbA workshop attracted 14 submissions. After careful multiple reviews by the workshop's program committee members, seven were accepted for presentation at the workshop and inclusion in the proceedings. This year's papers cover three main topics: visualization of behavior, the discovery of behavior from event data, and event abstraction.

The organizers wish to thank all the people who submitted papers to the EdbA'20 workshop, the many participants creating fruitful discussion and sharing insights, and the EdbA'20 Program Committee members for their valuable work in reviewing the submissions. A final word of thanks goes out to the organizers of ICPM 2020 for making this workshop possible.

Organization

Workshop Chairs

Benoît Depaire   Hasselt University, Belgium
Dirk Fahland   Eindhoven University of Technology,
           The Netherlands
Massimo Mecella   Sapienza University of Rome, Italy
Arik Senderovich   University of Toronto, Canada

Program Committee

Jan AertsHasselt   University, Belgium
Jochen De Weerdt   Catholic University of Leuven, Belgium
Claudio di Ciccio   Sapienza University of Rome, Italy
Bettina Fazinga   ICAR - National Research Council, Italy
Marwan Hassani   Eindhoven University of Technology,
           The Netherlands
Francesco Leotta   Sapienza University of Rome, Italy
Xixi Lu   Utrecht University, The Netherlands
Felix Mannhardt   Eindhoven University of Technology,
           The Netherlands
Niels Martin   Hasselt University, Belgium
Jan Mendling   Vienna University of Economics and Business,
           Austria
Marco Montali   Free University of Bozen-Bolzano, Italy

Visually Representing History Dependencies in Event Logs

Manuel Wetzel[1]([✉]), Agnes Koschmider[1], and Thomas Wilke[2]

[1] Group Process Analytics, Kiel University, Kiel, Germany
ak@informatik.uni-kiel.de
[2] Group Theory of Computing, Kiel University, Kiel, Germany
thomas.wilke@email.uni-kiel.de
https://www.pa.informatik.uni-kiel.de/en/

Abstract. Many process mining tools produce directly-follows graphs (DFG) as visual representations of event logs. While the "directly follows" relation is a good starting point for visualizations, there are simple phenomena it does not capture, for instance, when whether or not an event directly follows another event depends on the event directly preceding it. We call this a *history* dependency. This paper presents an empirical study of preferences for visualizing history dependencies: plain DFGs and two enhanced variants of DFGs (with additional arcs or rectangles) are evaluated. Our empirical study provides strong support for making an effort (to discover and) to explicitly visualize history dependencies. A ProM plug-in generating such explicit visualization is described in this paper.

Keywords: Directly-Follows Graph · Process visualization · Process discovery · History dependency · Empirical study

1 Introduction

Event log files are used as input to any process mining algorithm aiming to discover an as-is process model, to analyze processes or to identify bottlenecks. To reduce inappropriate conclusions from the discovered process model, it is essential that this model reflects the reality found in an event log as best as it can. Mostly, available commercial process mining tools produce a visualization of a directly-follows graph (DFG) as a representation of event logs. While the "directly follows" relation is a good starting point for a visualization, there are simple phenomena it does not capture, for instance, when whether or not an event directly follows another event depends on the event directly preceding it. Figure 1 illustrates this phenomenon in terms of traces.

According to Fig. 1a) the execution of activity D right after C is only allowed when A was executed directly before C. When activity B was executed directly before C, then only E may directly follow, but D must not. We call this event

© Springer Nature Switzerland AG 2021
S. Leemans and H. Leopold (Eds.): ICPM 2020 Workshops, LNBIP 406, pp. 5–16, 2021.
https://doi.org/10.1007/978-3-030-72693-5_1

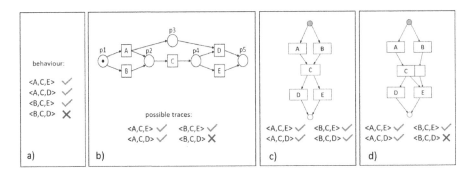

Fig. 1. a) example of a history dependency; b) discovered Petri net; c) discovered DFG; d) proposed visualization

dependency a *history dependency* (more precisely, a history-1 dependency). Figure 1b) shows a Petri net modeling the dependencies exactly.

A process mining algorithm based on the directly follows relation would produce the DFG in Fig. 1c). This DFG allows behavior that is not reflected in reality, namely the trace $\langle B, C, D \rangle$. Although limitations of DFGs have been demonstrated [9], the graphical visualization as DFG is still a common practice for available commercial process mining tools. The motivation behind our research, therefore, is not to resort to a completely different type of visualization, but rather to study visual enhancements of DFGs being suitable to visualize history dependencies. For the example given in Fig. 1a), we propose the visualization in Fig. 1d). Another example is given in Fig. 2. Mining an event log as given in Fig. 2a) leads to the DFG in Fig. 2c), which has an unintended cycle; we propose the visualization in Fig. 2d), which reflects reality exactly. A process model exactly modeling the event log is given in Fig. 2b).

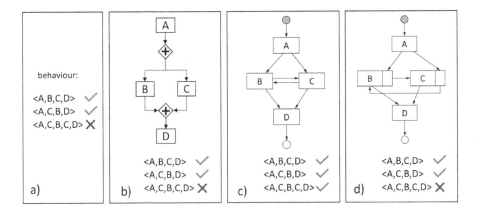

Fig. 2. a) a second example of a history dependency; b) BPMN model; c) discovered DFG; d) proposed visualization

To learn more about history dependencies and enhancements of the DFG, we conducted a user study. We compared plain DFGs to two enhanced visualizations, one with additional arcs (AA) and another one with additional rectangles (AR). Our findings show strong support for enhancing DFGs to visualize history dependencies. Thus, our finding does not correspond to common practical implementations (where history dependencies are not explicitly visualized); this is why we developed a ProM plug-in that produces enhanced DFGs, more precisely, it produces AR visualizations.

This paper is structured as follows. The next section compares our work with related works. Section 3 discusses, beside the plain DFG, visualization variants for history dependencies. Section 4 describes the design of a study to provide evidence about the visualization strategies for history dependency. The results of the study are discussed in Sects. 5 and 6. The ProM plug-in implementation is presented in Sect. 7. The paper concludes with a summary and an outlook.

2   Related Work

Process discovery algorithms generally distinguish between two types of dependencies [10]: explicit and implicit ones. An explicit dependency, which is also called direct or causal dependency [1], exists when an activity is directly followed by another activity in a considerable number of cases. An implicit dependency refers to various types of indirect (causal) relationships between activities, for instance, that an activity is eventually followed by another activity in a considerable number of cases. Dependency measures are used to determine whether or not and which kind of dependency is present. A history dependency in our sense takes into account causal dependencies that are not only concerned with two consecutive activities but a small number of consecutive activities [8]. Process models visualizing history dependency prevent, in some case, unintended cycles as demonstrated in Fig. 2 and thus overcome a limitation of the DFG discussed in [9].

We subsume our approach to techniques explicitly visualizing history- dependent information as in [4,5]. Compared to these approaches we do not introduce a new visualization technique nor label the process model with additional information but rather enhance the DFG, which is commonly used in commercial process mining tools, aiming to reflect the reality found in an event log better than a plain DFG.

3   Visualization Techniques

To understand the usefulness of the Directly-Follows Graph in terms of visualizing history dependencies we compared it with two other visualization variants as discussed in this section. The visualization techniques lay the foundation for the empirical validation of visualization preferences for history dependencies presented in Sect. 4.

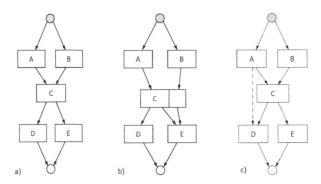

Fig. 3. Examples for a) DFG-based visualization, b) Visualization Additional Rectangle c) Visualization Additional Arc

3.1   Visualization Directly-Follows Graph

The DFG is widely used to visualize behavior between process activities. It gives information on a similar level of abstraction as the process modeling notations BPMN, EPK or Petri nets. Usually, commercial process mining tools also attach time and frequency based performance measures on the arcs for process monitoring reasons. Although the semantics of the DFG is easily understandable, the DFG does not represent a precise process model since it allows more behavior than has actually been recorded in the event log and can reasonably be expected as has shown in the introductory example. In this way, a history dependency is wrongly visualized and is even hardly spotted, which may lead to inappropriate conclusions inferred from the DFG. The following two visualization variants aim to overcome this issue.

3.2   Visualization Additional Rectangle

To visualize history dependency the plain DFG is enhanced with an additional rectangle. We insert to the activity routing the history dependency. We call this visualization technique "Additional Rectangle", see Fig. 3b) for an example. Additional Rectangle (AR) is inspired by hyperedges, a well-known concept from graph theory, see [2]. Incoming and outgoing arcs of the additional empty rectangle indicate the allowed behavior between two activities. Trace replays on this model would be: $\langle A, C, D\rangle$, $\langle A, C, E\rangle$, $\langle B, C, E\rangle$, which corresponds to the allowed behavior in Fig. 1a). The trace $\langle B, C, D\rangle$ cannot be replayed by this process model due to a missing empty rectangle at activity C. A more complex visualization of AR for large process models can be seen in Fig. 4b).

3.3   Visualization Additional Arc

A second technique to visualize history dependencies is "Additional Arc" (AA), see Fig. 3c). Here, the plain DFG is enhanced with dashed arcs. This visualization, like AR, explicitly illustrates a history dependency. The semantics of the dashed arc is "activity D can only be executed if activity A has been executed previously". Thus, the process model visualized through AA also does not allow to replay the trace ⟨B, C, D⟩ on the model. A complex example process model is shown in 4a) visualizing a model with a total of seven history dependencies.

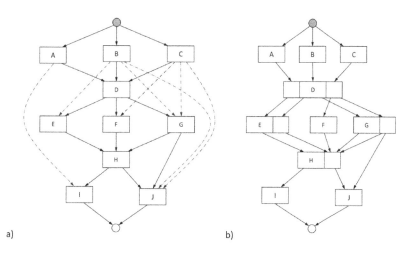

Fig. 4. Examples for visualizations a) Additional Arc b) Additional Rectangle for large process models.

4   Design Setting

To evaluate whether the proposed visualizations are suited to recognize history dependencies, a questionnaire was designed. The main focus of the study was to investigate the visualization preferences of the three visualization techniques presented in the previous section. Therefore, a web-based questionnaire was set up. Participants were free to answer the questions and could withdraw the completion of the questionnaire at any time and collection of data was anonymous. The following section describes the design of the questionnaire.

Objects. The objects evaluated by each participant were a set of traces and a total of eleven process models. Five of the process models were visualized according to the AR visualization, five with the AA visualization and one as a Directly-Follows Graph. Small process models (see Fig. 3) had six activities and one history-1 dependency. Large process models (see Fig. 4) were designed through ten activities having seven history dependencies including history-3 dependencies.

Response Variable. The response variable in our study is the level of understanding that the respondents displayed with respect to a visualization technique recognizing history dependencies. Understandability is measured as follows:

- the perceived ease of understanding (PEOU),
- preference between the visualization techniques,
- the number of correct traces spotted for a visualization technique,
- degree of agreement for comprehension questions

Instrumentation. The questionnaire was constructed as follows. After a motivation in history dependency, we asked the participants to complete for each visualization technique a task and to answer comprehension questions. The semantics of each visualization technique was not formally introduced. The task was to evaluate against a set of traces whether traces can be replayed by the process model. We showed 4 traces for each small process model and 21 traces for each large process model. The description of a task was *"Evaluate for each trace if the trace can be replayed by the process model. Please tick the corresponding box."*. After having completed this task, we asked the participants to rank (on a 5 point Likert scale) visualization preferences to represent history dependencies and to rank the visualization techniques based on their ability to represent history dependencies in a process model. In the second part of the questionnaire the participants had to complete two further tasks for the visualization techniques AA and AR. The first task was to choose between true or false statements. We provided for each of the two visualization techniques the following statements *"Please consider the three process models and choose if one of these models describes the following behavior according to the Additional Arc visualization The model is able to replay the traces: $\langle A, D, E \rangle$, $\langle A, D, F \rangle$, $\langle B, D, F \rangle$, $\langle C, D, F \rangle$. and is not able to replay $\langle B, D, E \rangle$ and $\langle C, D, E \rangle$ Please select the correct representation of the wanted behavior"* also providing the option *"none of the shown models describes the desired behaviour."*. Subsequently, the participants had to complete tasks for the visualizations AA and AR for large process models and to rank their individual preferences for both visualization techniques.

Subjects. The survey was conducted in July 2020. The link to the survey was sent to the participants of the "Advanced Process Mining"[1] course at the Kiel University as well as a number of experts across different European universities.

Data Collection. Along with the questionnaire, we asked the participants about the amount of experience they have in the fields of process mining and business process management. Furthermore, we asked whether they (practically) worked with process mining and if they understood the motivation behind history dependency that we presented at the beginning of the questionnaire.

5   Evaluation Results

Finally, the questionnaire was answered by 13 persons. 62% of the respondents had more than one year of experience in process mining and 38% had more than three years of experience. All participants used process mining in research. One

---

[1] Students of the "Advanced Process Mining" course also attended the course "Process Mining" in the winter term and thus had advanced process mining knowledge.

person worked with process mining in industry projects. The average time to complete the questionnaire was 41 min.

The results for visualization preferences were analyzed with respect to frequency distribution. Table 1 shows the statistical results for each preference, its answer options, the frequency in numbers per option, the frequency (%), and the cumulative frequency (%) for each question. Cumulative frequency is determined by aggregating agreement (strong agree, agree) and disagreement (disagree, strongly disagree) with the preference. Table 1 summarizes the statistical results. The results investigate the PEOU measure for each visualization technique while Table 2 shows the preference between the visualization techniques and an order between the three visualization techniques with respect to understandability. According to this result DFG is ranked in average 2.85 out of 3 with standard deviation of 0.36 meaning that it is less understandable as representation for history dependency. The Directly-Follows Graph is being perceived as not helpful to understand history dependencies. Visualization preferences for DFG received very low agreements (8% and 15%). In the ranking of individual preferences the DFG received the last position.

For small process models a slight preference exists for AR against AA visualization. 69% agreed that the AR visualization helped them better to recognize history dependencies, while 46% voted for AA as first choice. Related to individual visualization preferences AR was in average ranked (1.54 out of 3) compared to 1.62 for AA. The high standard deviation, however, might implicate that the participants are undecided.

A contrary individual preference is observed for large process models. The statement "*Additional Rectangle helped better than Additional Arc to recognize implicit dependency*" received an 85% approval, while the contrary statement only received a 23% approval. The preference for AR for large process models is also confirmed by the results in Table 2. The AR visualization was ranked 1.15 (out of 3), while the preference for AA declined to 1.85. So there is a significant preference for Additional Rectangle visualization over Additional Arc for larger process models when many implicit dependencies exist.

The results of user tasks (correct traces spotted for trace replay) are shown in Table 3. The tasks were not evaluated with respect to the correctness of an answer since no explanation of the semantics of each visualization technique was introduced to the users. Instead we compared the visualization techniques according to the overall consensus for each selection.

For each task multiple measures were calculated. The first three measures point to the consensus on a decision. This was measured by the relative amount of participants with identical answers. A consensus of 100% means that all participants made the identical decision, while a consensus of 50% means that half of the participants had the opinion that a trace can be replayed by the model while the other half had the opposite opinion. As for the trace replay tasks multiple decisions existed we determined the average [AVG] consensus of all traces within a task, the corresponding standard deviation [STD] and its minimum [MIN] which is the consensus on the most controverse trace selection.

Table 1. Results of preferences

Preference	Options	Freq	Freq. (%)	Cum. Freq. (%)
The visualization Directly-Follows Graph helped me better to recognize the implicit dependency over the visualization Additional Rectangle	s. agree	1	8 %	8 %
	agree	0	0 %	
	undecided	4	31 %	
	disagree	3	23 %	62 %
	s. disagree	5	38 %	
The visualization Directly-Follows Graph helped me better to recognize the implicit dependency over the visualization Additional Arc	s. agree	2	15 %	15 %
	agree	0	0 %	
	undecided	3	23 %	
	disagree	2	15 %	62 %
	s. disagree	6	46 %	
The visualization Additional Rectangle helped me better to recognize the implicit dependency over the visualization Directly-Follows Graph	s. agree	5	38 %	69 %
	agree	4	31 %	
	undecided	2	15 %	
	disagree	1	8 %	15 %
	s. disagree	1	8 %	
The visualization Additional Rectangle helped me better to recognize the implicit dependency over the visualization Additional Arc	s. agree	3	23 %	62 %
	agree	5	38 %	
	undecided	2	15 %	
	disagree	1	8 %	23 %
	s. disagree	2	15 %	
The visualization Additional Arc helped me better to recognize the implicit dependency over the visualization Directly-Follow Graph	s. agree	7	54 %	85 %
	agree	4	31 %	
	undecided	1	8 %	
	disagree	1	8 %	8 %
	s. disagree	0	0 %	
The visualization Additional Arc helped me better to recognize the implicit dependency over the visualization Additional Rectangle	s. agree	1	8 %	46 %
	agree	5	38 %	
	undecided	1	8 %	
	disagree	5	38 %	46 %
	s. disagree	1	8 %	
The visualization Additional Rectangle helped me better to recognize the implicit dependency over the visualization Additional Arc in large process models	s. agree	6	46 %	85 %
	agree	5	38 %	
	undecided	0	0 %	
	disagree	1	8 %	15 %
	s. disagree	1	8 %	
The visualization Additional Arc helped me better to recognize the implicit dependency over the visualization Additional Rectangle in large process models	s. agree	1	8 %	23 %
	agree	2	15 %	
	undecided	1	8 %	
	disagree	5	38 %	69 %
	s. disagree	4	31 %	

Table 2. Results of ranking the proposed visualizations

	Rank (AVG)	Rank (SD)
Additional Rectangle	1.54	0.63
Additional Arc	1.62	0.62
Directly-follows graph	2.85	0.36
Additional Rectangle for large models	1.15	0.36
Additional Arc for large models	1.85	0.36

The aggregated results show a consensus of 85% vs. 92% on average. Recall that for small process models users had to evaluate four trace replays. A value of 92% means that almost all participants had the same understanding of the AR visualization. Note, that it was a binary decision so the expected value of a random distribution would have been 50%. But not only the average overall decisions were better for the Rectangle visualization. The worst consensus was between 77% to 69% and also both calculated standard deviation metrics are smaller for AR. When evaluating trace replay for large process models (i.e., *"Please rate your visualization preferences to represent implicit dependencies in large process models"*), the consensus declines for both visualization techniques. But still, Additional Rectangle with a consensus of 90% is superior to Additional Arc. The last task (i.e., *"Please rank the presented visualization based on their ability to represent implicit dependencies in a process model"*) shows again a clear preference for AR (ten out of 13 choose it), while AA was ranked the second with 54% of the participants.

Table 3. Aggregated results of the tasks

Semantic	Task name	Cons. [AVG]	Cons. [STD]	Cons. [MIN]
DFG	Trace verification (small example)	100%	0%	100%
AR	Trace verification (small example)	92%	9%	77%
	Trace verification (large example)	90%	8%	69%
	Choose correct Model	77%		
AA	Trace verification (small example)	85%	15%	69%
	Trace verification (large example)	77%	17%	54%
	Choose correct Model	54%		

6   Discussion

Interpretation: The empirical study provides strong support for another visualization for history dependencies than the Directly-Follows Graph. Additional Rectangle, as well as Additional Arc, are better suited to visualize history dependencies. This finding does not directly correspond to common practical implementations. The DFG-based visualization, which is the common practice for available commercial process discovery tools, does not explicitly visualize history dependencies. Our statistical results show that AR and AA are easily understandable for users for small process models, while a strong support for AR was observed for large process models. Therefore, Additional Rectangle is a suitable alternative to current visualizations.

Implications: the design of a visual notation is a challenging task [7]. It requires a balance between *symbol deficit* (i.e., no constructs representing a graphical symbol), *symbol overload* (i.e., same graphical symbol for different representations), *symbol redundancy* (i.e., alternative graphical symbols for same representation) and *symbol excess* (i.e., showing all constructs on a diagram). The rejection of a DFG-based visualization is in line with the postulation of symbol deficit. When no construct is used to represent a graphical symbol then understandability decreases. Process discovery tools should implement an explicit visualization for history dependencies.

Limitations: The similar preference for Additional Rectangle vs. Additional Arc for small process models might be explained due to a weakness of understanding the semantics of the AA visualization. There are two contrary ways to interpret the dashed arc in 3 c). Either it is understood as "If A has been executed then it must be followed by D" or it can mean "activity D can only be executed if activity A has been executed before". Apparently, four out of twelve participants have intuitively interpreted the arc in the first way.

7   Implementation as ProM Plugin

In response to the evaluation results (and also the limitation of AA visualization discussed previously) we implemented a ProM plug-in, which experiments with visualizing history dependencies by enhancing the nodes in a DFG with additional rectangles. An overview of the components of the plug-in, called Dependent Directly Follows Model Miner (DDFM Miner), is shown in Fig. 5. The plug-in carries out three sequential steps.

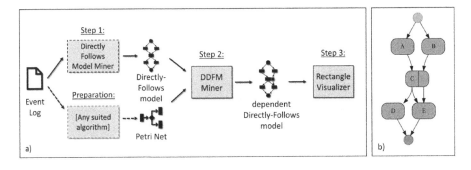

Fig. 5. a) Steps of the DDFM miner and b) its output

Step 1. The plain DFG is mined from an event log in XES format using the "DirectlyFollowsModelMiner" plugin [6], which produces a directly-follows matrix.

Step 2. History dependencies are being identified. For this purpose, the algorithm takes as input the directly-follows matrix from Step 1 and compares it with a Petri net manually mined from the same event log using a suited ProM plug-in miner (see "Preparation" in Fig. 5). The DDFM Miner then analyses every transition-place-transition triplet in the Petri net and determines whether this relation is already in the directly-follows matrix. If this is not the case, a history dependency is added to the history-dependency matrix.

Step 3. A visualizer plug-in relying on the Graphviz environment [3] translates the two matrices into a graphical representation. This includes four steps:

1. Calculate all *dependent paths* that connect two history dependent activities.
2. Pre-process node names to create the additional rectangles with Graphviz.
3. Generate nodes and *directly-follows edges*.
4. Iterate through *dependent paths* to remove *directly-follows edges* and add *dependency edges*.

Figure 5b) shows the final result of the DDFM miner.

8   Summary and Outlook

The simple understandablity of the DFG might be one reason for its high popularity. With regard to history dependencies, however, the DFG-based visualization fails. The objective of this work was to study how to visualize history dependencies. For this purpose, we compared the plain DFG with visualization variants in a user study. The results of the study provide strong support to enhance the plain DFG with additional rectangles to visualize history dependencies.

In response to our finding, the DDFM miner has been implemented, but this is only a first step. Future tasks are: (1) to evaluate the DDFM miner on large event logs, (2) to visualize history dependencies for multiple interfering dependent paths, as, for instance, present in Fig. 4: consider $\langle C, G, H, J \rangle$ and $\langle B, D, E, H, I \rangle$, (3) To enhance the semantics of AR visualization (see Appendix) with quantitative aspects such as frequency and time since our definition of history dependency is a "discrete" one.

A   Rigorous Definitions

Let $\mathcal{A}$ be a finite set of *activities* without $\triangleright$ and $\square$. A *trace* is a finite sequence $\triangleright A_0 \ldots A_n \square$ which satisfies $A_i \in \mathcal{A}$ for every $i \leq n$. An *event log* is a finite non-empty set of traces.

A *directly-follows graph with multiplicities (DFG+)* consists of a finite set V of vertices, a finite set $E \subseteq V \times V$ of edges, and a labeling function $\lambda \colon V \to \mathcal{A} \cup \{\triangleright, \square\}$ such that V together with E is a directed acyclic graph (DAG), there is exactly one vertex labeled $\triangleright$, and this vertex is the only vertex with no incoming edge, there is exactly one vertex labeled $\square$, and this vertex is the only vertex with no outgoing edge. A DFG+ represents the event log which consists of the labelings of all paths through it that are traces.

A DFG+ is a *directly-follows graph (DFG)* when every event is the label of at most one vertex. A DFG+ is *reduced* if every vertex is on some path from the vertex labeled $\triangleright$ to the vertex labeled $\square$ and there are no two distinct vertices with the same set of successor vertices.

Fact. For every event log there is, up to isomorphism, exactly one reduced DFG+ that represents it. Such a DFG+ is called *canonical* for the event log.

Definition. Let $\mathcal{D}$ be a canonical DFG+ for an event log. 1. The event log is *history-free* if $\mathcal{D}$ is a DFG.

2. Let k be a non-negative integer and $A \in \mathcal{A}$. There is a *k-conflict* for A if there are paths $v_0 v_1 \ldots v_k$ and $w_0 w_1 \ldots w_k$ in $\mathcal{D}$ such that $v_k \neq w_k$, $\lambda(v_i) = \lambda(w_i)$ for all $i < k$, and $A = \lambda(v_k) = \lambda(w_k)$. There is a *$k$-history dependency* $(k > 0)$ for A if there are no k-conflicts for it, but a $(k-1)$-conflict.

References

1. van der Aalst, W., Günther, C.: Finding structure in unstructured processes: the case for process mining, pp. 3–12, August 2007
2. Berge, C.: Graphs and Hypergraphs. North-Holland, New York (1973)
3. Ellson, J., Gansner, E.R., Koutsofios, E., North, S.C., Woodhull, G.: Graphviz and dynagraph - static and dynamic graph drawing tools. In: Jünger, M., Mutzel, P. (eds.) Graph Drawing Software. Mathematics and Visualization, pp. 127–148. Springer, Heidelberg (2003). https://doi.org/10.1007/978-3-642-18638-7_6
4. Goedertier, S., Martens, D., Baesens, B., Haesen, R., Vanthienen, J.: Process mining as first-order classification learning on logs with negative events. In: ter Hofstede, A., Benatallah, B., Paik, H.-Y. (eds.) BPM 2007. LNCS, vol. 4928, pp. 42–53. Springer, Heidelberg (2008). https://doi.org/10.1007/978-3-540-78238-4_6
5. van Hee, K., Oanea, O., Serebrenik, A., Sidorova, N., Voorhoeve, M.: History-based joins: semantics, soundness and implementation. In: Dustdar, S., Fiadeiro, J.L., Sheth, A.P. (eds.) BPM 2006. LNCS, vol. 4102, pp. 225–240. Springer, Heidelberg (2006). https://doi.org/10.1007/11841760_16
6. Leemans, S.J., Poppe, E., Wynn, M.T.: Directly follows-based process mining: a tool. ICPM **2019**, 9–12 (2019)
7. Moody, D.: The "physics" of notations: toward a scientific basis for constructing visual notations in software engineering. IEEE Trans. Softw. Eng. **35**, 756–779 (2010)
8. Sun, H., Du, Y., Qi, L., He, Z.: A method for mining process models with indirect dependencies via petri nets. IEEE Access **7**, 81211–81226 (2019)
9. van der Aalst, W.M.: A practitioner's guide to process mining: limitations of the directly-follows graph. Procedia Comput. Sci. **164**, 321–328 (2019)
10. Wen, L., Wang, J., Sun, J.: Detecting implicit dependencies between tasks from event logs. In: Zhou, X., Li, J., Shen, H.T., Kitsuregawa, M., Zhang, Y. (eds.) APWeb 2006. LNCS, vol. 3841, pp. 591–603. Springer, Heidelberg (2006). https://doi.org/10.1007/11610113_52

Analysis of Business Process Batching Using Causal Event Models

Philipp Waibel$^{(\boxtimes)}$ ⓘ, Christian Novak ⓘ, Saimir Bala ⓘ, Kate Revoredo ⓘ,
and Jan Mendling ⓘ

Institute for Information Business, Vienna University of Economics and Business
(WU), Vienna, Austria
{philipp.waibel,saimir.bala,kate.revoredo,jan.mendling}@wu.ac.at,
christian.novak@s.wu.ac.at

Abstract. Process mining supports business process management with operational insights extracted from event logs. A key challenge for process mining is that operational processes in production and logistics often include batching and unbatching, e.g., to delivery several packages using one truck tour. Such n:m relations blur the notion of a process instance and make the causality between events difficult to trace. In this paper, we address this research problem by introducing causal event models that capture batching behavior accurately. To this end, we construct conflict-free prime event structures for event instances of the event log, and devise various analysis techniques on top of them. We implemented the techniques in a tool and run in real data of a manufacturing company with various 1:n and n:1 relations in their production process showing the potential of our approach.

Keywords: Process mining · Business process modeling · Batching · Causality

1 Introduction

Business Process Management (BPM) comprises the various management activities that help organizations to discover, analyze, implement and monitor their processes [10]. Recently, BPM has become increasingly evidence-based thanks to advancements of process mining [1]. The availability of event log data from enterprise systems for various business processes is one of the key drivers of these developments as much as the commercial tool support.

Various algorithms have been proposed that support automatic process discovery, conformance checking, enhancement, or analysis of variants [1,10]. One aspect of specific interest to the process analyst is the batching behavior of processes, i.e., the merge of different objects either within the same case or between different cases. This practice contributes to both, more cost-efficient processing as well as delays. In order to grasp the batching behavior precisely, the knowledge about causality between the events is necessary. Without this knowledge spurious batches may occur.

ⓒ Springer Nature Switzerland AG 2021
S. Leemans and H. Leopold (Eds.): ICPM 2020 Workshops, LNBIP 406, pp. 17–29, 2021.
https://doi.org/10.1007/978-3-030-72693-5_2

In this paper, we address this research problem by introducing causal event models that capture batching behavior accurately. To this end, we construct conflict-free prime event structures, inspired by [3], for process instances stored in event logs, before devising various analysis techniques on top of them. Our use case demonstrates the benefits of our technique for the case of a manufacturing company with various 1:n and n:1 relations in their production process.

The rest of the paper is structured as follows. Section 2 discusses a motivational scenario for our work. Subsequently, Sect. 3 presents prior work related to our research problem with a focus on n:m relations and batching in business processes. Section 4 presents the conceptual foundations of our technique. Section 5 describes findings from applying our technique for a production process and discusses the lessons learned. Finally, Sect. 6 concludes the paper outlining future research.

2   Motivational Scenario

To motivate the presented work, we refer to an order-to-cash process example of one of our industry partners called Pastamaker (a pseudonym). Pastamaker's business is producing and delivering pasta to major supermarket chains. Their order-to-cash process is triggered by direct orders of a supermarket. These orders contain a list of items, where each of these items needs to be picked separately from the warehouse. In the next step, each item is packaged and sent as one or multiple deliveries. Each order generates an invoice that is settled and closed by a payment. In addition, packaging and delivering steps have sub-steps that are creating packaging notes and group delivery information.

Pastamaker uses batching at various stages of its production process, most importantly for bundling deliveries. For instance, one order can trigger different deliveries, and one delivery can include items from different orders. Such delivery batches are of central importance for keeping the operational costs of the process low. To further optimize the batching of the orders, Pastamaker would like to analyze them. In particular, Pastamaker would like to get information, like how many batching events took place, what events caused batches, or which steps are bottlenecks or caused delays.

3   Related Work

Perspectives on this problem have been discussed in two main streams of research: *i)* process mining of causal events with n:m relationships, which are common in database schemas; and *ii)* batching, which includes modeling batches and extracting knowledge about batches from event logs.

For what concerns stream *i)*, various recent publications consider the problem of n:m relationships of events. Lu et al. [19] present an approach that maintains complex event relationships based on database schemas and domain-knowledge to construct artifact-centric models with causal relationships. As we will discuss in the work at hand, we follow a similar approach of defining the causal relationships between events. The idea of modeling the causal relationships was further used in [11,12] to capture the concept of one event being part of multiple cases by using *labeled property graphs*. In these publications, Esser and Fahland transform event logs into graphs to store structural and temporal relationships between events. They discuss how edges between events define a causal relationship, based on the assumption that events are related to each other if there is an underlying entity to which both events belong. Moreover, the authors show that their approach provides the means for fast querying of the data. Berti and van der Aalst [6] provide support for exploring event logs stored in databases from multiple viewpoints. González López de Murillas et al. [22] identify interesting case notions from databases, while Bala et al. evaluate heuristics for finding suitable case identifiers. [4]. Li et al. [17] follow another approach and create an object-centric event log format that does not require a case notion as it is required for the XES format. Li et al. argue that this object-oriented event log format helps to store relationships in the form of 1:n and n:m as it is common in databases. The problem involved in the usage of classic "flattened" event logs is also discussed in [2] in which an object-centric process mining approach is presented. Lu et al. [18], Diamantini et al. [8] and Genga et al. [14] consider the causality between events by modeling the traces as a partial order of its events.

Dumas and García-Bañuelos [9] discuss a process mining approach based on prime event structures. This publication transforms the cases in an event log into prime event structures and then use the concept of asymmetric event structures to create a process model. This approach is further used in [3] to diagnose behavioral differences between business process models. Ponce de León et al. choose a similar approach in [16] that uses event structures together with the occurrence nets to create process models. In [5], the author also uses prime event structures as an intermediate step to create a process model. In comparison to [3] and [16], the author uses a different approach for creating the process model out of the separated prime event structures.

The approach presented in the work at hand is using the concept of prime event structures to represent the causal relationship between different events stored in a database with 1:n and n:m relationships. Moreover, we use labeled property graphs to store and querying the event structures for batching events.

For what concerns stream *ii)*, Fahland [13] presents the concept of event synchronization, which is related to the concept of batching. He emphasizes that proper semantics for processes with n:m interactions require, among others, cardinality constraints and synchronization of transitions. Research on modeling batch behavior in a business process addresses this point at the type level. Pufahl et al. [23] extend BPMN with a specific batch activity type that considers an activation rule, a grouping attribute, a maximum batch size, and an execution order along with the definition of corresponding operational semantics. Martin et al. [21] present batching metrics for identifying patterns in event logs that point to batches. These include frequency of batch processing, batch size, instances per batch, duration and waiting times of instances in batches and temporal overlaps of batches. In [20], Martin et al. define a mining technique for discovering batch activation rules in event logs assuming that any observation of events done by the same resource doing the same activity for different cases represents a batch. Without knowledge about causality between events, this assumption may lead to spurious batches. Klijn and Fahland [15] present an approach to detect batches from event logs by analyzing the performance spectrum discussed in [7]. Their approach provides different metrics to quantify the batches, e.g., the batch size, different time metrics, or the batching frequency. A current limitation of this approach is a clear-cut between when disjoint cases are batches and when not.

As we will see in the following, we provide the means to identify batching events and to analyze them in great detail by using the causal event models.

4   Batch Analysis Based on Causal Event Models

This section proposes our approach to discover specific batching behavior, which explains characteristics in batching behavior from an event log. Section 4.1 presents the underlying formal concept used to capture causally related data from an ERP system. Section 4.2 discusses how all relevant batching nodes are identified, how batches are visualized and insights into batching behavior is presented. Finally, Sect. 4.3 discusses the implementation of our approach.

4.1   Determine Causal Event Models for Event Log

The first step of our approach is concerned with identifying the causal relations between event instances of the event log. To this end, we make use of foreign key relationships between entities of the database schema. Based on these relationships, we are able to reconstruct which events have triggered each other in passages of the process that exhibits 1:n or n:1 relationships like in order:delivery (n:1). As a formal structure for representing causal event models of the event log, we build on conflict-free prime event structures.

Definition 1 (Conflict-free prime event structure). *A* labeled *conflict-free prime event structure* is defined by the tuple $cf\_PES = \langle E, \leqslant, \lambda \rangle$, where E is *a set of events,* $\leqslant$ *defines the causality relation as a partial order on* E *and* $\lambda : E \to \Lambda$ *is a labeling function.*

A cf_PES is based on the *prime event structure (PES)* as defined in [3] excluding conflict relations. The latter are excluded from cf_PES, because conflicts and decisions that where made during the process execution are not visible in the event log, and therefore they can also not be represented in our causal event model. A cf_PES is equivalent to the notion of labeled partial order. As an example for a cf_PES, Fig. 1 depicts the order-to-cash processes from our industry partner discussed in Sect. 2. The example in Fig. 1 shows one order with three separate items that got picked individually from the warehouse and then delivered in one package.

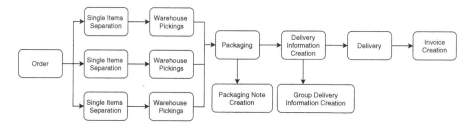

Fig. 1. Example cf_PES for the motivational scenario.

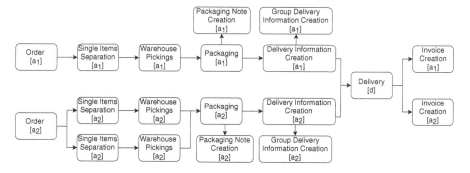

Fig. 2. Example DB_PES for the motivational scenario. The numbers in square brackets depicts the order affiliation.

If the whole event log is described by using a cf_PES, we call it a database of conflict-free prime event structure (DB_PES) or a *causal event log*. Such a DB_PES can contain several separated cf_PES, but also cf_PESs that share one or several events. We call these shared events *batch nodes*, since they are bundling together several process instances.

For example, in the running order-to-cash process example, the orders a_1, and a_2 might be bundled into one delivery d. A corresponding $cf\_PES = \langle E, \leqslant, \lambda \rangle$ would then be composed of $E = \{a_1, a_2, d\}$, $\leqslant = \{(a_1, d), (a_2, d)\}$, and $\lambda = \{(a_1, order), (a_2, order), (d, delivery)\}$, when we omit the other events. Figure 2 depicts these two order-to-cash cf_PES. As can be seen in Fig. 2 the two individual orders (the text in square brackets represents the event affiliation to E) share a common delivery event.

4.2   Batching Analysis

Once the DB_PES is created, various analysis operations can be performed to analyze the data. In the work at hand, the functionality to analyze the batching behavior is presented. In the following, we will first define some preliminaries and then the analysis approach.

Definition 2 (Event Type, Case Identifier, Node Identifier, Preset and Postset Nodes). *Given E is a set of events and ET is a set of event types, we define $\tau : E \rightarrow ET$ as the function that returns the type of an event. We define*

$e^\tau = \tau(e)$. *Furthermore, we define* $id : E \to I$ *as an index function for the node ID, and* $c\_ids : E \to I$ *as an index function that defines the case IDs of the cases that are using an event* $e \in E$. *Furthermore, for a relation* $R \subseteq E \times E$, *we define for an* $e \in E$ *the preset of nodes* $\bullet e = \{x \mid (x, e) \in R\}$ *and the postset of nodes* $e\bullet = \{x \mid (e, x) \in R\}$.

To perform an analysis of the batching operations, we devise an algorithm that first identifies all batching nodes $E_{batches}$ of a DB_PES, i.e., $E_{batches} = \{e | e \in E \wedge |c\_ids(e)| > 1\}$. In a second step the algorithm identifies for each $e \in E_{batches}$ the cf_PES that contains the batching node, defined by $cf\_PES^e_{batch}$. This is done by iterating through the nodes of the DB_PES with $\bullet e$ and $e\bullet$, starting from the batch node $e \in E_{batches}$, until the start and end nodes, i.e., $|\bullet e| = 0$, respectively $|e \bullet| = 0$, are reached. The start and end nodes are stored in E^e_{start} and E^e_{end}.

Depending on the size of the $cf\_PES^e_{batch}$, the visualization can be too crowded for a clear visualization. To overcome this problem, the algorithm aggregates all nodes with the same event type, i.e., e^τ, of the $cf\_PES^e_{batch}$ together. This aggregation step also counts the quantity of the aggregated nodes and relationships, and the cardinality in a form of $\{1{:}1, 1{:}N, N{:}1, N{:}M\}$ of each relationship. Additionally, the algorithm gets all preceding nodes of the batching events, i.e., $E^e_{prec} = \{\bullet e | e \in E_{batches}\}$. Eventually, the collected information, i.e., $cf\_PES^e_{batch}$, $E^e_{batches}$, E^e_{start}, and E^e_{end}, is returned by the algorithm.

The returned information can then be used to visualize and analyze the batches. To this end, we use a visualization based on the aggregated $cf\_PES^e_{batch}$ in a way that all cf_PESs are shown and the corresponding batch nodes (stored in $E^e_{batches}$) are highlighted. Furthermore, we provide for each $e \in E^e_{batches}$: event type e^τ, the batching factor $|c\_ids(e)|$ (i.e., the size of the batch), the execution start and end time of e, the average duration of the start of the preceding nodes E^e_{prec} to the start of the batch node e, the earliest start time in E^e_{start}, and the latest end time in E^e_{end}.

4.3   Implementation

Our approach has been implemented as a proof-of-concept, called Causal Miner. The prototype is developed in Java and uses the Spring Framework (vers. 2.3.0). We store the cf_PESs built from the event log in the graph database Neo4j (vers. 4.0.4) and use Cypher as query language. At the current state, the event log data can be read from Oracle DB as well as from Microsoft SQL Server. The visualizations are provided in a Web UI. The source code of the proof-of-concept prototype can be accessed on GitHub (https://github.com/piwa/causal-miner).

For the analysis of the batching operations, several Cypher queries are used. Two of them are presented in Listing 1.1 and Listing 1.2. The former returns the $E_{batches}$'s ordered by the batching factor. To limit the result, the query returns only the three $E_{batches}$'s with the highest batching factor, this limitation is configurable. Listing 1.2 returns the $cf\_PES^e_{batch}$ for the $e \in E_{batches}$ with the node $ID(e) = 5647$. For better readability, we replaced some configuration parameters in line 5 with "$\cdots$".

Listing 1.1. Neo4j Cypher Query to Find all Batching Nodes.

```
1   MATCH (n:InstanceActivity) WHERE size(n.instanceIds) > 1
2   RETURN DISTINCT ID(n) AS batchNodeId, size(n.instanceIds) AS batchSize,
        n.instanceIds AS batchInstanceIds
3   ORDER BY batchSize DESC LIMIT 3;
```

Listing 1.2. Neo4j Cypher Query to get all *cf_PES* that Share a Common Batching Node.

```
1   MATCH (n:InstanceActivity) WHERE ID(n) = 5647
2   MATCH p=(:InstanceStartActivity)-[*1..10]->(n)-[*1..10]->(:
        InstanceEndActivity)
3   UNWIND nodes(p) AS unwindedNodes
4   WITH collect(distinct unwindedNodes) AS collectedNodes
5   CALL at.ac.wuwien.extendedGroup(...) YIELD node, relationship
6   RETURN collect(distinct node) AS modelActivityList, collect(distinct
        relationship) AS modelRelationshipList
```

5   Results and Evaluation

In this section, we evaluate our prototypical implementation illustrating how it enables both visual and quantitative analyses of batching.

5.1   Setup and Dataset

The evaluation uses a real-world dataset from our industry partner, the previously mentioned food production company Pastamaker. The dataset is composed of the order-to-cash processes of the company. In total the dataset contains nearly 70,000 orders with more than 8,500,000 events. As discussed in Sect. 2, the process is composed of the following steps: A new instance is triggered by a supermarket order. The order is then broken down into single order items. These items are then picked from the warehouse separately. The items are then packed, and sent as one or multiple deliveries. As the last step, the invoice is created. Some of these steps also contain substeps and can be shared by several orders. For instance, it is often the case that several orders are delivered together, i.e., sharing the same delivery event, as depicted in Fig. 2.

The company uses an ERP system, which is built on an SQL database. For the evaluation, we first import the data from this SQL database into the Neo4j graph database according to the approach presented in Sect. 4. Subsequently, we perform different analysis steps on the Neo4j data.

5.2   Visualization of the Results

The Causal Miner offers different ways to work with the data. Figure 3 shows a screenshot of the Causal Miner. This screenshot exhibits an aggregated view of the *DB_PES*, which brings together all event nodes of the same type. It also shows the node and relationship quantity (in parenthesis) and the cardinality.

Along with the aggregated view presented in Fig. 3, the Causal Miner offers several other visualization methods, such as methods to validate the *cf_PESs*

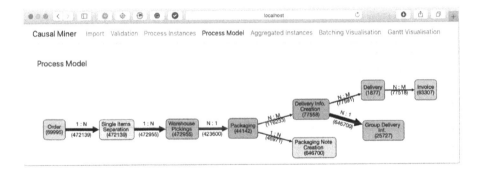

Fig. 3. Aggregated view of the DB_PES.

according to a given process structure, to visualize single cf_PESs, and to represent the activity durations in GANTT charts. The work at hand focuses on the visualization of batching. A screenshot of this view is presented in Fig. 4. The main UI consists of two tabs. In the Filter tab, several filtering functionalities can be selected. In the Upper Batches tab, the results of the batching queries are visualized. The view depicted in Fig. 4 is showing the batching node with the highest batching factor, together with the corresponding $cf\_PES^e_{batch}$. The tables on the right side show the information about the batches that are gathered by the algorithm discussed in Sect. 4.2. If a deeper analysis of the $cf\_PES^e_{batch}$ is required, an analyst can click on the *Show Instances* link under the table. This link opens the view depicted in Fig. 5 that shows all cf_PESs that are involved in the current batching node, together with information about the single events.

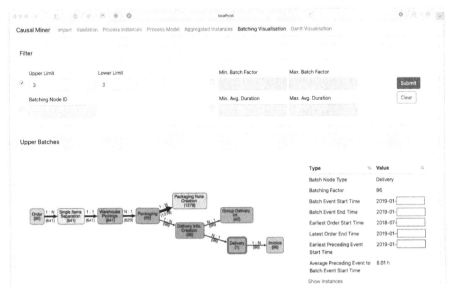

Fig. 4. Filtering options and the batching node, including the aggregated cf_PES, with the biggest batch factor (exact times obfuscated due to privacy).

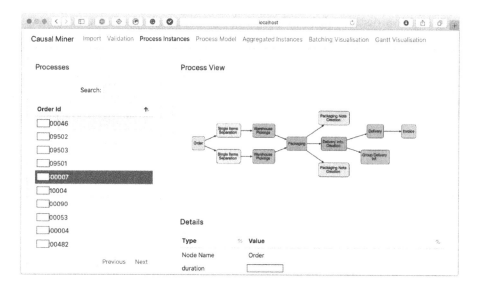

Fig. 5. Depiction of all cf_PES that share a particular batch node (note: some information was anonymized due to privacy concerns).

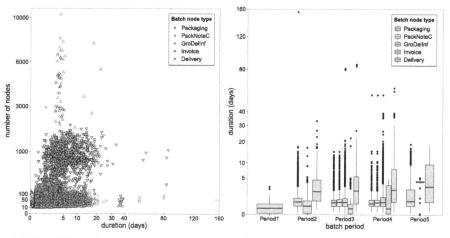

(a) Total Node Count versus Duration.   (b) Batching in Different Time Periods.

Fig. 6. Data analyses of batching.

5.3   Data Analyses

Besides the visualization possibilities shown in Sect. 5.2, the approach presented in Sect. 4, can be used for more detailed quantitative data analyses.

Figure 6 provides two types of analyses that allow us to gather interesting insights into the batching. Figure 6a presents a scatter plot that shows the total amount of nodes before the batching node versus the duration between the earliest start time in E^e_{start} and the latest end time in E^e_{end}. Each event type is plotted with a distinct shape. Figure 6a helps to visualize at least three main clusters of batching. Especially, it is possible to observe that *Invoices* are typically batched in five days. Figure 6b depicts a boxplot that shows the duration between the earliest start time in E^e_{start} and the latest end time in E^e_{end} in different time periods for each event type. Moreover, it can be observed that most of the time, the orders that are batched in the *Delivery* event need more time than the orders that are batched in different events. The shorter durations in the other case are partially due to corrections in the bookings without logistic activities and internal orders without deliveries.

The presented charts enable a process analyst to quickly identify outlier instances of the process. These outliers can then be analyzed in greater detail with the help of the Causal Miner by analyzing the $cf\_PES^e_{batch}$ and the involved cf_PES. Figure 7 presents an outlier that was identified from the scatter plot.

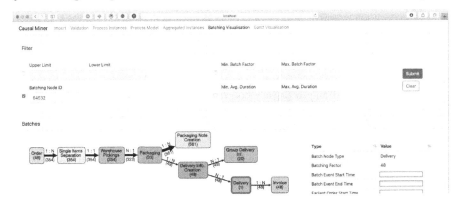

Fig. 7. Identified outlier (Exact Times Obfuscated due to Privacy).

5.4 Discussion

As shown in the evaluation, our approach provides different ways to analyze the batching behavior: First, the Causal Miner provides a way to visualize the processes as a whole, together with the event quantities and relationship cardinalities, by using the aggregated view of the *DB_PES*. Second, the Causal Miner provides a view to analyze the batching behavior of the processes. This view further provides a filter functionality to analyze the batching behavior regarding different aspects. At the current state, this filtering provides, e.g., the means to filter for the batching factor. Third, a separate view allows the analysis of the batched processes on the instance and single events level. Fourth, the approach provides different ways to analyze the batching by using plots. As shown, these plots can be used to analyze the batching behavior and to detect outliers. These outliers can then be analyzed further by using the process visualizations.

These functionalities provide an analyst a way to start with a high-level analysis of the batching, using the plots and the filters, and then dig deeper into the batching behavior by analyzing the aggregated and the single process instances. Moreover, an analyst can even go to the level of single events. In addition to the presented features, the Causal Miner provides different other views, like representing the process instances as a GANTT chart.

Since our approach considers all events stored in a graph-based database, in our case Neo4j, query performance also plays an important role. In our evaluation with the 8,500,000 events, queries that are searching for specific events or process instances need less than 200 ms. Only queries like the one presented in Listing 1.1 needs around 7.5 s since this query searches for all batching nodes, and Listing 1.2 needs around 1.6 s. The evaluation was done on a server with eight cores with 2,6 GHz and 47 GB RAM.

6   Conclusion

This paper introduces a technique to use causal event models to capture batching behavior. Our approach can be used to identify batches and determine its most important attributes, which helps to retrieve further insights of the batch processing. The algorithm is evaluated on real-world event logs, showing the practicability and usability of the approach. The resulting data can be used to discover batches and understand their context.

The implemented approach shows that there are important factors for batching. The processes are batched by different node types that have different characteristics. Important differences can be seen in the complexity of the process and total duration time of batches. Process analysts can use the data to conduct further performance analysis and trigger process improvements.

Future work will aim at empirically comparing alternative batches in processes and further improve batching analysis methods. This includes automatic identification of outliers and automatic evaluation of its causing process instances. Another direction for future research involves the automatic suggestions for process improvement based on the batch analysis.

References

1. van der Aalst, W.M.P.: Process Mining - Data Science in Action, 2nd edn. Springer, Heidelberg (2016). https://doi.org/10.1007/978-3-662-49851-4
2. Aalst, W.M.P.: Object-centric process mining: dealing with divergence and convergence in event data. In: Ölveczky, P.C., Salaün, G. (eds.) SEFM 2019. LNCS, vol. 11724, pp. 3–25. Springer, Cham (2019). https://doi.org/10.1007/978-3-030-30446-1_1
3. Armas-Cervantes, A., Baldan, P., Dumas, M., García-Bañuelos, L.: Diagnosing behavioral differences between business process models: an approach based on event structures. Inf. Syst. **56**, 304–325 (2016)
4. Bala, S., Mendling, J., Schimak, M., Queteschiner, P.: Case and activity identification for mining process models from middleware. In: Buchmann, R.A., Karagiannis, D., Kirikova, M. (eds.) PoEM 2018. LNBIP, vol. 335, pp. 86–102. Springer, Cham (2018). https://doi.org/10.1007/978-3-030-02302-7_6

5. Bergenthum, R.: Prime miner - process discovery using prime event structures. In: International Conference on Process Mining, ICPM 2019, Aachen, Germany, 24–26 June 2019, pp. 41–48. IEEE (2019)

6. Berti, A., van der Aalst, W.: Extracting multiple viewpoint models from relational databases. In: Ceravolo, P., van Keulen, M., Gómez-López, M.T. (eds.) SIMPDA 2018-2019. LNBIP, vol. 379, pp. 24–51. Springer, Cham (2020). https://doi.org/10.1007/978-3-030-46633-6_2

7. Denisov, V., Fahland, D., van der Aalst, W.M.P.: Unbiased, fine-grained description of processes performance from event data. In: Weske, M., Montali, M., Weber, I., vom Brocke, J. (eds.) BPM 2018. LNCS, vol. 11080, pp. 139–157. Springer, Cham (2018). https://doi.org/10.1007/978-3-319-98648-7_9

8. Diamantini, C., Genga, L., Potena, D., van der Aalst, W.M.P.: Building instance graphs for highly variable processes. Expert Syst. Appl. **59**, 101–118 (2016)

9. Dumas, M., García-Bañuelos, L.: Process mining reloaded: event structures as a unified representation of process models and event logs. In: Devillers, R., Valmari, A. (eds.) PETRI NETS 2015. LNCS, vol. 9115, pp. 33–48. Springer, Cham (2015). https://doi.org/10.1007/978-3-319-19488-2_2

10. Dumas, M., Rosa, M.L., Mendling, J., Reijers, H.A.: Fundamentals of Business Process Management, 2nd edn. Springer, Heidelberg (2018). https://doi.org/10.1007/978-3-662-56509-4

11. Esser, S., Fahland, D.: Storing and querying multi-dimensional process event logs using graph databases. In: Di Francescomarino, C., Dijkman, R., Zdun, U. (eds.) BPM 2019. LNBIP, vol. 362, pp. 632–644. Springer, Cham (2019). https://doi.org/10.1007/978-3-030-37453-2_51

12. Esser, S., Fahland, D.: Multi-dimensional event data in graph databases. CoRR abs/2005.14552 (2020)

13. Fahland, D.: Describing behavior of processes with many-to-many interactions. In: Donatelli, S., Haar, S. (eds.) PETRI NETS 2019. LNCS, vol. 11522, pp. 3–24. Springer, Cham (2019). https://doi.org/10.1007/978-3-030-21571-2_1

14. Genga, L., Alizadeh, M., Potena, D., Diamantini, C., Zannone, N.: Discovering anomalous frequent patterns from partially ordered event logs. J. Intell. Inf. Syst. **51**(2), 257–300 (2018). https://doi.org/10.1007/s10844-018-0501-z

15. Klijn, E.L., Fahland, D.: Performance mining for batch processing using the performance spectrum. In: Di Francescomarino, C., Dijkman, R., Zdun, U. (eds.) BPM 2019. LNBIP, vol. 362, pp. 172–185. Springer, Cham (2019). https://doi.org/10.1007/978-3-030-37453-2_15

16. de León, H.P., Rodríguez, C., Carmona, J., Heljanko, K., Haar, S.: Unfolding-based process discovery. CoRR abs/1507.02744 (2015)

17. Li, G., de Murillas, E.G.L., de Carvalho, R.M., van der Aalst, W.M.P.: Extracting object-centric event logs to support process mining on databases. In: Mendling, J., Mouratidis, H. (eds.) CAiSE 2018. LNBIP, vol. 317, pp. 182–199. Springer, Cham (2018). https://doi.org/10.1007/978-3-319-92901-9_16

18. Lu, X., et al.: Semi-supervised log pattern detection and exploration using event concurrence and contextual information. In: Panetto, H., et al. (eds.) OTM 2017. LNCS, vol. 10573, pp. 154–174. Springer, Cham (2017). https://doi.org/10.1007/978-3-319-69462-7_11

19. Lu, X., Nagelkerke, M., van de Wiel, D., Fahland, D.: Discovering interacting artifacts from ERP systems. IEEE Trans. Serv. Comput. **8**(6), 861–873 (2015)

20. Martin, N., Solti, A., Mendling, J., Depaire, B., Caris, A.: Mining batch activation rules from event logs. IEEE Trans. Serv. Comput. (2019, early access)

21. Martin, N., Swennen, M., Depaire, B., Jans, M., Caris, A., Vanhoof, K.: Retrieving batch organisation of work insights from event logs. Decis. Support Syst. **100**, 119–128 (2017)
22. de Murillas, E.G.L., Reijers, H.A., van der Aalst, W.M.P.: Case notion discovery and recommendation: automated event log building on databases. Knowl. Inf. Syst. **62**(7), 2539–2575 (2019). https://doi.org/10.1007/s10115-019-01430-6
23. Pufahl, L., Weske, M.: Batch activity: enhancing business process modeling and enactment with batch processing. Computing **101**(12), 1909–1933 (2019). https://doi.org/10.1007/s00607-019-00717-4

Process "Prospecting" to Improve Renewable Energy Interconnection Queues: A Case Study

Gerry Murphy[⊠]

PGMoDE, LLC, Boulder, CO 80302, USA
gerry.murphy@pgmode.om

Abstract. Globally, interconnecting a new solar or wind generation project to the grid involves navigating a queue requiring financial deposits, engineering studies, and fees to upgrade the electric grid. The process can take years, during which time changes to regulatory regimes, tax incentives, financial markets, or competitive pressures can make a project suddenly nonviable for an investor. For grid operators, the increasing saturation of intermittent generation concurrent with retiring fossil fuel generation makes every new project increasingly complex to assess. This paper provides a case study of applying process mining techniques to address the question of whether the options proposed by Duke Energy Carolinas (DEC) to reform its generation interconnection queue process are warranted. Two options for reform have been proposed: creating study clusters based on concurrency or creating them based on locational proximity. Results indicate support for aspects of both options, although some causes may prove uncontrollable due to their origin in external factors such as market competition and power systems engineering decision making.

Keywords: Interconnection queue · Process discovery · Conformance analysis

1 Background

This paper provides a case study of applying process mining techniques to "prospect" options for reforming generation interconnection queue procedures operated by Duke Energy Carolinas (DEC), an electric utility in the United States with 2.6M customers [1]. *Interconnection queue* refers to the process for new power plants to get approval for connecting to the electric grid. DEC interconnection workflows consists of the following steps: 1.) application and review, 2.) system impact study, 3.) potential restudy loops as needed including feasibility and facilities studies, 4.) interconnection agreement, 5.) construction, 6.) commissioning, and 7.) commercial operations.

Driven by rapid equipment cost declines, government incentives, and a favorable economic environment, the quantity of solar generation capacity installed in North Carolina (NC) grew from ~1,000 MW of installed capacity in 2015 to 6,435 MW of installed capacity as of Q1 2020 [2, 3]. This $9B cumulative investment has made NC the second ranked state in the United States for solar generation capacity [3].

© Springer Nature Switzerland AG 2021
S. Leemans and H. Leopold (Eds.): ICPM 2020 Workshops, LNBIP 406, pp. 30–42, 2021.
https://doi.org/10.1007/978-3-030-72693-5_3

Investment has been influenced by periodic expiration of the Federal Production Tax Credit (PTC). Created in 1992, PTC has been renewed 12 times since expiring in 1999 [4]. In early 2016, PTC was extended until 12/31/2019. This created unprecedented market stability for solar project developers. In late 2019, PTC was extended only through 12/31/2020 [4]. For this study, 12/31/2015 and 12/31/2019 are key dates.

Investors risk developing and operating solar generation plants in exchange for selling their electricity to an electric utility at a fixed price that guarantees a positive return on investment. Because these investments are speculative in nature until final stages, economic and competitive forces can have a large effect on investor behavior.

In 2019, North Carolina Utilities Commission (NCUC), which regulates the electricity market in NC, required DEC to expedite the interconnection queue [5]. DEC proposed two approaches: clustering new solar projects based on a temporal basis versus doing so on a locational basis. The proposed changes would involve the same activity sequence, but activities would be coordinated across multiple projects. Grid upgrade costs would be shared across multiple projects as opposed to having the single project that triggers an upgrade bearing the full upgrade cost [6].

Quarterly DEC regulatory filings were used as source data for an event log of interconnection milestones for new solar projects. Due to their legal nature, the filings were assumed accurate. Filings had quarterly intervals so trace alignment would not have tied to daily workflow activities; a standard approach to identify factors driving activity bottlenecks was not viable. Summarized below, a "prospecting" approach was taken to compare process performance of multiple study groups filtered within a single data set (Table 1).

Table 1. The process "prospecting" approach taken for this study.

Research question	Data limitation	"Prospecting" adaptations
Is there evidence to support temporally grouped cluster studies? • H_a: there is seasonal variation in project activities • H_0: there is no seasonal variation	Data describe queue performance in quarterly snapshots but do not capture daily activities	• Define two seasonal study groups of cases based on their queue entry date • Compare Petri net behavior and event log conformance for the seasonal groups
Is there evidence to support locationally grouped cluster studies? • H_a: locational clustering can be seen in a higher count of projects per substation where interconnection occurs • H_0: there is no locational clustering	Data lack project developer activities that describe how they select locations or how they manage projects through the queue	• Define a third study group of cases located at top quartile substations in terms of project volume. • Compare Petri net behavior and event log conformance for top quartile substation projects vs. Seasonal groups

(continued)

Table 1. (*continued*)

Research question	Data limitation	"Prospecting" adaptations
Is there evidence to explain how developers navigated the key date effect on their investments? • H_a: PTC expiration dates influence activities • H_0: PTC expiration dates do not influence activities	The effect of external factors on queue dynamics was not directly measured	• Define a fourth study group of top quartile installers based on their number of cases • Compare project cancellations or sales in key date years vs. Other years

2   Methodology

2.1   Gather DEC Regulatory Filing Data and Convert It into MS Excel Format

An event log was assembled by collecting documents provided by DEC to NCUC on a quarterly basis from Oct 2015–Apr 2020 [7]. The documents were converted to a spreadsheet, standardized, and prepared for process mining in a relational database.

The original data set contained 18,560 events for 4,868 cases. It included these attributes: Queue Number, Queue Issued Date, Installer Account Name (5,424 null entries), Energy Source Type (1 nulls), Installed Capacity (no nulls), Facility County (919 nulls), Substation Name (752 nulls), and Feeder Number (1,331 nulls). The Installed Capacity attribute was removed due to unit of measure variations. The following activity types were discovered: Additional Field Work Required, Cancelled, Construction - In Progress, Construction – Pending, Engineering Design - In Progress, Engineering Design – Pending, Facility Study - In Progress, Facility Study – Pending, Feasibility Study – Pending, Interconnection Agreement Execution – Pending, IR Review - In Progress, Open, Request Incomplete, and Superseded.

2.2   Assess Process Performance and Generate an Event Log CSV File

Using SQL queries, start and completion times were calculated for project activities. Completion times were not given, so they were imputed by determining the quarterly filing report date in which that activity or project disappeared: for example, if a project was listed as "construction – in progress" in one report but then the project was no longer listed in the subsequent report, construction was assumed to have completed and thus given a completion date of the report when it first disappeared. This reduced total events from 18,560 to 6,659 as events without date information were removed.

Activity types reported by DEC changed over time. Newer activity types that appeared from 2018 were filtered because they had low occurrence and incomplete data. Only these events were analyzed: "Open," "Cancelled," "Superseded," "Construction – In Progress," and "Construction – Pending." Total events reduced from 6,659 to 6,456. A CSV file was created in event log structure and exported.

2.3   Conduct Petri Net Behavior and Event Log Conformance Analysis

The CSV file was uploaded to PROM 6.9, converted to XES format, and filtered into the following groups: a.) projects initiated between July 1 and December 31 ("Ones") b.) projects initiated between January 1 and June 30 ("Zeros"), and c.) projects initiated for interconnection at a substation within the top quartile of interconnection requests ("TopQS"). Filter groups a.) and b.) addressed the temporal clustering question. Filter group c.) addressed the locational clustering question. The sequence of analysis was as follows: 1.) Petri net analysis, 2.) Conformance analysis using Multi-perspective Process Explorer and Replay A Log On Petri Net for Conformance packages.

Using PROM 6.9, Petri nets were created using Mine Petri Net With Inductive Miner package utilizing default settings. Inductive Miner was chosen because it produced a Petri net model with sequential activities most resembling the actual DEC interconnection process, unlike Alpha Miner and ILP packages. A key assumption was that Inductive Minor can correctly identify the main process behavior in this event log; all deviations identified in conformance analysis are true process deviations.

2.4   Conduct Event Log Visualization and Directly Follows Graph Analysis

Log analysis was conducted using Explore Event Log (Track Variants), Log Pattern Explorer, and Dotted Chart visualizations. Lastly, Mine Matrix package was run to generate event causality data for comparison.

2.5   Analyze Key Date Behavior of Project Developers (*Installers*)

In its regulatory filing data, DEC uses the term *installer* to refer to project developers. To assess their key date behavior a fourth study group was created for *installers* within the top quartile of solar project volume – TopQI. The project events of TopQI were compared to the remaining 75% of *installers* (Rest), focusing on events surrounding key legislative dates: 12/31/15 and 12/31/19. A key assumption was that larger developers have more engineers and resources compared to smaller *installers*. Consequently, TopQI were expected to be savvier in their responses to key dates.

3   Results

3.1   Assess Process Performance

Figure 1 below is a forward-looking chart created in MS Excel that counts projects in the queue on a quarterly basis in terms of what their future end state will become.

DEC performance in processing interconnection requests improved after 2017. If you were a developer entering the queue in August 2018, you would have had 3X the likelihood of completing your project as someone entering just 10 months before.

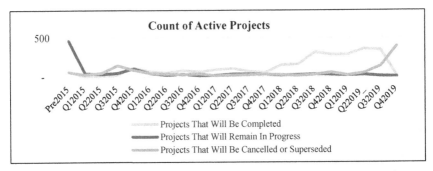

Fig. 1. Quarterly count of active projects by future disposition in the DEC interconnection queue.

Above, the 12/31/15 key date effect is visible in the precipitous decline of projects that will remain in progress following Q1 2015. Of this cohort of 527 projects, 20% were cancelled in Q3 2015, 64% went on to be cancelled or sold ("superseded") in 2016–2019, 11% were in pre-construction state as of April, 2020, and only 6% passed onto the construction stage. The 12/31/19 key date effect is again visible above in the increase in projects that will be canceled or superseded as of Q4 2019.

Calculated in MS Excel, Figs. 2 and 3 below show the average and standard deviation of project duration for projects in the DEC interconnection queue.

Fig. 2. Quarterly average duration (days) of projects in DEC interconnection queue.

Fig. 3. Quarterly standard deviation duration (days) of projects in DEC interconnection queue.

For projects that will be completed, the average project duration for projects declined from 1,112 days in Q1 2015 to 161.93 days in Q4 2019. Standard deviation of for projects that will be completed declined from 482 days in Q1 2015 to 107 days in Q4 2019.

Perhaps the best metric of interconnection queue performance success is whether a project gets constructed. Figure 4 below gives an overview of project success trends:

Fig. 4. Success rate for new projects initiated each quarter: the count of projects that eventually start construction/count of all projects initiated.

Figure 4 provides further evidence of improving queue performance between 2017 and 2019. Yet as the key date of 12/31/2019 approached, construction starts fell.

Table 2 below is a summary of queue performance across the three study groups:

Table 2. 2015–2020 interconnection queue performance for Zeros, Ones, and TopQS groups

Filter group	Total projects initiated	Avg elapsed time/Case (Days)	Success rate
Zeros: Q1–Q2 start	1,671	167.41	89%
Ones: Q3–Q4 start	1,838	154.51	80%
TopQS	890	161.25	88%

Between July and December, Ones group initiated more projects (1,838) vs. Zeros group from January to June (1,671). Ones group projects had shorter elapsed time compared to Zeros group but had a lower success rate. TopQS located projects had lower elapsed time compared to Zero group while having a similar success rate to Ones group.

3.2   Petri Net Behavioral Analysis

Petri nets for the Zeros, Ones, and TopQS groups were created using the Mine Petri Net Using Inductive Miner package and assessed using Analyze Behavioral Property of Petri Net and Analyze Structural Property of Petri Net packages. Results were compared using the Show Petri Net Metrics package. Table 3 below summarizes results.

Every Petri net was found to be a sound workflow net. All three groups had Extended Cardoso values of 14 - 20, which per Cardoso places them in the "easy to understand" complexity category [8]. Extended Cyclomatic metrics show wider variation than Extended Cardoso metrics: there is a wider difference in the number of possible linear paths across the three groups compared to the number/type of splits. The Structuredness metric results align more with Extended Cardoso results: TopQS has the most complex model, followed by Zeros and then Ones. The three filtered study groups each have greater workflow complexity than the event log as a whole (All).

Table 3. Petri analysis results for Zeros, Ones, and TopQS groups.

	Ones	Zeros	TopQS	All
Density metric	0.08824	0.08824	0.06875	0.16667
|F|	36	36	44	22
|P X T|	204	204	230	66
Extended Cardoso metric	15	17	20	9
Extended Cyclomatic metric	26	22	35	9
Number of arcs	36	36	44	22
Number of places	12	12	16	6
Number of transitions	17	17	20	11
Structuredness metric	66	100	117	22

3.3 Conformance Analysis

Conformance analysis was conducted on each filter group and the results were compared side by side. The results of Multi-perspective Process Explorer and Replay A Log On Petri Net are shown in Table 4 below.

Table 4. Conformance analysis results for Zeros, Ones, and TopQS groups.

	Ones	Zeros	TopQS	All
Avg activity precision	79.2%	86.4%	72.6%	94.7%
# Moves observed	29,897	27,488	15,089	32,342
# Moves possible	37,726	31,806	20,776	34,143
Avg. fitness	63.7%	58.6%	58.2%	68.2%
% Violations	33.4%	37.9%	39.3%	41.8%
# Correct events	3,983	3,946	1,619	6,973
# Wrong events	1,895	2,160	975	5,011
# Missing events	98	251	75	-
# Traces	2,408	2,548	1,130	4,671
# Events	5,878	6,106	2,594	11,984
# Event classes	5	5	5	5

The Ones group had higher average fitness compared to Zeros and TopQS. All other conformance metrics were lower for Zeros versus Ones. Combined with the lower Structuredness and Extended Cardoso metrics for Ones in Table 3 above, projects initiated from July to December (Ones) have better performing models than projects initiated from January to June (Zeros). TopQS group covers the entire year with a locational focus, and it has the lowest overall precision and fitness. Despite its successful balance of low case duration, high success rate, and lowest raw fitness cost, the TopQS model has more violations and fewer correct events than Ones and Zeros.

3.4 Event Log Visualizations

The figures below summarize log visualizations for Zeros, Ones, and TopQS groups:

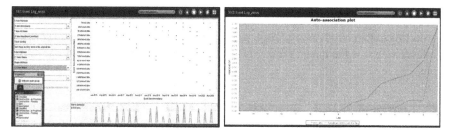

Fig. 5. Zeros group Dotted Chart and Auto-association visualizations.

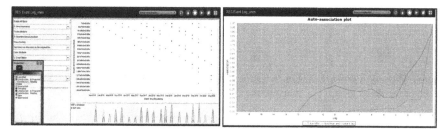

Fig. 6. Ones group Dotted Chart and Auto-association visualizations.

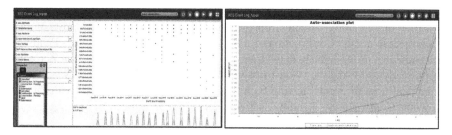

Fig. 7. TopQS group Dotted Chart and Auto-association visualizations.

The Dotted Charts for Zeros and Ones groups are similar. In the upper right corner of the chart, which displays recent, lowest duration events, Ones have better performance than Zeros. For Ones, this corner is denser with events and there are more blue "Construction – In Progress" and green "Construction – Pending" dots that indicate success. Ones also have more tan "Superseded" dots and fewer pink "Cancelled" dots compared to Zeros. TopQS shows strongest overall performance in the dotted chart upper right corner: it is densest and has highest number of blue "Construction – In Progress" and green "Construction – Pending" dots.

Auto-association plots vary most in terms of the Goodman and Kruskal's tau values. For Ones, association values decline overall in the lag range of 9 down to 4 before increasing again. Zeros and TopQS do not have this mid-range decline. All three groups rapidly increase their association values at the lowest lag values.

3.5 Directly Follows Graph

Convert Log to Directly Follows Graph package was next run on each group to compare high-level views of the queue process. Figures 8, 9 and 10 below show the resulting graphs.

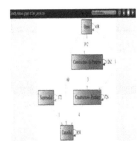 

Fig. 8. Directly follows graph of the Zeros group

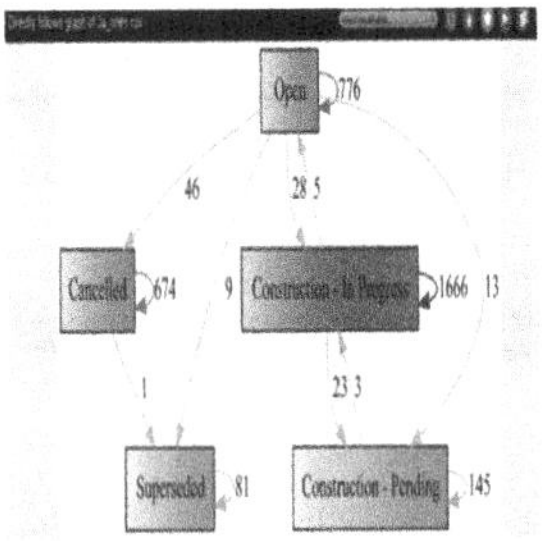

Fig. 9. Directly follows graph of the Ones group

Fig. 10. Directly follows graph of the TopQS group

Comparing Figs. 8, 9 and 10 to the interconnection queue in Fig. 1 above, the directly follows graphs differ in the flow of events through the "Construction – Pending" activity. The DEC proposed process shows a series of engineering studies of a new power plant and the milestone payments that lead to a facilities study and then an interconnection agreement prior to construction start. In Figs. 8, 9 and 10, a project can proceed to "Construction – Pending" directly from the initial "Open" activity, but it is most common for a project to attain "Construction – In Progress" prior to "Construction – Pending." Ones and Zeros differ in the path to "Cancelled" activity, which in the case of Zeros can occur directly following the "Construction – Pending" activity. TopQS shows the most hierarchical flow of events, having both "Cancelled" and "Superseded" both directly following "Construction – Pending" activity.

3.6 Key Date Behavioral Analysis: Solar Project Developers

Key date behavioral analysis focused on comparing *installers* (developers) accounting for the top quartile of solar project volume (TopQI) group to the rest of *installers* (Rest). Comparing the October to January period for key dates of 12/31/15 and 12/31/19, when PTC was set to expire, versus 12/31/2016, 12/31/17, and 12/31/2018, a sharper picture of queue dynamics emerges. Table 5 below shows this comparison.

Table 5. Comparison of key project activities that start between October to January for PTC expiration dates (2015, 2019) versus other years (2016, 2017, 2018).

Time period	Cancelled	Construction - in progress	Construction - pending	Queue issued	Superseded
PTC expiration 2015	134	0	0	304	0
PTC expiration 2019	166	0	0	501	47
2016, 2017, 2018	192	296	49	948	60

PTC expiration years of 2015 and 2019 had a disproportionate number of projects cancelled or superseded ($347 = 134 + 166 + 0 + 47$) versus the other years ($252 = 192 + 60$). No projects were committed to construction in 2015 or 2019. More projects were initiated in the 2019 end of year period (501) versus 2015 (304), although this was far below 2016–2018 levels (948).

The Tables 6 and 7 below compare top quartile installers (TopQI) to the Rest of installers.

Table 6. Comparison of key project activities of TopQI group during end of year time intervals.

Time Period	Cancelled	Construction - in progress	Construction - pending	Queue issued	Superseded
PTC expiration 2015	77	0	0	218	0
PTC expiration 2019	72	0	0	8	23
2016, 2017, 2018	103	58	13	306	21

Table 7. Comparison of key project activities of rest group during end of year time intervals.

Time period	Cancelled	Construction - In progress	Construction - pending	Queue issued	Superseded
PTC expiration 2015	57	0	0	86	0
PTC expiration 2019	94	0	0	493	24
2016, 2017, 2018	89	238	36	642	39

For *Cancelled* projects, TopQI did not change practices from 2015 to 2019 key dates (77 vs. 72), but the Rest increased cancellations by 65% in 2019 (57 vs. 94). For project starts (*Queue Issued*), TopQI decreased theirs by 96% from 2015 to 2019 key dates (218 vs. 8) and the Rest increased *Queue Issued* by 83% (86 vs. 493). For *Superseded*, TopQI and the Rest both increased this activity in 2019 (0 vs. 23 and 24, respectively).

4   Discussion

4.1   Petri Net Behavioral Analysis

Variation in complexity metrics observed across groups and scenarios based on the same business workflow could point to both anomalies in the data set and opportunities to streamline the workflow and standardize its data model, simply from the perspective of reducing errors. Since all groups derive from the same event log which has been simplified by removing low occurrence events, these complexity differences could reflect real variances in the process. Additionally, better Petri net performance observed in the Ones model could explain its advantage in average case time assuming poor process performance is reflected in project delays. However, the observations rely on Inductive Minor's ability to correctly portray the main process model in Petri net outputs.

4.2   Conformance and Event Log Analysis

Seasonality is a factor in model performance. There is seasonal variance in case duration, in the number of events and traces per case, and in model conformance.

Assuming Inductive Minor did accurately capture the main process model in its generated Petri nets so that variance across study groups reflects real process variation, Zeros had worse model performance than Ones but also had a higher success rate. Projects initiated from July to December (Ones) had more low-occurrence events that were removed during data preparation and this could have advantaged its model performance compared to projects that initiated from January to June (Zeros). Having a lower project workload did not make Zeros group interconnection requests get processed faster than Ones group. It is possible that Zeros having extra time helped resolve issues blocking construction starts, but the key date effect could just as likely have increased cancellations for Ones during the end of the year. The higher share of project cancellations and supersessions (sales) during the end of year period could have been a factor on Ones having a lower success rate versus Zeros. Despite having the lowest density model, TopQS had the lowest model fitness and precision. The simpler model for TopQS did not advantage its model conformance. From a model conformance perspective, it is not clear what drove TopQS success.

Dotted Chart results of Figs. 5, 6 and 7 above show differentiation across groups in shortest duration events occurring in the most recent time intervals. In this upper right quadrant area, Ones have higher success versus Zeros and TopQS have highest success overall. Relative success and short duration of TopQS projects could be a recent trend.

Overall, if Inductive Minor did not accurately portray the main process flows then the validity of conformance analysis results is questionable. However, the study group differentiation observed in Dotted Chart results could support the differentiation observed in conformance analysis.

4.3 Key Date Behavioral Analysis

End of year project activities differed between the TopQI and Rest of installers group. Comparing TopQI to the Rest, there is a relative increase in Q4 cancellations across all years, but not on the specific key dates of 2015 and 2019. TopQI also showed a relative increase in superseded projects in Q4 and on key dates compared to the Rest. For project initiations, TopQI had a greater volume in Q4 across all years including key dates compared to the Rest, but the volume of project initiations was still lower during this time of year. Overall, it appears that TopQI were reducing DEC area investment at the end of 2019 while the Rest were still ramping up DEC area investment. TopQI were aggressive investors in 2015 but cautious or exiting the DEC area in 2019.

4.4 Approach Viability

Converting regulatory filings into an event log for process "prospecting" analysis was a novel approach. Because the main questions in the queue reform debate had already been framed within publicly available documents, process "prospecting" was able to provide valuable context despite limitations of the data set.

Filtering study groups from a common data source met the objective of finding process variations relative across the groups. Going further to benchmark this interconnection queue data set against that for queues in other regions would be a more challenging topic which would require more robust data.

Addressing more detailed questions about root cause would have required alignment analysis at the trace level. Alignment analysis was not viable because the end dates of many activities were imputed on a quarterly basis, which may have created concurrency that did not really exist.

5 Conclusions

Based on results, the null hypotheses for seasonal and locational cluster patterns can be rejected. Neither the process performance nor process "prospecting" results indicate seasonal or locational uniformity in the DEC interconnection queue. In terms of project developer behavior, the null hypothesis that PTC expiration dates do not influence project events can also be rejected. Addressing the influence of key dates on project events will be an important consideration to finalizing the proposed DEC cluster study process.

Process "prospecting" played a useful role in addressing whether the proposed queue clustering approaches were warranted. Its insights into model performance could be useful to design the optimal cluster study workflow. As a followup, process mining on a more robust DEC interconnection queue data set could determine whether the process variances observed by Inductive Minor are accurate. In addition to supporting the business and regulatory mandate for queue reform at DEC, this could help improve the design of the cluster study workflow.

More broadly, this study confirmed that process mining can be incorporated to benefit process-focused business scenario analysis. PROM 6.9 offered a vast array of analytical options, which was advantageous. Since it is a research tool, the downside of PROM 6.9

is that each plugin has its own documentation and accompanying research papers. This caused some confusion around interpreting process mining results.

From the industry perspective, process "prospecting" is a common scenario; businesses are likely to begin a process transformation initiative with a small pilot and limited data. There is an opportunity in the process mining community to craft a "prospecting" interface that allows practitioners to assess their data and recommend plugin options that meet their study objectives. Long term, meta-research studies across the suite of PROM 6.9 plugins may be useful to cultivate the process mining body of knowledge.

References

1. Duke Energy website. https://www.duke-energy.com/our-company/about-us. Accessed 27 July 2020
2. US Energy Information Agency website. https://www.eia.gov/state/?sid=NC. Accessed 27 July 2020
3. Solar Energy International website. https://seia.org/state-solar-policy/north-carolina-solar. Accessed 27 July 2020
4. Congressional Research Service. https://crsreports.congress.gov/product/pdf/R/R43453/17. Accessed 27 July 2020
5. NCUC website. https://starw1.ncuc.net/NCUC/PSC/PSCDocumentDetailsPageNCUC.aspx? DocumentId=6d82de14-71de-492f-a6d4-f9a14099642a&Class=Order. Accessed 27 July 2020
6. NCUC website. DEC and DEP Queue Reform Update, 15 October 2019. https://starw1.ncuc. net/NCUC/ViewFile.aspx?Id=0ca6866e-dcaf-4622-b543-486ab37cc34a. Accessed 27 July 2020
7. NCUC website, Document Search. https://starw1.ncuc.net/NCUC/page/DocumentsTextSe arch/portal.aspx. Search term: "Interconnection Queue". Accessed 27 July 2020
8. Sanchez-Gonzalez, L., et al.: Towards Thresholds of Control Flow Complexity Measures for BPMN Models (2011). https://jorge-cardoso.github.io/publications/Papers/CP-2011-060-SAC-Towards-thresholds-of-control-flow.pdf. Accessed 27 July 2020

Automated Discovery of Process Models with True Concurrency and Inclusive Choices

Adriano Augusto[1]([✉]), Marlon Dumas[2], and Marcello La Rosa[1]

[1] University of Melbourne, Melbourne, Australia
{a.augusto,marcello.larosa}@unimelb.edu.au
[2] University of Tartu, Tartu, Estonia
marlon.dumas@ut.ee

Abstract. Enterprise information systems allow companies to maintain detailed records of their business process executions. These records can be extracted in the form of event logs, which capture the execution of activities across multiple instances of a business process. Event logs may be used to analyze business processes at a fine level of detail using process mining techniques. Among other things, process mining techniques allow us to discover a process model from an event log – an operation known as automated process discovery. Despite a rich body of research in the field, existing automated process discovery techniques do not fully capture the concurrency inherent in a business process. Specifically, the bulk of these techniques treat two activities A and B as concurrent if sometimes A completes before B and other times B completes before A. Typically though, activities in a business process are executed in a true concurrency setting, meaning that two or more activity executions overlap temporally. This paper addresses this gap by presenting a refined version of an automated process discovery technique, namely Split Miner, that discovers true concurrency relations from event logs containing start and end timestamps for each activity. The proposed technique is also able to differentiate between exclusive and inclusive choices. We evaluate the proposed technique relative to existing baselines using 11 real-life logs drawn from different industries.

1 Introduction

Enterprise information systems, such as Enterprise Resource Planning (ERP) systems, maintain detailed records of each execution of the business processes they support. These records can be extracted in the form of event logs. An event log is a set of event records capturing the execution of activities across a set of instances of a process.

Process mining techniques allow us to exploit event logs in order to analyze business processes at a fine level of detail. Among other things, process mining techniques allow us to discover a process model from an event log – an operation known as *automated process discovery*. Despite a rich body of research in the field, existing automated process discovery techniques do not fully capture the concurrency inherent in business processes. Indeed, the bulk of automated process discovery techniques operate under an interleaved concurrency model – a model of concurrency where two events are concurrent if they occur in either order. Specifically, existing techniques treat two activities

S. Leemans and H. Leopold (Eds.): ICPM 2020 Workshops, LNBIP 406, pp. 43–56, 2021.
https://doi.org/10.1007/978-3-030-72693-5_4

A and B as concurrent if sometimes A completes before B and other times B completes before A. The interleaved concurrency model is suitable in systems where actions are atomic. However, in a business process, activities have a duration and the execution of two or more activities may overlap temporally. In other words, business processes contain *true concurrency*. The failure of existing automated process discovery techniques to take into account this true concurrency leads them to miss certain concurrency relations. For example, when an activity A always completes before activity B (because B takes longer) even though A and B overlap, existing techniques treat A and B as sequential. If A is then followed by C and C usually completes after B (but overlaps with it), they conclude that A, B and C are sequential, thus missing the observed concurrency.

This paper addresses this gap by presenting a refined version of an automated process discovery algorithm, namely Split Miner [6], capable of discovering true concurrency relations from event logs that record both the start and end timestamps of activity executions. The proposed technique, namely Split Miner 2.0, is also able to differentiate between exclusive and inclusive choices. The paper reports on an empirical evaluation that compares Split Miner 2.0 against existing baselines in terms of accuracy and model complexity measures.

The rest of the paper is structured as follows. Section 2 briefly reviews existing automated process discovery techniques. Section 3 introduces the approach to exploit true concurrency for automated process discovery. Section 4 presents the empirical evaluation while Sect. 5 summarizes the findings and further possible extensions.

2   Background and Related Work

An *event log* records information about a set of executions of a business processes (a.k.a. cases). Concretely, an event log is a chronological sequence of events, each one capturing a state change in the execution of an activity. As a minimum, each event in a log has three attributes: the identifier of the process execution (a.k.a. *case ID*); the label (i.e. the process activity the event refers to); and the timestamp (e.g. 10/07/2020 10.43). Optionally, an event may have other attributes such as the resource who triggered the event, their department, etc. In this paper, we require that at least one fourth attribute is attached to each event, namely the *life-cycle transition*. For a given event, this attribute indicates what state-change the referenced activity has undergone. The life-cycle of an activity captures all the states in an activity execution and their possible transitions. In general, one could observe very complex life-cycles, including states such as created, assigned, started, suspended, etc. In this paper, we adopt a simple life-cycle model wherein an activity execution can be in one of two states: *start* (i.e. the activity execution started); and *end* (i.e. the activity execution ended).

Event logs can be exploited for different types of analysis including conformance checking, process performance mining, and automated process discovery [16]. In this paper, we focus on the latter. The goal of automated process discovery is to discover a process model (such as the one in Fig. 1) by analysing an event log such as the one in Table 1 (the latter is just an extract and not a full log).

The quality of an automatically discovered process model is traditionally assessed over four dimensions: fitness – the amount of process behaviour recorded in the event

Table 1. Event log example.

Case-ID	Activity	Life-cycle	Timestamp
1	a	start	2020-07-08 10.03
2	a	start	2020-07-08 10.42
1	a	end	2020-07-08 10.57
2	a	end	2020-07-08 11.21
1	b	start	2020-07-08 13.29
1	c	start	2020-07-08 14.13
2	b	start	2020-07-08 15.22
2	b	end	2020-07-09 10.24
1	b	end	2020-07-09 10.37
2	d	start	2020-07-09 11.13
2	d	end	2020-07-09 12.28
1	c	end	2020-07-09 12.53

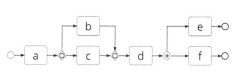

Fig. 1. Process model example.

log that can be replayed by the process model; precision – the amount of behaviour captured by the process model that can be found in the event log; generalization – the amount of behaviour captured by the process model that even not being observed in the event log is likely to belong to the original process; and simplicity – quantifying how difficult is to understand the process model. Furthermore, a process model should be sound. The notion of *soundness* has been defined on Workflow nets [17] as a correctness criterion, and is also applicable to BPMN models. Formulated on BPMN models, soundness encompasses three properties: i) every process instance eventually reaches the end event (no deadlocks); ii) no end event is reached more than once during a process execution (proper completion); iii) each process activity is triggered in at least one process execution (no dead activities).

A recent literature review of automated process discovery algorithms [5] showed that only few algorithms stand out for accuracy and performance among those outputting procedural process models. Specifically, Inductive Miner (IM) [10], Evolutionary Tree Miner (ETM) [7], and Split Miner (SM) [6]. IM and ETM are known to discover process models that are either highly fitting or precise, discovering simple, block-structured and sound process models, while SM focuses on maximizing both fitness and precision at the cost of simplicity, structuredness, and in rare cases compromising the soundness of the process models [5,6]. However, of these three automated process discovery algorithms, only IM provides a variant that takes into account the activities' life-cycle when discovering a process model. IM life-cycle variant [11] analyses the activities' life-cycles to distinguish between concurrency and interleaving relations.

Past studies that investigated the problem of discovering control-flow relations between activities by leveraging life-cycle information or execution times include: (1) a simple algorithm [13] for discovering block-structured process models from complete and noise-free event logs; (2) an extension of the α-algorithm, i.e. the β algorithm [19]; (3) an extension of Heuristics Miner [8]; and (4) the work of Senderovich et al. which explores process performance modelling via temporal network representation [14]. The first one is limited to noise-free log. The second and third are based on underlying algorithms that produce unsound and inaccurate models when applied to real-life event logs, as shown in [5]. The fourth approach is not geared to discovering process models but rather targets the problem of performance mining.

In this paper, we extend the SM algorithm, which has been shown to produce accurate and (generally) sound process models over real-life logs. Figure 2 shows an overview of how SM discovers a process model from an event log. Given an input event log, SM operates over five steps: i) discover the directly-follows graph (DFG) and loops from the event log; ii) analyse the DFG for discovering concurrency relations; iii) filter the DFG by removing the infrequent behaviour; iv) discover the split gateways; v) discover the join gateways. Each step is a standalone operation based on tailored algorithms [6], such a modular approach allows the replacement of any step with alternative methods. In this paper, we show how we updated the first, second, and fifth steps to discover true concurrency and inclusive choices, and reduce the chances of producing unsound process models via heuristics.

Fig. 2. Overview of the Split Miner approach [6].

3   Approach

In this section, we describe how we redesigned the first two steps of the Split Miner original approach [6] and integrated in the last step two heuristics to repair models that are unsound due to improper completion and identify inclusive relations between activities, enabling the discovery of OR-splits.

3.1   Refined Directly-Follows Graph Discovery

Given an event log, the first step performed by Split Miner is to sequentially read the events and build the directly-follows graph (DFG). Although this operation is straightforward, its output strictly depends on how the event log and the DFG are defined. Definitions 1, 2, and 3 capture the notion of DFG used in the original Split Miner.

Definition 1. *[Event Log as in [6]] Given a set of process activity labels $\mathscr{A}$, an event log $\mathscr{L}$ is a multiset of traces, where a trace $t \in \mathscr{L}$ is a sequence of activity labels $t = \langle a_1, a_2, \ldots, a_k \rangle$, with $a_i \in \mathscr{A}, 1 \leq i \leq k$. In addition, we use the notation $a \in \mathscr{A}$ to refer an activity a that belongs to a generic trace $t \in \mathscr{L}$.[1]*

Definition 2. *[Directly-Follows Relation as in [6]] Given an event log $\mathscr{L}$ and two process activities $a_x, a_y \in \mathscr{A}$, we say that activity a_y directly-follows activity a_x, with notation $a_x \rightsquigarrow a_y$, if and only if (iff) $\exists \langle a_1, a_2, \ldots, a_k \rangle \in \mathscr{L} \mid a_i = a_x \land a_j = a_y \land j = i+1 \land 0 < i < n$.*

---
[1] For simplicity, we use the term activity to refer to its label.

Definition 3. *[Directly-Follows Graph as in [6]] Given an event log $\mathscr{L}$, its Directly-Follows Graph (DFG) is a directed graph $\mathscr{G} = (N, E)$, where N is the non-empty set of nodes, where each node represents a unique activity $a \in \mathscr{L}$ and there exists a bijective function $\lambda : N \mapsto \mathscr{A}$ such that $\lambda(n)$ retrieves the activity n refers to; and E is the set of edges capturing the directly-follows relations of the activities observed in $\mathscr{L}$, $E = \{(n, m) \in N \times N \mid \lambda(n) \rightsquigarrow \lambda(m)\}$.*

To capture the activities' lifecycle information, we refine the concept of event log.

Definition 4. *[Refined Event Log] Given a set of events $\mathscr{E}$, a refined event log $\mathscr{L}_\rho$ is a multiset of traces, where a trace $t \in \mathscr{L}_\rho$ is a sequence of events $t = \langle e_1, e_2, \ldots, e_k \rangle$, with $e_i \in \mathscr{E}, 1 \leq i \leq k$. Each event $e \in \mathscr{E}_\rho$ is a tuple $e = (l, p, t)$, where $l \in \mathscr{A}$ is the process activity the event refers to, retrieved with the notation e^l; $p \in \{start, end\}$ is the state of the life-cycle of activity l, retrieved with the notation e^p; and t is the timestamp of the event, retrieved with the notation e^t.*

While redefining the event log to capture the activities' life-cycle information is intuitive and follows from its original definition [16], the same does not apply for the DFG. Indeed, more than one approach could be used to generate a DFG from a refined event log. The simplest approach would be to disregard all the events of a specific state of an activity life-cycle, for example, we could remove from $\mathscr{L}_\rho$ all the events $e \in \mathscr{L}_\rho \mid e^p = start$ or all the events $e \in \mathscr{L}_\rho \mid e^p = end$. Then, the refined event log would turn into an event log (Definition 1) and the DFG would be constructed according to Definition 3, but this would be equivalent to discarding the activities' lifecycle information.

An alternative approach was proposed by Leemans et al. [11] and incorporated into a variant of the Inductive Miner that takes into account lifecycle transitions, herein called Inductive Miner Lifecycle (IM-lc). According to [11], an activity a_y directly-follows an activity a_x if any of the life-cycle states of activity a_y is observed after any of the life-cycle states of activity a_x in the same trace and between the two observations no activity completes the execution of its full life-cycle (see Definition 5).

Definition 5. *[Directly-Follows Relation as in [11]] Given a refined event log $\mathscr{L}_\rho$ and two process activities $a_x, a_y \in \mathscr{A}$, the relation $a_x \rightsquigarrow a_y$ holds iff $\exists \langle e_1, e_2, \ldots, e_k \rangle \in \mathscr{L}_\rho \mid e^l_i = a_x \wedge e^l_j = a_y \wedge i < j \wedge \nexists n, m \in \,]\, i, j\, [\, \mid n < m \wedge e^p_n = start \wedge e^p_m = end \wedge e^l_n = e^l_m$.*

According to Definition 5, a directly-follows relation would hold between two activities whose life-cycles overlap (i.e. the start-state of an activity is observed between

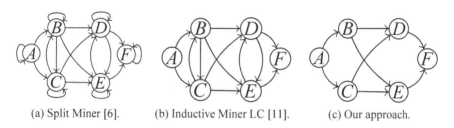

| (a) Split Miner [6]. | (b) Inductive Miner LC [11]. | (c) Our approach. |

Fig. 3. Examples of discovered DFGs by applying Definition 2, 5, and 6 (left to right).

the start-state and the end-state of another activity). While this is important and useful for IM-lc to discover concurrency relations [10], it would not be beneficial for Split Miner, since Split Miner requires to remove the directly-follows relations between activities that are considered concurrent [6]. Consequently, we are interested in discarding directly-follows relations of activities whose life-cycles overlap. We redefine the directly-follows relation of activities observed in a refined event log as follows. An activity a_y directly-follows an activity a_x if the start-state of the life-cycle of activity a_y is observed after the end-state of the life-cycle of activity a_x and no end-state of other activities are observed in between (see Definition 6).

Definition 6. *[Directly-Follows Relation] Given a refined event log $\mathscr{L}_\rho$ and two process activities $a_x, a_y \in \mathscr{A}$, the relation $a_x \leadsto_r a_y$ holds iff $\exists \langle e_1, e_2, \ldots, e_k \rangle \in \mathscr{L}_\rho \mid e^l_i = a_x \wedge e^l_j = a_y \wedge e^p_i = end \wedge e^l_j = start \wedge 1 \leq i < j \leq k \wedge \nexists n \in \,]\,i,j\,[\mid e^p_n = end.$*

The new version of Split Miner we propose in this paper relies on Definition 6. Depending on the definition of directly-follows relation that one adopts when generating the DFG, one may discover very different DFGs. As an example, let us consider the following refined event log (captured as a collection of traces, where each event is represented as the activity it refers to – including its life-cycle state as subscript, s standing for *start* and e standing for *end*): $\mathscr{L}_{\rho_x} =$
$\{\langle A_s, A_e, B_s, C_s, C_e, B_e, E_s, D_s, D_e, E_e, F_s, F_e \rangle, \langle A_s, A_e, B_s, C_s, B_e, C_e, E_s, D_s, E_e, D_e, F_s, F_e \rangle,$
$\langle A_s, A_e, C_s, B_s, B_e, C_e, D_s, E_s, D_e, E_e, F_s, F_e \rangle, \langle A_s, A_e, C_s, B_s, C_e, B_e, D_s, E_s, E_e, D_e, F_s, F_e \rangle\};$
Fig. 3 shows the DFGs discovered from the $\mathscr{L}_{\rho_x}$ by applying Definition 2 (original Split Miner approach), Definition 5 (Inductive Miner life-cycle approach), and Definition 6 (this paper approach).

3.2 Refined Concurrency Discovery

The second step of the original Split Miner that we redesigned is the concurrency discovery. Split Miner relies on a simple heuristic to discover concurrency, precisely, given a DFG and two activities $A, B \in \mathscr{A}$ such that neither A nor B is a self-loop, A and B are assumed concurrent iff three conditions are true: A directly-follows B and B directly-follows A (Relation 1); A and B do not form a short-loop (Relations 2 and 3); the frequency of the two directly-follows relations $A \leadsto B$ and $B \leadsto A$ is similar (Relation 2).[2]

$$A \leadsto B \wedge B \leadsto A \tag{1}$$

$$\nexists \langle a_1, a_2, \ldots, a_k \rangle \in \mathscr{L} \mid a_i = A \wedge a_{i+1} = B \wedge a_{i+2} = A \ \wedge \ i \in [1, k-2] \tag{2}$$

$$\nexists \langle a_1, a_2, \ldots, a_k \rangle \in \mathscr{L} \mid a_i = B \wedge a_{i+1} = A \wedge a_{i+2} = B \ \wedge \ i \in [1, k-2] \tag{3}$$

$$\frac{||A \leadsto B| - |B \leadsto A||}{|A \leadsto B| + |B \leadsto A|} < \varepsilon \quad (\varepsilon \in [0,1]) \tag{4}$$

The simplicity of the concurrency oracle of Split Miner derives from the simplicity of the input event log (see Definition 1). However, when receiving as input a refined

---

[2] The frequency of a directly-follows relation is the number of times the relation is observed.

event log (Definition 4), it is possible to identify true concurrency by focusing on activities whose life-cycles overlap and are hence truly executed concurrently (e.g. by different process resources). Consequently, we redefine the concurrency discovery oracle as follows. Given two activities $A, B \in \mathscr{A}$ and a refined event log $\mathscr{L}_\rho$, we say A and B are concurrent if the following relation holds:

$$2 \cdot \frac{|A \asymp B|}{|A| + |B|} \geq \varepsilon \quad (\varepsilon \in [0,1]) \tag{5}$$

where $|A \asymp B|$ is the total number of observations of overlapping life-cycles of A and B in $\mathscr{L}_\rho$; $|A|$ and $|B|$ are respectively the total number of complete life-cycle[3] observations of activity A and activity B in $\mathscr{L}_\rho$; and ε is an arbitrary variable (given as input parameter) defining the minimum percentage of times that the two activities' life-cycles are required to overlap to assume the two activities concurrent. In particular, when $\varepsilon = 1$ our notion of concurrency is equivalent to the notion of *strong simultaneousness* defined by Van der Werf et al. [18] as well as Allen's interval relations [3] of *overlaps, contains, starts,* and *is finished by*. While for any other value of $\varepsilon > 0$ it is equivalent to a parametrized notion of *weak simultaneousness* [18]. Given that real-life event logs often contain noise and infrequent process behaviour, requiring $\varepsilon = 1$ would be very restrictive and may lead to the discovery of no concurrent activities.

Although both our approach and IM-lc infer concurrency relations between activities from the observation of overlapping life-cycles, we rely on an heuristic before validating the concurrency relations (i.e. Eq. 5) – in-line with the original Split Miner; while IM-lc assumes the information contained in the log to be valid a priori (this is mitigated by another extension of IM-lc that embeds a filtering technique [11]).

3.3 Heuristic Improvement

Although Split Miner guarantees to discover sound acyclic process models and *deadlock-free* cyclic process models with *no dead activities*, for cyclic process models it does not guarantee *proper completion*. However, it is possible to reduce the chances to discover process models exhibiting improper completion by applying the following heuristic: for each AND-split gateway in a process model with an outgoing edge that is a loop-edge (leading to a topologically deeper node of the process model), we create a preceding XOR-split gateway and set this latter as source of the loop-edge. Figure 4 intuitively show how the heuristic operates, the loop-edge is highlighted in blue and, in general, activities could be present in the loop-edge.

Lastly, we integrated an heuristic to discern between concurrency and inclusive relations, in other words identifying when an AND-split gateway is a candidate OR-split gateway. This second heuristic operates as follows. For each AND-split gateway in a process model, we consider all the successor activities and we check pairwise whether there exist traces where the pair of activities are mutually exclusive (i.e. one of the two activities is executed but not the other). Then, if the majority of the pairs of activities are both mutually exclusive and concurrent in different traces,[4] we turn

---

[3] E.g. including start and end states.

[4] With at least one observation of mutual exclusiveness every two observations of concurrency or vice-versa.

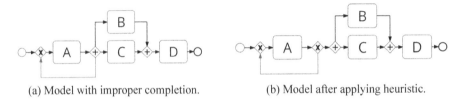

(a) Model with improper completion.   (b) Model after applying heuristic.

Fig. 4. Heuristic removal of improper completion generated by loops.

the AND-split gateway into an OR-split gateway and we update accordingly the OR-join gateway. As an example, let us consider the model in Fig. 5a and the event log $\mathscr{L}_{\rho_y} = \{\langle A_s, A_e, B_s, C_s, D_s, B_e, D_e, C_e, E_s, E_e\rangle^3, \langle A_s, A_e, C_s, D_s, C_e, D_e, E_s, E_e\rangle^2,$ $\langle A_s, A_e, B_s, D_s, D_e, B_e, E_s, E_e\rangle, \}$; B and C are observed three times concurrently and three times are mutually exclusive, B and D are observed four times concurrently and two times mutually exclusive, C and D are observed five times concurrently and one mutually exclusive. Given that two pairs of activities out of three (B,D and B,C) are eligible for inclusiveness, we turn the AND gateways into OR gateways (Fig. 5b).

4   Evaluation

In this section, we present an empirical evaluation that compares Split Miner 2.0 (SM$_{2.0}$) with three state-of-the-art automated process discovery algorithms: the original Split Miner [6] (SM), the Inductive Miner Lifecycle (IM-lc) [10] including its infrequent behaviour filter [11], and the most recent version of IM, namely IMfa [9].

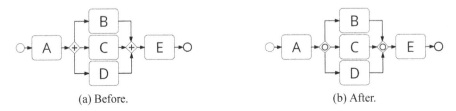

(a) Before.   (b) After.

Fig. 5. Heuristic identification of OR-split gateways.

4.1   Dataset and Setup

As testing dataset, we selected eleven real-life event logs (L1–L11) containing activity lifecycle information. The logs were sourced from companies operating in different fields (e.g. insurance, manufacturing, banking) and geographic areas (i.e. Europe and Australia). Given that these logs are not publicly available, we added a publicly available simulated event log known as the "Repair example" (R-Log),[5] which also contains

---

[5] http://www.promtools.org/prom6/downloads/example-logs.zip.

Table 2. Descriptive statistics of the logs.

Event log	Total traces	Distinct traces	Total events	Distinct events	Trace length		
					MIN	AVG	MAX
L1	28,504	2.64%	443,862	23	4	15	1230
L2	3,885	9.11%	15,096	6	2	3	60
L3	954	10.80%	13,740	18	6	14	46
L4	37	86.49%	1,156	18	22	31	36
L5	146	78.08%	3,764	18	2	25	84
L6	551	96.55%	19,174	80	2	34	126
L7	70,512	0.28%	830,522	8	4	11	40
L8	9,906	2.19%	9,906	26	6	44	354
L9	1,182	92.81%	46,282	9	12	39	276
L10	608	11.51%	18,238	21	4	2	88
L11	1,214	20.18%	11,226	12	4	9	58
R-Log	1,104	5.53%	15,468	8	6	14	30

activity lifecycle information. We did not include the BPIC12 and BPIC17 logs simply because the former does not have any overlapping lifecycle, and for the latter both SM and $SM_{2.0}$ produced the same model, which was analysed in previous studies [4–6].

Table 2 displays the characteristics of the event logs, highlighting their variety, with logs containing short to long traces (length 2 to 1,230), a wide range of distinct traces (0.28% to 96.55%) and distinct events (6 to 80), as well as notable differences in the total number of traces (37 to 70,512) and events (1,156 to 830,522). The lifecycle information for each activity recorded in these event logs was complete, i.e. the start and end events were recorded for each activity.

From each log, we discovered a process model with $SM_{2.0}$, SM, and IM-lc, and compared the quality of the discovered models over three quality measures: fitness, precision, and simplicity. Several methods have been proposed for measuring fitness and precision of an automatically discovered process model [15]. In this paper we use two methods, the one proposed by Adriansyah et al. [1,2] (alignment-based accuracy) and the one proposed by Augusto et al. [4] (Markovian accuracy). As proxy for simplicity we use the following three metrics [12]: *Size* – the total number of nodes of a process model; *Control-flow complexity* (CFC) – the amount of branching induced by the split gateways in the process model; *Structuredness* – the percentage of nodes located inside a single-entry single-exit fragment of the process model.

We implemented $SM_{2.0}$ as a Java command-line application,[6] and we ran the experiments on an Intel Core i7-8565U @ 1.80 GHz with 32 GB RAM running Windows 10 Pro (64-bit) and JVM 8 with 14 GB of allocated RAM (10 GB Stack and 4 GB Heap). All the discovery algorithms (SM, $SM_{2.0}$, and IM-lc) were executed using their default

---

[6] Available as "Split Miner 2.0" at http://apromore.org/platform/tools.

input parameters, and we set a timeout of 30 min for each algorithm execution and for each measurement.

4.2 Results

Table 3 reports the fitness, precision, and simplicity measurements. Due to space limits, the table does not show the measurements for IMfa because they were either equal or worse than those for IM-lc, with the exception of those obtained on L9 (which reported a slight improvement).

We can observe that $SM_{2.0}$ is less prone to discovering unsound models than SM, with the latter discovering an unsound model every three and the former only discovering sound models. This achievement reflects the effectiveness of our heuristics for removing improper completion.

In terms of accuracy, the results obtained with the alignment-based accuracy and the Markovian accuracy are partially consistent in line with previous findings [4]. In fact, the two measuring methods agree on the best models in terms of fitness, precision, and F-score, respectively 100%, 66%, and 75% of the times.

As for fitness, IM-lc outperforms both SM and $SM_{2.0}$ as expected [5]. In terms of precision and F-score $SM_{2.0}$ and SM achieve the highest scores, with $SM_{2.0}$ performing better than SM, most of the times discovering more precise and fitting process models ultimately achieving a higher F-score. In fact, $SM_{2.0}$ accuracy scores are either higher than or equal to those of SM, the latter outperforming the former in fitness or precision only two times according to the alignment-based accuracy, and only three times according to the Markovian accuracy. Compared to IM-lc, $SM_{2.0}$ discovers eleven times more precise process models, independently of the measurement method.

As for simplicity, $SM_{2.0}$ stands out by producing smaller models than those discovered by both SM and IM-lc (9 times out of 12) and with a lower CFC (10 times out of 12). However, $SM_{2.0}$ and SM cannot systematically produce fully-structured process models as opposed to IM-lc which achieves this by design. Lastly, the execution times of IM-lc, SM, and $SM_{2.0}$ are negligible: all the process models were discovered within a minute (except for log L7, where IM-lc timed out).

Figure 6 shows two qualitative examples of the improvements yielded by $SM_{2.0}$. Considering the models from the L6 log (Figs. 6a and 6b), $SM_{2.0}$ discovered the inclusive-OR relations between several activities of the process and removed the improper completion, while SM produced an unsound model. In the specific case of the L6 log, we also had the chance to validate the discovered model with the process analysts of the organization this log was extracted from, who confirmed that the activities were indeed in an inclusive-OR relation. Considering the models from the R-log (Figs. 6c and 6d), only $SM_{2.0}$ discovers the concurrency relations between its activities, while SM mixes us the concurrency relations with loops.

Table 3. Experiment results.

Event log	Discovery approach	Alignment accuracy [1,2]			Markovian accuracy [4]			Simplicity		
		Fitness	Precision	F-score	Fitness	Precision	F-score	Size	CFC	Struct.
L1	IM-lc	0.88	0.78	0.83	0.82	0.15	0.25	**40**	26	**1.00**
	SM	**0.98**	0.94	**0.96**	**0.96**	**0.44**	**0.60**	47	32	0.47
	SM$_{2.0}$	0.83	**0.97**	0.90	0.44	0.34	0.38	45	**25**	0.56
L2	IM-lc	0.87	0.44	0.59	0.53	0.14	0.22	20	11	**1.00**
	SM	**0.92**	**1.00**	**0.96**	**0.69**	**0.88**	**0.77**	**14**	**6**	**1.00**
	SM$_{2.0}$	**0.92**	**1.00**	**0.96**	**0.69**	**0.88**	**0.77**	**14**	**6**	**1.00**
L3	IM-lc	**0.98**	0.71	0.82	**0.88**	0.08	0.14	49	27	**1.00**
	SM	0.96	0.97	**0.96**	0.72	**0.41**	**0.52**	36	16	0.58
	SM$_{2.0}$	0.93	**0.99**	**0.96**	0.40	0.07	0.12	**31**	**10**	0.77
L4	IM-lc	**0.98**	0.41	0.57	**1.00**	0.06	0.12	35	12	**1.00**
	SM	0.84	**1.00**	**0.91**	0.45	**0.79**	**0.57**	26	6	0.46
	SM$_{2.0}$	0.94	0.66	0.78	0.93	0.08	0.15	**25**	**3**	**1.00**
L5	IM-lc	**0.83**	0.53	0.65	**0.90**	0.17	0.29	33	12	**1.00**
	SM	*Unsound*			*Unsound*			31	11	0.45
	SM$_{2.0}$	0.76	**0.79**	**0.78**	0.86	**0.19**	**0.31**	**27**	**3**	0.59
L6	IM-lc	*Measurements timeout*			0.10	0.00	0.01	**126**	**78**	**1.00**
	SM	*Unsound*			*unsound*			161	98	0.14
	SM$_{2.0}$	**0.70**	**0.66**	**0.68**	**0.31**	**0.23**	**0.26**	138	80	0.50
L7	IM-lc	*Discovery timeout*			*Discovery timeout*			*Discovery timeout*		
	SM	**0.88**	**1.00**	**0.94**	**0.73**	**0.90**	**0.81**	**12**	**2**	**1.00**
	SM$_{2.0}$	**0.88**	**1.00**	**0.94**	**0.73**	**0.90**	**0.81**	**12**	**2**	**1.00**
L8	IM-lc	**0.85**	0.40	0.55	**0.87**	0.03	0.06	61	39	**1.00**
	SM	*Unsound*			*Unsound*			160	118	0.02
	SM$_{2.0}$	0.77	**0.76**	**0.77**	0.38	**0.33**	**0.35**	**46**	26	0.70
L9	IM-lc	**0.94**	0.26	0.41	**0.92**	0.43	**0.58**	23	11	**1.00**
	SM	*Unsound*			*Unsound*			17	5	0.53
	SM$_{2.0}$	0.57	**0.91**	**0.70**	0.28	**0.45**	0.35	17	5	0.47
L10	IM-lc	**0.95**	0.75	0.84	**0.98**	0.15	0.26	31	8	**1.00**
	SM	0.77	**1.00**	**0.87**	0.95	**0.93**	**0.94**	**29**	**6**	**1.00**
	SM$_{2.0}$	0.77	**1.00**	**0.87**	0.95	**0.93**	**0.94**	**29**	**6**	**1.00**
L11	IM-lc	**0.91**	0.75	0.82	**0.45**	0.14	0.22	36	21	**1.00**
	SM	0.83	**0.90**	**0.87**	0.29	0.26	**0.28**	44	28	0.16
	SM$_{2.0}$	0.84	**0.90**	**0.87**	0.06	**0.33**	0.10	**22**	**11**	0.59
R-Log	IM-lc	**0.99**	0.98	**0.99**	**1.00**	0.96	**0.98**	16	5	**1.00**
	SM	0.91	**0.99**	0.95	0.45	0.83	0.59	**14**	**4**	0.36
	SM$_{2.0}$	0.98	0.97	0.98	0.94	**0.98**	0.96	16	5	0.50

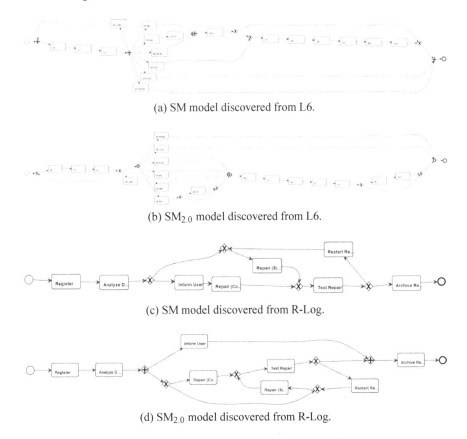

(a) SM model discovered from L6.

(b) $SM_{2.0}$ model discovered from L6.

(c) SM model discovered from R-Log.

(d) $SM_{2.0}$ model discovered from R-Log.

Fig. 6. Models discovered by SM and $SM_{2.0}$ from the L6 and R-Log.

5 Conclusion

In this paper, we presented Split Miner 2.0 ($SM_{2.0}$), an extension of Split Miner (SM) [6] that exploits the activities' start and end timestamps recorded in an event log to discover true concurrency and inclusive choice relations between activities. This is achieved by redesigning the discovery of a directly-follows graph from an event log, adapting the concurrency notion of Van der Werf et al. [18], and introducing an intuitive heuristic to identify inclusive relations. Furthermore, given that SM cannot guarantee sound process models, we designed an heuristic that reduces the chances of discovering process models exhibiting improper completion. The empirical evaluation shows that $SM_{2.0}$ can discover more concurrent relations than SM, remove improper completion, and identify OR-splits, while preserving SM's model accuracy and reducing the complexity.

Although several studies have investigated the problem of automated process discovery from event logs [5], most of them operate on simple event logs with only three attributes: case id, timestamp, and activity label. Future research work in this area may focus on designing more sophisticated automated process discovery algorithms that can discover more complex BPMN process models by leveraging additional information that may be available in real-life event logs. Another direction for future work is to design accuracy measures such as fitness and precision that go beyond simple control-flow relations and include support for inclusive gateways, including the OR-join.

Acknowledgments. Research funded by the Australian Research Council (grant DP180102839) and the Estonian Research Council (grant PRG887).

References

1. Adriansyah, A., Munoz-Gama, J., Carmona, J., van Dongen, B.F., van der Aalst, W.M.P.: Alignment based precision checking. In: La Rosa, M., Soffer, P. (eds.) BPM 2012. LNBIP, vol. 132, pp. 137–149. Springer, Heidelberg (2013). https://doi.org/10.1007/978-3-642-36285-9_15
2. Adriansyah, A., Munoz-Gama, J., Carmona, J., van Dongen, B., van der Aalst, W.M.P.: Measuring precision of modeled behavior. ISeB **13**(1), 37–67 (2015)
3. Allen, J.F.: Maintaining knowledge about temporal intervals. Commun. ACM **26**(11), 832–843 (1983)
4. Augusto, A., Armas Cervantes, A., Conforti, R., Dumas, M., Rosa, L.: Measuring fitness and precision of automatically discovered process models: a principled and scalable approach. IEEE TKDE (2020, to appear)
5. Augusto, A., et al.: Automated discovery of process models from event logs: review and benchmark. IEEE TKDE **31**(4), 686–705 (2019)
6. Augusto, A., Conforti, R., Dumas, M., La Rosa, M., Polyvyanyy, A.: Split miner: automated discovery of accurate and simple business process models from event logs. Knowl. Inf. Syst. **59**(2), 251–284 (2018). https://doi.org/10.1007/s10115-018-1214-x
7. Buijs, J.C.A.M., van Dongen, B.F., van der Aalst, W.M.P.: On the role of fitness, precision, generalization and simplicity in process discovery. In: Meersman, R., et al. (eds.) OTM 2012. LNCS, vol. 7565, pp. 305–322. Springer, Heidelberg (2012). https://doi.org/10.1007/978-3-642-33606-5_19
8. Burattin, A., Sperduti, A.: Heuristics miner for time intervals. In: ESANN (2010)
9. Leemans, S.J.J., Fahland, D.: Information-preserving abstractions of event data in process mining. Knowl. Inf. Syst. **62**(3), 1143–1197 (2019). https://doi.org/10.1007/s10115-019-01376-9
10. Leemans, S.J.J., Fahland, D., van der Aalst, W.M.P.: Discovering block-structured process models from event logs containing infrequent behaviour. In: Lohmann, N., Song, M., Wohed, P. (eds.) BPM 2013. LNBIP, vol. 171, pp. 66–78. Springer, Cham (2014). https://doi.org/10.1007/978-3-319-06257-0_6
11. Leemans, S.J.J., Fahland, D., van der Aalst, W.M.P.: Using life cycle information in process discovery. In: Reichert, M., Reijers, H.A. (eds.) BPM 2015. LNBIP, vol. 256, pp. 204–217. Springer, Cham (2016). https://doi.org/10.1007/978-3-319-42887-1_17
12. Mendling, J.: Metrics for Process Models: Empirical Foundations of Verification, Error Prediction, and Guidelines for Correctness. LNBIP, vol. 6. Springer, Heidelberg (2008). https://doi.org/10.1007/978-3-540-89224-3
13. Schimm, G.: Mining exact models of concurrent workflows. Comput. Ind. **53**(3), 265–281 (2004)
14. Senderovich, A., Weidlich, M., Gal, A.: Temporal network representation of event logs for improved performance modelling in business processes. In: Carmona, J., Engels, G., Kumar, A. (eds.) BPM 2017. LNCS, vol. 10445, pp. 3–21. Springer, Cham (2017). https://doi.org/10.1007/978-3-319-65000-5_1

15. Syring, A.F., Tax, N., van der Aalst, W.M.P.: Evaluating conformance measures in process mining using conformance propositions. In: Koutny, M., Pomello, L., Kristensen, L.M. (eds.) Transactions on Petri Nets and Other Models of Concurrency XIV. LNCS, vol. 11790, pp. 192–221. Springer, Heidelberg (2019). https://doi.org/10.1007/978-3-662-60651-3_8
16. van der Aalst, W.M.P.: Process Mining - Data Science in Action. Springer, Heidelberg (2016). https://doi.org/10.1007/978-3-662-49851-4
17. van der Aalst, W.M.P., et al.: Soundness of workflow nets: classification, decidability, and analysis. Formal Asp. Comput. 23(3), 333–363 (2011)
18. van der Werf, J.M.E.M., Mans, R., van der Aalst, W.M.P.: Mining declarative models using time intervals. In: PNSE+ ModPE, pp. 313–331. Citeseer (2013)
19. Wen, L., Wang, J., van der Aalst, W.M.P., Huang, B., Sun, J.: A novel approach for process mining based on event types. J. Intell. Inf. Syst. 32(2), 163–190 (2009)

A Novel Approach to Discover Switch Behaviours in Process Mining

Yang Lu$^{(\boxtimes)}$ ⓘ, Qifan Chen ⓘ, and Simon Poon ⓘ

School of Computer Science, The University of Sydney, 2006 Sydney, NSW, Australia
{yalu8986,qche8411}@uni.sydney.edu.au, simon.poon@sydney.edu.au

Abstract. Process mining is a relatively new subject which builds a bridge between process modelling and data mining. An exclusive choice in a process model usually splits the process into different branches. However, in some processes, it is possible to switch from one branch to another. The inductive miner guarantees to return sound process models, but fails to return a precise model when there are switch behaviours between different exclusive choice branches due to the limitation of process trees. In this paper, we present a novel extension to the process tree model to support switch behaviours between different branches of the exclusive choice operator and propose a novel extension to the inductive miner to discover sound process models with switch behaviours. The proposed discovery technique utilizes the theory of a previous study to detect possible switch behaviours. We apply both artificial and publicly-available datasets to evaluate our approach. Our results show that our approach can improve the precision of discovered models by 36% while maintaining high fitness values compared to the original inductive miner.

Keywords: Process discovery · Complex behaviours detection · Switch behaviours · Inductive miner · Process trees

1 Introduction

Process mining is useful for analyzing business processes along with improving and predicting which contains three parts – process discovery, conformance checking and process enhancement [1]. The most critical part of process mining is process discovery, which aims at extracting insight of the system workflow from real data. The resulting process model should not only have a high fitness value, but also be an accurate representation of the real process [2]. The inductive miner is one of the leading process discovery algorithms which can guarantee to produce sound process models within finite time [1, 3]. Given the direct outcome of the inductive miner is a process tree [3], the behaviours being represented are limited. When giving complex event logs as input, the inductive miner often returns so-called "flower models" which preserve high fitness but have very low precision values [3, 4]. Although we can still replay the majority of traces on the process model, "flower models" fail to represent real processes accurately and precisely [1, 2].

© Springer Nature Switzerland AG 2021
S. Leemans and H. Leopold (Eds.): ICPM 2020 Workshops, LNBIP 406, pp. 57–68, 2021.
https://doi.org/10.1007/978-3-030-72693-5_5

When dealing with an exclusive decision choice in a process model, the decision point is split into multiple branches [5]. However, in many real-life processes, it can be possible to switch between branches after a decision has been made. Although the inductive miner is known to be useful in generating sound models from data, it fails to discover an accurate model when switch behaviours exist.

In this paper, we propose a novel extension to the process tree model to handle switch behaviours between different exclusive choice branches. We then develop a novel extension to the inductive miner to discover sound process models with switch behaviours based on the theory in [6]. From a broader perspective, our proposed method not only guarantees to produce sound process models but also not being limited to produce block-structured process models. We apply both artificial and publicly-available datasets to evaluate our approach. Fitness, precision and F-score [4, 7] are used to measure the accuracy of resulting models, size (the number of nodes) and CFC (the number of branching caused by split gateways) [8] are adopted to measure the model complexity.

The rest of the paper is structured as follows: Sect. 2 is a literature review of related work. Section 3 introduces formal definitions of some terms. Section 4 introduces the extension to the process tree model and how to translate it into a workflow net. In Sect. 5, we describe our process discovery technique. The approach is evaluated in Sect. 6. We finally conclude our paper in Sect. 7.

2 Background

When modelling switch behaviours between different exclusive choice branches using Petri-nets, a hidden transition is needed since we cannot connect two places directly [6]. The classical alpha algorithm [9] cannot discover any hidden transitions. [6, 10] improve the classical alpha algorithm to allow the detection of invisible tasks. Although the alpha algorithms are not robust to noises and cannot guarantee to produce sound models. [6] proposes a heuristic for detecting invisible transitions between activities directly from event logs. If there is a hidden transition between two activities on different exclusive choice branches, a switch behaviour is detected.

In reviewing other process discovery algorithms which can discover switch behaviours between different exclusive choice branches including the alpha algorithms with invisible tasks [6, 10], heuristics miners [11], genetic miners [12] and the ILP algorithm [13], none of them can guarantee to produce a sound process model. In addition, some of them cannot handle noises, thus, not suitable to be applied to real data. Although the split miner [7] can discover switch behaviours and guarantee to produce deadlock-free models. It still cannot guarantee to produce sound models as defined in [9], which defines soundness as (a) safeness, (b) proper completion, (c) option to complete, (d) absence of dead tasks.

The inductive miners are a family of process discovery algorithms which utilize the divide-and-conquer approach in the field of process discovery [3, 14–18]. The inductive miners recursively divide the activities into different partitions and split event logs until base cases are touched. The direct outcomes of the inductive miners are process trees, which can be easily translated into equivalent block-structured workflow nets [3]. An important feature of the inductive miner family is that the resulting model is always sound regardless of the input log. However, process trees also limit the behaviours which can be represented. For example, they fail to represent switch behaviours between exclusive choice branches.

When the given event log is complex, the inductive miner [3] can easily return a "flower model" with high fitness but low precision. [14] removes infrequent relations between activities before partitioning the activities. However, according to the benchmark in both [4] and [7]. The inductive miner infrequent (IMf) in [14] still returns models with low precision values compared with other algorithms. [19] tries to solve the problem by giving duplicate labels to the same activity when a local "flower model" is returned. The algorithm successfully improves the precision of the outcome models but leads to longer execution time. Besides, if we apply the algorithm in [19] with the inductive miners, the outcome models are still block-structured workflow nets.

The process mining framework in the original inductive miner [3] allows researchers to define their ways to partition activities and customized process tree semantics. For example, [17] puts lifecycle information into the process discovery to distinguish "interleaving" behaviours from "parallel" behaviours. [18] defines new operators on the process tree and uses the inductive miner to discover cancellation behaviours.

3 Preliminaries

In this section, we present some formal definitions which will be used in this paper. For process trees and block-structured workflow nets, we refer to [3], for soundness of Petri-nets, we refer to [9]. Besides, for IWF-net (workflow nets with invisible tasks), DIWF-net (a subset of IWF-nets) and log completeness, we refer to [6]. For clarification, in this paper, we use "X" to represent the exclusive choice operator, "$\rightarrow$" to represent the sequence operator, "$\wedge$" to represent the parallel operator and "$\circlearrowright$" to represent the loop operator in the process tree [3].

Definition 1 (Relations between activities). Let L be an event log of a workflow net N, let a, b be two activities in L. Then:

1. $a >_L b$ if there is a trace $t \in L$ where t $= < \ldots\ldots, a, b, \ldots\ldots >$,
2. $a \sim_L b$ if there is a trace $t \in L$ where t $= < \ldots\ldots, a, b, a, \ldots\ldots >$, and there is a trace $t \in L$ where t $= < \ldots\ldots, b, a, b, \ldots\ldots >$,
3. $a \rightarrow_L b$ if $a >_L b \wedge (b \not>_L a \vee a \sim_L b)$,
4. $a ||_L b$ if $a >_L b \wedge b >_L a \wedge a \not\sim_L b$.

Definition 2 (Mendacious dependency) [6]. Let N $= (P, T_v \cup T_{iv}, F)$ be a potential sound IWF-net, T_v is the set of visible tasks, T_{iv} is the set of invisible tasks. There is a mendacious dependency between activities a, b in event log L, denoted as $a \rightsquigarrow_L b$, iff $a \rightarrow_L b \wedge \exists x, y \in T_v : a \rightarrow_L x \wedge y \rightarrow_L b \wedge y \not>_L x \wedge x \not\rightarrow_L b \wedge a \not\rightarrow_L y$.

4 The Switch Process Tree

In this section, we formally define the switch behaviour and its corresponding representation on the process tree. The switch process tree is a novel extension based on the process tree model described in [3].

Definition 3 (First, Path function). Let n be a leaf in a process tree, tp be an arbitrary operator type. First (n, tp) refers to the first ancestor node of n with operator type tp. For example, in the process tree shown in Fig. 1, First (Node 3, X) refers to the root node. First $(n, tp) = \varnothing$ if none of the ancestor nodes of n has type tp. Let n_1, n_2 be two arbitrary nodes of a process tree, Path (n_1, n_2) refers to the path from n_1 to n_2 (excluding n_1 and n_2). Path $(n_1, n_2) = \varnothing$ if n_2 is not reachable from n_1 or $n_1 = n_2$. Referring back to Fig. 1, Path (Node 0, Node 8) = <Node 2>, Path (Node 1, Node 8) = $\varnothing$.

Definition 4 (Switch process tree and switch behaviour). Assume a finite alphabet A of activities. A switch process tree is a normal process tree with switch leaf operators $a \Rightarrow B$ where $a \in A$, $B \subset A$. Combined with an exclusive choice operator X, the novel leaf node denotes the place we execute activity a, and have an option to switch to one of the activities in set B on another branch of an exclusive choice operator. $a \Rightarrow b$ is a switch behavior if there exists $a \Rightarrow B$ such that $b \in B$, we call a the source of the switch behavior, b the destination of the switch behaviour. To ensure the process model is still sound, we define the constraints below:

1. The activities on different sides of a switch leaf node must be put on different branches of an exclusive choice operator X, i.e. we can only switch execution rights from one exclusive choice branch to another.
2. If there exists a leaf operator $a \Rightarrow B$ in the process tree, then $\forall b \in B$, First $(a \Rightarrow B, \bigwedge)$ = First $(b, \bigwedge)$, and if First $(a \Rightarrow B, \bigwedge)$ = First $(b, \bigwedge) \neq \varnothing$, then Path (First $(a \Rightarrow B, \bigwedge), a \Rightarrow B) \bigcap$ Path (First $(b, \bigwedge), b) \neq \varnothing$. i.e. we cannot switch out of a parallel branch.

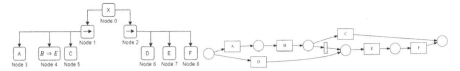

Fig. 1. An example switch process tree and its corresponding workflow net

Figure 1 shows an example switch process tree and its corresponding workflow net. There are three possible traces in the model, which are <A, B, C>, <D, E, F> and <A, B, E, F>.

Definition 5 (Translating switch process trees into workflow nets). Translating a switch process tree into a workflow net is straightforward. We first ignore the switch leaf nodes and translate the process tree into a block-structured workflow net. Then we connect the activities on different sides of the switch operators using hidden transitions. Suppose Tr is a switch process tree, S is the set of all the switch leaf nodes in Tr, Tr^* is an equivalent process tree of Tr but all the switch behaviours are removed, i.e., for all the $a \Rightarrow B \in S$ in Tr, we convert them into a in Tr^*. $N = (P, T_v \cup T_{iv}, F)$ is a block-structured workflow net corresponding to Tr^*. For each $a \Rightarrow B \in S$ and $b \in B$, we create a new invisible task t_{switch} into set T_{switch}, then:

1. If $|a\cdot| = 1$ in N, $p_{a-out} = a\cdot$, $|\cdot p_{a-out} \backslash T_{switch}| = 1$, then we add a new arc f into N, $f = (p_{a-out}, t_{switch})$.
2. If $|\cdot b| = 1$, $p_{b-in} = \cdot b$, $|p_{b-in} \cdot \backslash T_{switch}| = 1$, then we add a new arc f into N, $f = (t_{switch}, p_{b-in})$.
3. If $|a\cdot| = 1$ in N, $p_{a-out} = a\cdot$, $|\cdot p_{a-out} \backslash T_{switch}| > 1$, we first delete the arc $f_1 = (a, p_{a-out})$ from N, then we create another new invisible task t_{bridge} and place p_{bridge} into N. We finally add arcs $f_2 = (a, p_{bridge}), f_3 = (p_{bridge}, t_{bridge}), f_4 = (t_{bridge}, p_{a-out})$ and $f_5 = (p_{bridge}, t_{switch})$.
4. If $|\cdot b| = 1$, $p_{b-in} = \cdot b$, $|p_{b-in} \cdot \backslash T_{switch}| > 1$, we first delete the arc $f_1 = (p_{b-in}, b)$ from N, then we create another new invisible task t_{bridge} and place p_{bridge} into N. We finally add arcs $f_2 = (p_{bridge}, b), f_3 = (t_{bridge}, p_{bridge}), f_4 = (p_{b-in}, t_{bridge})$ and $f_5 = (t_{switch}, p_{bridge})$.
5. If $|a\cdot| > 1$, there is a "and split" after a. We add a new invisible task t_{bridge} after a as the split point and then go back to step 1, i.e., $|a\cdot| = 1$, $|t_{bridge}\cdot| > 1$, $a \cdot \cap \cdot t_{bridge} \neq \emptyset$.
6. If $|\cdot b| > 1$, there is a "and join" before b. We add a new invisible task t_{bridge} before b as the joining point and then go back to step 1, i.e., $|\cdot b| = 1$, $|\cdot t_{bridge}| > 1$, $\cdot b \cap t_{bridge} \cdot \neq \emptyset$.

To illustrate the translation process, we use three examples translated from the above different scenarios in Fig. 2, Fig. 3 and Fig. 4:

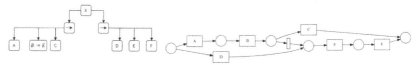

Fig. 2. An example translation from a switch process tree to a workflow net (Definition 5, case 1, 2).

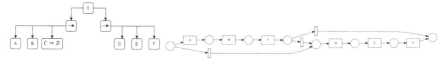

Fig. 3. An example translation from a switch process tree to a workflow net (Definition 5, case 3, 4).

Theorem 1. If we translate a switch process tree into a workflow net, the resulting workflow net is always sound if the constraints in Definition 4 are all satisfied.

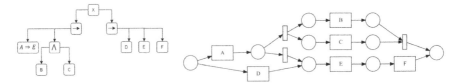

Fig. 4. An example translation from a switch process tree to a workflow net (Definition 5, case 5).

Proof. Assume we ignore all the switch behaviours in the process tree during the translation, according to [3], we can get an equivalent sound block-structured workflow net. According to Definition 5, each switch invisible transition is always connected to one single input place and one single output place. The translation process does not increase the number of input/output places of any transitions. As a result, it is free for a token in the model to choose whether firing a switch invisible transition or not. Thus, the resulting process model will not contain dead tasks and is always safe. In addition, since the original block-structured workflow net is sound, if we move a token from one exclusive choice branch to another one, the process can still be completed properly as long as we don't move the token out of a parallel branch. Thus, if the constraints in Definition 4 are all satisfied, the resulting workflow net is always sound.

5 Discovering Switch Process Trees

In [6], researchers define the prime invisible tasks into SKIP, REDO, SWITCH, INITIALIZE and FINALIZE where SWITCH refers to switching execution rights between alternative branches. Thus, the SWITCH invisible tasks can be used to represent the switch behaviours we define in Sect. 4. Researchers in [6] prove that given L is a complete event log of a sound DIWF-net $N = (P, T_v \cup T_{iv}, F)$, if $a, b \in T_v$ are two visible tasks, then there is a prime invisible task $t \in T_{iv}$ between a and b, i.e., $a \cdot \cap \cdot t \neq \emptyset$ and $t \cdot \cap \cdot b \neq \emptyset$ iff $a \leadsto_L b$. Although the scope of the proof is limited, the evaluation of [6] shows that the power of the theory is not limited to complete logs of DIWF-nets. More importantly, [6] provides us with a heuristic to predetermine possible invisible tasks between activities from event logs directly. Suppose we know two activities are on two different exclusive choice branches and there is an invisible task between them, then we know there is a switch behaviour between the two activities.

To discover switch process trees using the inductive miner, we extend the normal exclusive choice cut of the inductive miner framework to a switch exclusive choice cut. In this section, we show the switch exclusive choice cut step by step. To illustrate the process, we use a complete log of the example model presented in Fig. 2 $L_1 = \langle A, B, C \rangle, \langle D, E, F \rangle, \langle A, B, E, F \rangle$ as a running example. To make sure we detect all the switch behaviours, we put the switch exclusive choice cut before the other three cuts in each iteration. The extended IM framework is shown in Algorithm 1.

Algorithm 1. The extended IM framework augumented with switch behaviours
Input: An event log L
Output: A Switch Process Tree Tr
Discover (L)

1	**If** $BaseCase(L) ! = \phi$ **Then Return** $BaseCase(L)$
2	**Else**
3	$(cut, (\Sigma_1, \dots, \Sigma_k), switches) = SwitchExclusiveChoiceCut(G(L), L)$
4	**If** $k \leq 1$ **Then** $\big(cut, (\Sigma_1, \dots, \Sigma_k)\big) = SequenceCut(G(L))$ //cut in the original IM
5	**If** $k \leq 1$ **Then** $\big(cut, (\Sigma_1, \dots, \Sigma_k)\big) = ConcurrentCut(G(L))$ //cut in the original IM
6	**If** $k \leq 1$ **Then** $\big(cut, (\Sigma_1, \dots, \Sigma_k)\big) = LoopCut(G(L))$ //cut in the original IM
7	**If** $cut = \phi$ **Then Return** $Fallthrough(L)$ //function in the original IM
8	$NumOfActivitiesBefore = CountActivities(L)$
9	$(L_1, \dots, L_k) = SplitLog(L, (cut, (\Sigma_1, \dots, \Sigma_k)))$ //function in the original IM
10	**If** $cut = SwitchExclusiveChoiceCut$ **Then**
11	$NumOfActivitiesAfter = CountActivities(L_1, \dots, L_k)$
12	**If** $NumOfActivitiesBefore ! = NumOfActivitiesAfter$ **Then**
13	$(cut, (\Sigma_1, \dots, \Sigma_k)) = XORCut(G(L), L)$ //Exclusive choice cut of the original IM
14	**If** $k \leq 1$ **Then** $\big(cut, (\Sigma_1, \dots, \Sigma_k)\big) = SequenceCut(G(L))$ //cut in the original IM
15	**If** $k \leq 1$ **Then** $\big(cut, (\Sigma_1, \dots, \Sigma_k)\big) = ConcurrentCut(G(L))$ //cut in the original IM
16	**If** $k \leq 1$ **Then** $\big(cut, (\Sigma_1, \dots, \Sigma_k)\big) = LoopCut(G(L))$ //cut in the original IM
17	**If** $cut = \phi$ **Then Return** $Fallthrough(L)$ //function in the original IM
18	$(L_1, \dots, L_k) = SplitLog(L, (cut, (\Sigma_1, \dots, \Sigma_k)))$ //function in the original IM
19	**Return** $cut(Discover(L_1), \dots, Discover(L_k))$

5.1 The Switch Exclusive Choice Cut (Line 3)

Step 1: Adding Artificial Start and End Activities

According to Definition 5, if the source activity of a switch behaviour is at the end of an exclusive choice branch or if the destination activity of a switch behaviour is at the beginning of an exclusive choice branch, we need to add an extra invisible task before the destination activity or after the source activity to represent the process precisely. However, the process model is no longer a DIWF-net after adding the extra invisible task according to [6], so we may fail to detect an invisible task between the two activities using the mendacious dependency.

To solve the problem, before a switch cut, we first identify all the start and end activities in the event log and add a unique start and end activity to each of them. For example, after adding artificial activities into L_1, we get $L_1^* = \langle Start_A, A, B, C, End_C \rangle, \langle Start_D, D, E, F, End_F \rangle, \langle Start_A, A, B, E, F, End_F \rangle$.

Step 2: Calculating All the Mendacious Dependencies Between Activities

We then go through the event log and identify all the mendacious dependencies. Besides, we ignore all the mendacious dependencies containing artificial start and end activities. After the mendacious dependencies are identified, we delete all the artificial start and end activities.

In our example, we get one mendacious dependency, which is $B \rightsquigarrow_L E$.

Step 3: Finding Switch Exclusive Choice Cut and Switch Leaf Operators

Firstly, if there is a mendacious dependency between two activities in the directly-follows graph, we replace the edge between them with an invisible edge.

Definition 6 (Invisible edge). Given $G(L)$ is a directly-follows graph of event logs L, A is the set of activities in L. $a, b \in A$, there is an invisible edge from a to b in $G(L)$ iff $a \leadsto_L b$.

Definition 7 (Switch exclusive choice cut). Suppose E is the set of all edges in $G(L)$, E^* is the set of all invisible edges. A switch exclusive choice cut is a cut $\Sigma_1, \Sigma_2, \ldots, \Sigma_n$ of a directly-follows graph $G(L)$ such that (Fig. 5):

1. There are only invisible edges between $\Sigma_{1 \leq i \leq n}$ and $\Sigma_{1 \leq j \leq n}$.

$$\forall i \neq j \bigwedge a_i \in \Sigma_i \bigwedge a_j \in \Sigma_j : (a_i, a_j) \notin E \backslash E^*$$

Fig. 5. Switch exclusive choice cut for L_1

If an invisible edge is cut during the switch exclusive choice cut, i.e., $\exists a_i \in \Sigma_i, a_j \in \Sigma_j : (a_i, a_j) \in E^*$, then a new switch behaviour $a_i \Rightarrow a_j$ is discovered. By merging all the switch behaviours with the same source together in the end, we can get switch leaf operators.

Step 4: Removing Traces with Switch Behaviours
We use the same exclusive choice cut split function as the inductive miner infrequent [14] to split the event logs after an exclusive choice switch cut. A problem here is splitting the event log could cause extra "skip" behaviours. In our running example, since we partitioned the activities into two groups which are $\{A, B, C\}$ and $\{D, E, F\}$, the trace $<A, B, E, F>$ will be projected into either $< A, B >$ or $< E, F>$. The options will either produce an extra end activity B or an extra start activity E in the local sub-process. To resolve the issue, we consider deleting the traces with switch behaviours before splitting the log. For example, we delete $< A, B, E, F >$ from L_1 before splitting the log. However, deleting traces increases the requirement of log completeness, which may cause the loss of activities or behaviours when dealing with real-life data. We decide to make the option adjustable (Fig. 6).

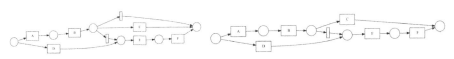

Fig. 6. The resulting model of L_1 with the deleting traces option off (left) and on (right), there is one more extra trace $< A, B >$ on the left graph.

5.2 Verifying the Exclusive Choice Switch Cut (Line 10–18)

Performing the switch exclusive choice cut and splitting the event log may cause unnecessary loss of activities. Every time after we perform the switch exclusive choice cut and split the event log, we check if the total number of activities changes. If there is a change in the number of activities (line 12), we abort the whole cut, redo the log split and disable the exclusive choice cut in the next iteration (line 13). We enable the exclusive choice cut again after the current log has been split into sub logs.

5.3 Removing Incorrect Switch Behaviours

Although we can identify switch behaviours during the exclusive choice cut, we are unable to determine if the constraints in Definition 4 are met before the whole process tree has been constructed. To ensure a sound model is returned, we iterate through the whole process tree at the end and delete any switch behaviours which violate the constraints defined in Definition 4.

6 Evaluation

We implement our approach on the inductive miner directly in the ProM framework [20]. Our code and evaluation results are available at https://github.com/bearlu1996/switch. We applied both artificial and publicly-available event data to evaluate our algorithm. Fitness and precision are used to evaluate the accuracy of our process models. Besides, we use the formula in [4] and [7] to calculate F-score, i.e., $F - Score = 2 * \frac{fitness * precision}{fitness + precision}$. CFC (the number of branching caused by split gateways) [8] and size (the number of nodes) are also used to evaluate the complexity of our process models. For replicable purposes, we use "Replay a log on Petri net for conformance analysis" in ProM to calculate fitness, "Check Precision based on Align-ETConformance" to calculate the precision. We use "Calculate BPMN Metrics" to calculate model complexity. In addition, the tools in the "BPMN Miner" are used to covert between Petri-nets and BPMNs. We use default settings for all the parameters.

6.1 Evaluation Using Artificial Data

We first use several artificial logs with switch behaviours to demonstrate the performance of our approach. When applying the original inductive miner on these logs, it fails to discover precise models. Instead, "flower" models with low precision are returned. We show that after using our extension, we can get precise models (Table 1).

6.2 Evaluation Using Publically-Available Data

We use a publicly-available dataset called "BPIC13-incident" from the "4TU Center for Research Data" to evaluate our algorithm. We use "Event name + lifecycle" as the activity classifier, the dataset contains 7554 traces, 2278 distinct traces, 65533 events and 13 distinct events. The average length of traces is 9 while the shortest length is 1 and

Table 1. Evaluation using artificial data.

Log: <A, B, C>, <D, E, F>, <A, F>	
IM	**IM augmented with switch behaviours**
Fitness: 1.0, Precision: 0.38	Fitness: 1.0, Precision: 1.0
Log: <A, B, C>, <D, E, F>, <A, E, F>, <D, E, C>, <A, E, C>	
IM	**IM augmented with switch behaviours**
Fitness: 1.0, Precision: 0.42	Fitness: 1.0, Precision: 1.0
Log: <A, B, C>, <D, E, F>, <A, B, D, E, F>	
IM	**IM augmented with switch behaviours**
Fitness: 1.0, Precision: 0.94	Fitness: 1.0, Precision: 1.0
Log: <A, B, C>, <D, E, F>, <D, E, C>, <A, F>, <A, C>	
IM	**IM augmented with switch behaviours**
Fitness: 1.0, Precision: 0.42	Fitness: 1.0, Precision: 0.97

the longest length is 123. We combine our approach with the inductive miner infrequent (IMf) [14] and switch off the "delete trace" option, we also compare our results with the split miner (SM) [7]. In addition, we use default settings for all the parameters.

Evaluation results are presented in Table 2. All three methods can produce a model with high fitness. However, the IMf returns a model with low precision. Our approach rises the precision of IMf by 36%. In addition, our approach returns a model with both higher precision and F-score than the split miner. For the model complexity, our approach also achieves both smaller size and CFC than the split miner.

Table 2. Evaluation results with the publicly-available dataset (IMs refers to our approach)

	Accuracy			Complexity	
	Fitness	Precision	F-Score	Size	CFC
IMf	0.95	0.59	0.73	35	**33**
SM	**0.98**	0.71	0.82	39	48
IMs	0.97	**0.80**	**0.88**	**33**	46

7   Discussion and Conclusion

In this paper, we present an extension to both the inductive miner and the process tree model. We allow the inductive miner to discover sound process models but not being limited to block-structured workflow nets. The evaluation results show that our approach can reduce the chance for the inductive miner to return flower models. Besides, in our evaluation, our approach can also discover models that are comparable in terms of both model accuracy and complexity to these produced by the split miner.

One limitation is that when performing the switch exclusive choice cut, we do not know if the switch behaviour is valid or not, thus we need to check the validity of the switch behaviours to make sure the model is still sound in the end. It has to be noted that the fitness of resulting models might be reduced if too many switch behaviours are removed. We aim to develop better algorithms to repair the models in the future. Besides, as shown in the artificial data evaluation, when the same place is both the input and output of two switch invisible transitions, there might be redundant hidden transitions in the model, future work is required to remove these redundant hidden transitions.

Finally, we also aim to conduct more experiments to evaluate the performance of our approach in the future, including the impacts of different orders of cuts.

References

1. van der Aalst, W.M.P.: Process Mining - Data Science in Action. Springer, Heidelberg (2016). https://doi.org/10.1007/978-3-662-49851-4_1
2. Buijs, J., van Dongen, B.F., van der Aalst, W.M.P.: Quality dimensions in process discovery: the importance of fitness, precision, generalization and simplicity. Int. J. Coop. Inf. Syst. **23**(1), 1440001 (2014)
3. Leemans, S.J.J., Fahland, D., van der Aalst, W.M.P.: Discovering block-structured process models from event logs - a constructive approach. In: Colom, J.-M., Desel, J. (eds.) PETRI NETS 2013. LNCS, vol. 7927, pp. 311–329. Springer, Heidelberg (2013). https://doi.org/10.1007/978-3-642-38697-8_17
4. Augusto, A., et al.: Automated discovery of process models from event logs: review and benchmark. IEEE Trans. Knowl. Data Eng. **31**, 686–705 (2018)
5. Dumas, M., La Rosa, M., Mendling, J., Reijers, H.: Fundamentals of Business Process Management. Springer, Heidelberg (2013).https://doi.org/10.1007/978-3-642-33143-5
6. Wen, L., Wang, J., van der Aalst, W.M.P., Huang, B., Sun, J.: Mining process models with prime invisible tasks. Data Knowl. Eng. **69**(10), 999–1021 (2010)

7. Augusto, A., Conforti, R., Dumas, M., La Rosa, M., Polyvyanyy, A.: Split miner: automated discovery of accurate and simple business process models from event logs. Knowl. Inf. Syst. **59**(2), 251–284 (2018). https://doi.org/10.1007/s10115-018-1214-x

8. Cardoso, J.S.: Business process control-flow complexity: metric, evaluation, and validation. Int. J. Web Serv. Res. **5**(2), 49–76 (2008)

9. van der Aalst, W.M.P., Weijters, A.J.M.M., Maruster, L.: Workflow mining: discovering process models from event logs. IEEE Trans. Knowl. Data Eng. **16**(9), 1128–1142 (2004)

10. Guo, Q., Wen, L., Wang, J., Yan, Z., Yu, P.S.: Mining invisible tasks in non-free-choice constructs. In: Motahari-Nezhad, H.R., Recker, J., Weidlich, M. (eds.) BPM 2015. LNCS, vol. 9253, pp. 109–125. Springer, Cham (2015). https://doi.org/10.1007/978-3-319-23063-4_7

11. Weijters, A., Ribeiro, J.: Flexible heuristics miner (FHM). In: CIDM, pp. 310–317. IEEE (2011)

12. de Medeiros, A., Weijters, A., van der Aalst, W.: Genetic process mining: an experimental evaluation. Data Min. Knowl. Discov. **14**(2), 245–304 (2007)

13. van Zelst, S.J., van Dongen, B.F., van der Aalst, W.M.P., Verbeek, H.M.W.: Discovering workflow nets using integer linear programming. Computing **100**(5), 529–556 (2017). https://doi.org/10.1007/s00607-017-0582-5

14. Leemans, S.J.J., Fahland, D., van der Aalst, W.M.P.: Discovering block-structured process models from event logs containing infrequent behaviour. In: Lohmann, N., Song, M., Wohed, P. (eds.) BPM 2013. LNBIP, vol. 171, pp. 66–78. Springer, Cham (2014). https://doi.org/10.1007/978-3-319-06257-0_6

15. Leemans, S.J.J., Fahland, D., van der Aalst, W.M.P.: Discovering block-structured process models from incomplete event logs. In: Ciardo, G., Kindler, E. (eds.) PETRI NETS 2014. LNCS, vol. 8489, pp. 91–110. Springer, Heidelberg (2014). https://doi.org/10.1007/978-3-319-07734-5_6

16. Leemans, S.J.J., Fahland, D., van der Aalst, W.M.P.: Scalable process discovery and conformance checking. Softw. Syst. Model. **17**(2), 599–631 (2016). https://doi.org/10.1007/s10270-016-0545-x

17. Leemans, S.J.J., Fahland, D., van der Aalst, W.M.P.: Using life cycle information in process discovery. In: Reichert, M., Reijers, H.A. (eds.) BPM 2015. LNBIP, vol. 256, pp. 204–217. Springer, Cham (2016). https://doi.org/10.1007/978-3-319-42887-1_17

18. Leemans, M., van der Aalst, W.M.P.: Modeling and discovering cancelation behavior. In: Panetto, H., et al. (eds.) OTM 2017. LNCS, vol. 10573, pp. 93–113. Springer, Cham (2017). https://doi.org/10.1007/978-3-319-69462-7_8

19. Lu, X., Fahland, D., van den Biggelaar, F.J.H.M., van der Aalst, W.M.P.: Handling duplicated tasks in process discovery by refining event labels. In: La Rosa, M., Loos, P., Pastor, O. (eds.) BPM 2016. LNCS, vol. 9850, pp. 90–107. Springer, Cham (2016). https://doi.org/10.1007/978-3-319-45348-4_6

20. van Dongen, B.F., de Medeiros, A.K.A., Verbeek, H.M.W., Weijters, A.J.M.M., van der Aalst, W.M.P.: The ProM framework: a new era in process mining tool support. In: Ciardo, G., Darondeau, P. (eds.) ICATPN 2005. LNCS, vol. 3536, pp. 444–454. Springer, Heidelberg (2005). https://doi.org/10.1007/11494744_25

Process Model Discovery from Sensor Event Data

Dominik Janssen[1]([✉])[iD], Felix Mannhardt[2,3][iD], Agnes Koschmider[1][iD],
and Sebastiaan J. van Zelst[4,5][iD]

[1] Group Process Analytics, Kiel University, Kiel, Germany
{dominik.janssen,ak}@informatik.uni-kiel.de
[2] Department of Technology Management, SINTEF Digital, Oslo, Norway
f.mannhardt@tue.nl
[3] Department of Computer Science, NTNU, Trondheim, Norway
[4] Fraunhofer Institute for Applied Information Technology, Fraunhofer Gesellschaft,
Sankt Augustin, Germany
sebastiaan.van.zelst@fit.fraunhofer.de
[5] Chair of Process and Data Science, RWTH Aachen University, Aachen, Germany

Abstract. Virtually all techniques, developed in the area of process mining, assume the input event data to be discrete, and, at a relatively high level (i.e., close to the business-level). However, in many cases, the event data generated during the execution of a process is at a much lower level of abstraction, e.g., sensor data. Hence, in this paper, we present a novel technique that allows us to translate sensor data into higher-level, discrete event data, thus enabling existing process mining techniques to work on data tracked at a sensory level. Our technique discretises the observed sensor data into activities by applying unsupervised learning in the form of clustering. Furthermore, we refine the observed sequences by deducing imperative sub-models for the observed discretised data, i.e., allowing us to identify concurrency and interleaving within the data. We evaluated the approach by comparing the obtained model quality for several clustering techniques on a publicly available data-set in a smart home scenario. Our results show that applying our framework combined with a clustering technique yields results on data that otherwise would not be suitable for process discovery.

Keywords: Process mining · Sensor data · Event correlation · IoT

1 Introduction

The rise of the Internet-of-Things (IoT), i.e., interconnected devices, mechanical and digital machines, gradually digitalises the day-to-day operations of modern-day enterprises. More-and-more devices are interconnected and store valuable traces of behavioural data, generated during their interaction with humans, as well as other interconnected devices. For example, consider the concepts of

© Springer Nature Switzerland AG 2021
S. Leemans and H. Leopold (Eds.): ICPM 2020 Workshops, LNBIP 406, pp. 69–81, 2021.
https://doi.org/10.1007/978-3-030-72693-5_6

autonomous production and the adoption of *robotics in healthcare*, in which operational processes are gradually digitised and automated, utilising interconnecting and communicating devices and machines.

Whereas the design of a single device, connected to a larger network of devices, remains manageable (though it is complex in its own right), deficiencies in inter-device communication or handover of work-packages, easily lead to global process under-performance. Hence, a clear understanding of the general flow of work, as well as an understanding of bottlenecks and synchronisation points is of utmost importance to further improve the efficiency of the executed processes. *Process mining* techniques aim to exploit behavioural data, stored in the information systems to support the execution of processes and to distil process models [1]. In particular, they can derive-and-construct process models based on tracked event data, i.e., in a *completely automated* fashion.

In general, process mining relies on discrete event data, typically assumed to be tracked at *the business level*, i.e., the event data directly relates to high-level business process concepts. However, often, the level at which the event data is tracked within information systems is at a much lower level.

Possible application scenarios are settings where the movement of objects or people (entities) is tracked by motion sensors, light barriers or similar types of sensors that only detect absence and presence of a person or object and cannot distinguish between different observed entities. Those sensors can be found in smart home settings, smart factories and healthcare-related applications. If in these possible settings, it is of interest to discover frequent behaviour patterns or abnormal behaviour, our proposed method provides a novel approach that translates sensory data, into a process model. In particular, unlabelled raw sensor events are aggregated and clustered by an unsupervised learning technique to identify activities through clusters of related event sequences. To identify the activities, we discover a process model for each identified cluster. The activities, labelled by a domain expert, serve as input for process mining-based model discovery, which allows to identify concurrent and interleaving behaviour in sensor event data. We evaluate our approach on the publicly available CASAS dataset [2] and compare two clustering methods. The obtained results show promising results, hinting towards a better result by using clustering based on a self-organising map (SOM) in comparison to basic k-means in this context based on our methodology.

To the best of our knowledge, this paper suggests the first activity and process discovery technique for unlabelled sensor event data using SOM as model and addressing the challenges of concurrent behaviour between activities and multiple residents.

The remainder of this paper is structured as follows. The next section presents related work. Subsequently, Sect. 3 presents our approach, which has been evaluated using a real-life data-set. The evaluation is summarised in Sect. 4. The paper concludes with an outlook on future work in Sect. 5.

2   Related Work

A large body of research exists that partially addresses the discovery of events and activities at different levels. In the following we consider related approaches that use sensor data aiming to translate it into higher-level, discrete event data or applying process mining on raw sensor data. Our focus of related approaches also lays in smart homes as we used data of smart home sensors for evaluation.

Activity recognition in smart home has been widely addressed relying the recognition on different sensor types like motion or video [3–5] or analysing data from wearables [6] or reference sensors [7]. Recently, Deep Learning (DL) methods for detecting and predicting activities in IoT environments have been increasingly explored [8]. Unlike classical machine learning techniques, DL networks automatically derive features from the data and produce promising results in different domains. Particularly in the field of smart homes or ambient assisted living, there are first approaches that recognise activities based on sensor event data [9–12]. Activity recognition is predominantly used for a situational prognosis [13]. Also these kinds of approaches identify simple activities [14,15]. Complex activities like people's daily activities can only be identified using extra sensors [14,16]. Although our method for process model discovery from raw location sensor data also requires a manual labelling of clusters of high-level events, we believe that the process model view on raw sensor data advances existing approaches and is beneficial in terms of evaluating the quality of data aggregations, which DL-based approaches are not capable of.

Mapping low-level events to activities for process mining is still a challenge [17]. Leotta et al.[18] envision to use similar techniques as we employ: however, only discuss challenges. The current status-quo is that approaches indicate only likelihoods of mappings, since there is often more than one possible solution [19]. Our approach for event aggregation in combination with unsupervised learning aims to bridge this gap. Related literature for activity discovery for process mining either use supervised techniques [6,20] or visualise human habits [21] in order to accurately identify activities. Some works exist that detect activities from high-level events through unsupervised techniques [20,22,23], which have been compared in this paper. These related works [20,22,23] use patterns or local process models to aggregate event data towards higher abstraction levels. But they did not allow to discover meaningful activities for our data set. For unlabelled training sets, related approaches suggest to use a time-based label refinement [24] or locations [25] as characteristics in order to segment the event log and to abstract activities out of it. However, the methods already expects particular representations of traces. Given our scenario, the application of local process models did not allow to identify useful process fragments.

3   Translating Sensor Data to High-Level Traces

Our method for process model discovery from raw location sensor data assumes a location sensor event log E_L as input derived from a set of sensors S e.g.,

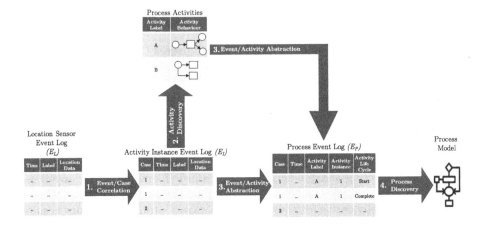

Fig. 1. Process discovery approach for location sensor event data.

networks of WiFi-access points, or motion sensors in smart homes. We expect events $e \in E_L$ to satisfy some minimal requirements: For each event we can retrieve a timestamp $time(e)$ inducing a partial order on the events, a sensor label $sensor(e) \in S$ indicating which sensor was activated and some form of information that either implicitly or explicitly refers to a location (i.e., $location(e) \in L$). The location information can be explicit in the form of coordinates (e.g., latitude, longitude) or implicit by providing labelled locations together with a distance function providing pairwise distances between them. Throughout the paper we assume that E_L was generated by one or more *entities* $n \in N$. An entity may be a person or an object in the observed area.

Events in a location sensor log do not necessarily have a unique identifier attached to identify by which entity they were triggered. Often data contains overlapping and concurrent activities by multiple entities. In smart homes or factories, multiple entities can be present at the same time. It has to be ensured that the analysed activities are all associated with the correct entity, to obtain a meaningful process model on a by-entity-level. Our method targets such scenarios where a sensor cannot identify entities utilising a unique identifier. Figure 1 gives an overview of the proposed approach, which consists of the following four steps that are explained in the following sections:

1. *Event Correlation:* Correlation of events from a location sensor event log E_L to (unlabelled) activity instances yielding an instance log E_I.
2. *Activity Discovery:* Discovery of process activities A together with their labels and sensor-level process models describing the expected behaviour on a sensor level.
3. *Event Abstraction:* Abstraction of the instance log E_I to a process event log E_P where events are directly related to the *start* or *completion* of process activities A.

4. *Process Discovery:* Process discovery based on the process event log E_P resulting in an activity-level process model defined over activities A.

3.1   Event Correlation

The first step towards process mining on raw sensor events is to group the input data according to a set of numbered activity instances by correlating each individual location sensor event $e \in E_L$ to an activity instance $i \in \mathbb{N}$. This results in an *instance log* E_I in which, beyond the requirements for E_L, each event $e_i \in E_I$ is additionally assigned an activity instance that can be retrieved with *instance*$(e) \in \mathbb{N}$. The main goal of this step is to produce traces such that each trace can be associated with an activity instance. We assume every recorded event in the raw data is caused by an entity. In order to determine which entity $n \in N$ caused which event, the raw event location data is assigned to the respective entities. Eventually the trace of each entity is divided into smaller sub-traces (cases) that contain only one single activity: we denote this as *sensor case slicing*. Here, also an approach for *entity detection* is required.

Entity Detection. In a setting with sensors providing only information whether an object is present or absent, a distinction between entities is not possible. However, if we know the relative location of the sensors to each other, our weighted average distance approach can be implemented and distinguish between multiple entities. The very first time any of the sensors detects the presence of an entity is the beginning of the first entity's trace. For every subsequent sensor activation, we have to decide which entity caused the activation of a sensor. Each time a sensor is activated, we calculate which already registered entity is closest to the current sensor activation, based on the entities' last known position. If no entity is close enough, the algorithm assumes that a new entity has entered the observed area and creates a new trace for this new entity. Both the proximity threshold and the maximum number of entities are parameters that can be manually adjusted based on the scenario. This straightforward implementation works well if entities keep their distance to each other. But as soon as entities cross paths in a spot that is only covered by a single sensor, this method will not be able to correctly assign the sensor activations after the entities moved on, since the newly activated sensor has the same distance to every entity in that single spot. This limitation can be overcome, by assuming, entities will preserve their direction of motion and predict where entities are headed by also considering the entities' previous locations combined with a decay function in the distance function.

Sensor Case Slicing. During its presence in the observed area, the entity executes most likely more than one activity. To identify meaningful activities from the continuous recording (what is called a "long trace"), an appropriate separation into smaller sub-traces, called cases, is required. We have to divide the traces here, because we are identifying and clustering activities by their sensor-activation-signature, therefore the sub-traces can only contain one single activity.

In concrete terms, in our approach, a long trace is cut into sub-traces of a predefined fixed length. Depending on the application, the optimal fixed length might be different. Our implementation incorporates a grid search, comparing the results for different sub-trace lengths, to maintain flexibility. The challenge is to avoid sub-traces that are too short and contain too little sensor-data to extract meaningful activities. But at the same time, the sub-traces cannot be too long, as a too-long sub-trace may consist of multiple activities.

3.2 Activity Discovery

Having obtained the instance log E_I, we aim to infer a set of process activities A that are likely to have generated the raw sensor events. The outputs of this step in our approach are a set of activities A. Each activity $a \in A$ has both an activity label $label(a)$ as well as a process model describing the low-level behaviour of that activity a in terms of events on the sensor-level. The main challenge in this part of the approach is to determine a good division of activity instances into clusters, i.e., an *activity clustering* where each of the clusters should represent a distinct activity on the process level. This refers not only to the clustering itself but also to finding a good number of clusters. Furthermore, a suitable *activity labelling* needs to be found.

Activity Clustering. Independent of the implemented clustering technique, the objective remains the same: Find similar sub-traces and group them. For this, we used a *Self-Organising Map* (SOM) clustering and k-means. The challenge with the discovery of similarities is to find a criterion to define the similarity between sub-traces. Usually, in SOM this is achieved by calculating the euclidean distance between vectors. However, this is challenging if sensors have arbitrary label names. We experimented with three alternative representations of the traces: First, we counted how often each sensor is activated in a trace. Second, we counted for how long each sensor is activated for in a trace. And third, we combined both the quantity method and time method in one vector. The third representation retains the most information of the original trace and is, therefore, the preferred choice.

Activity Labelling and Validation. Having discovered clusters of similar traces corresponding to distinct activities, we still lack insights into the kind of activity that are represented by each cluster. Also, it may be challenging to judge the quality of the obtained clustering. We assume that activity labelling generally requires a human-in-the-loop with appropriate domain knowledge. Thus, we propose to discover a process model based on the events of each cluster by using the Inductive Miner. The quality of the process model is evaluated based on the F1-score combination of the common *fitness* and *precision* measure. These interpretable process models make our method suitable for complex processes and the quality measure can be used to validate the clustering result. Having access to the process models and their quality evaluation a domain expert can interpret, validate and label each cluster with an appropriate activity label.

3.3   Event Abstraction

The third step of our approach combines the sub-traces yielded by the *event correlation* step (Sect. 3.1) and the activity clusters detected in the *Activity Discovery* step (Sect. 3.2). This results in a process event log E_P that groups together events from the original location sensor event log E_L to process events $e_p \in E_P$. For each process event e_p we can obtain the following attributes: $time(e_P) \in \mathbb{R}$, $activity(e_P) \in A$, $entity(e_P) \in N$, and $transition(e_P) \in \{start, completed\}$. Thus, each process event refers to a specific high-level process activity and indicates a transition in the transactional life-cycle, i.e., whether the activity instance has been started or completed.

3.4   Process Discovery

Having promoted the raw location sensor events E_L to the level of activity instances, our process event log E_P fulfills almost all requirements for high-level process discovery. Anyway, still missing are process cases that are meaningful to our analysis goal. Identifying process cases is highly dependent on the particular circumstance. In our application scenario, we propose to focus on re-occurring behaviour of an entity starting with a specific activity (e.g., entering the smart home). Based on our event correlation step (Sect. 3.1), we build a separate trace for each entity. Then, the potentially very long trace referring to a single entity is subdivided into multiple traces by dividing it into separate traces each time the activity of interest occurs. To discover a meaningful process model, we have to assume that regular and routine behaviour is observable. As a starting point, an activity has to be selected that most likely will be the origin of the routine behaviour such as *entering the observed area*. Finally, an overall process model is discovered using a standard technique, e.g., Inductive Miner [26]. The final output is a process model reflecting the observed behaviour of the entities aggregated only from raw sensor data.

4   Evaluation

4.1   Set-up

We evaluated our approach on the publicly available CASAS data-set, which contains raw sensor data from a smart home environment [2]. The CASAS data fulfils the two requirements of our approach: it contains the timestamps and location information of sensor events. The data was recorded in a smart home test-bed with two residents and a house equipped with 51 motion sensors. Figure 2 shows the house plan and the positions of the motion sensors. Each motion sensor generates low-level events, where each sensor entry is tagged with a timestamp, the sensor ID and the binary sensor value (active/not active). We extracted sensor data from 7 consecutive days (02/05–09/02/2010) from the 20-Kyoto-2-Daily life, 2010–2012 data set.

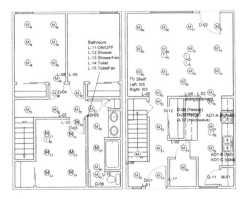

Fig. 2. Sensor layout of an apartment in the CASAS project [2].

We applied our method for different values of parameters such as sub-trace length, number of clusters and the similarity measure used. We used grid search to identify best parameter values for the clustering based on the average combined fitness [27] and precision [28] (F1-score) obtained for the process models discovered for each cluster of high-level events. We employ standard filtering techniques (most frequent traces and activities) used in process mining to focus on the dominant behaviour in each cluster. We compared the proposed SOM clustering with k-means clustering based on the same similarity measures. The implementation of step 1 and 2 is openly accessible[1] We used PM4Py 1.1.1 and heuristicmineR for the process discovery and evaluation.

Having discovered activities and obtained traces based on the idea to discover re-occurring behaviour starting with the same activity (Sect. 3.4), we applied Heuristics Miner to discover a process model of the behaviour. Based on the spatial layout of the smart home (Fig. 2), we choose to create traces that start with the activity *Walk entrance/stairs/storage* as the entry point into the house. Heuristics Miner was selected as we expect the inhabitants of the smart home environment to show a lot of infrequent behaviour, for which Heuristics Miner has shown to be appropriate [29].

4.2   Results and Discussion

Figure 3 shows the results of our grid search. We experimented with trace lengths ranging from four to twelve. Shorter trace lengths generally lead to a better F1-score. However, we need to impose a minimal trace length since traces consisting only of a single event would trivially lead to the discovery of process models with perfect fitness and precision. In our case, less than four events did not allow to infer a set of meaningful activities.

Evaluating sample data has shown that considering both the frequency of activation as well as the duration of the activations as a similarity measure (the

---

[1] https://github.com/d-o-m-i-n-i-k/Process-Model-Discovery-public.

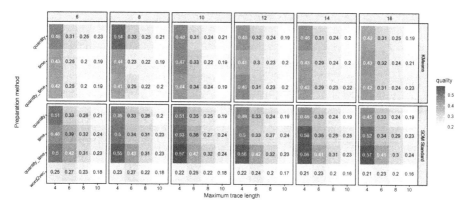

Fig. 3. Average F1-score for process models discovered for the clusters based on six different cluster sizes (6–16), five different maximum trace lengths (4–12), four vector preparation methods and two clustering algorithms.

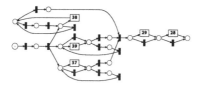

Fig. 4. Example of a Petri net discovered using Inductive Miner for a cluster in the *Activity Discovery* step.

method *quantity_time*) yields superior results, compared to only regarding one aspect. When choosing too few clusters or too many, the quality score decreases. In turn, choosing too many clusters may lead to several clusters representing the same activities, which should have been grouped. We also qualitatively evaluated the clustering by manually inspecting and labelling some of the results.

For example, the Petri net discovered by Inductive Miner on a cluster shown in Fig. 4 is a reasonable candidate. The three sensors that can be activated simultaneously are all located in the bathroom. The subsequent sensors M29 and M28 are located in the hall with M28, which is furthest from the bathroom. From this example process model, it is reasonable to infer that this cluster refers to activities where the entity spends some time in the bathroom and then left the room. Overall, the similar results are obtained for 10 and 16 clusters with a trace length of 4 and using our proposed *quantity&time* vectorisation approach.

We grouped the activity instances of the best clustering results (16 clusters) into traces at the level of process instances. Afterwards we filtered the resulting event log to only retain traces of a length in the range of 5 to 25 events. This yields a log with 5898 events grouped in 273 traces with an average length of 21.6. The application of Heuristics Miner with a dependency threshold of 0.8 and a frequency threshold of 10 returns the Causal net dependencies shown in Fig. 5. The activity in entrance area of the house marks the starting point of our

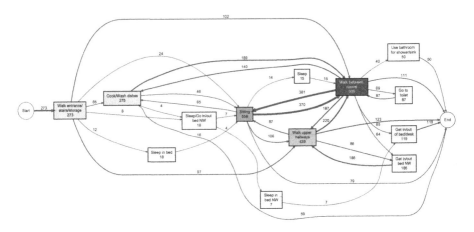

Fig. 5. Causal net discovered with the Heuristics Miner on the obtained process event log.

Causal net. The activities that can mostly be observed after the entry activity are *walking between the rooms, walking in the upper hallways* and *going to the kitchen to cook or wash the dishes.* After cooking the dishes it often occurs that the resident would walk between the rooms to sit down, presumably to eat in the living room.

4.3 Limitations

A drawback of our method is the assumption of continuous movement in the event correlation step (Sect. 3.1). As soon as the motion at the rendezvous location is more than just a mere passing by, our approach might not return the desired results. Additionally, the entity recognition could be improved by using more sophisticated methods, e.g. hidden Markov models that have already shown promising results in differentiating people from one another [30]. Moreover, the sensor case slicing mechanism could take variable sub-trace length into consideration, i.e., depending on the activity, the number of involved events, and therefore the sub-trace length may vary. For example, the activities *sleeping, cooking* and *washing hands* are activities with a distinctive difference in the number of involved events.

5 Conclusion

IoT environments generate a large amount of data, predestined for further analysis. Process mining can give valuable insights into how real-life activities perform when extracting meaningful activities instances from raw sensor events. This paper combined unsupervised learning in the form of clustering and process mining, to discover activities and process models from motion sensors. We evaluated our approach by comparing the obtained model quality for several clustering techniques on a publicly available data-set in a smart home scenario and found it to be superior. To fully relieve domain experts from process modelling and to automate the process of model discovery, we believe that an accurate approach for entity centricity is imperative. For this, future tasks are to fuse heterogeneous sensor events as input for high-level aggregation, to take into account other vectorisation methods such the shortest path distance between sensors (i.e., relational or pair-wise distances only) to better disambiguate between residents and to apply non-end-to-end process discovery methods such as Local Process Model discovery [22]. In further research, we plan to include spatial information, like room layouts in smart homes, into our approach as well as implement variable trace lengths and experiment with other machine-based learning techniques to further improve the discovered process models.

References

1. van der Aalst, W.M.P.: Process Mining - Data Science in Action, 2nd edn. Springer, Heidelberg (2016). https://doi.org/10.1007/978-3-662-49851-4
2. Rashidi, P., Youngblood, G.M., Cook, D.J., Das, S.K.: Inhabitant guidance of smart environments. In: Jacko, J.A. (ed.) HCI 2007. LNCS, vol. 4551, pp. 910–919. Springer, Heidelberg (2007). https://doi.org/10.1007/978-3-540-73107-8_100
3. Nentwig, M., Stamminger, M.: A method for the reproduction of vehicle test drives for the simulation based evaluation of image processing algorithms. In: 13th International IEEE Conference on Intelligent Transportation Systems, pp. 1307–1312. IEEE, September 2010
4. Zhang, M., Sawchuk, A.A.: Motion primitive-based human activity recognition using a bag-of-features approach. In: Proceedings of the 2nd ACM SIGHIT Symposium on IHI 2012. IHI 2012, ACM Press, pp. 631–640 (2012)
5. Diete, A., Sztyler, T., Weiland, L., Stuckenschmidt, H.: Improving motion-based activity recognition with ego-centric vision. In: PerCom Workshops 2018, pp. 488–491. IEEE Computer Society (2018)
6. Sztyler, T., Carmona, J., Völker, J., Stuckenschmidt, H.: Self-tracking reloaded: applying process mining to personalized health care from labeled sensor data. In: Koutny, M., Desel, J., Kleijn, J. (eds.) Transactions on Petri Nets and Other Models of Concurrency XI. LNCS, vol. 9930, pp. 160–180. Springer, Heidelberg (2016). https://doi.org/10.1007/978-3-662-53401-4_8
7. Larue, G.S., Rakotonirainy, A., Pettitt, A.N.: Predicting reduced driver alertness on monotonous highways. IEEE Pervasive Comput. **14**(2), 78–85 (2015)
8. Weerdt, J.D., van den Broucke, S., Vanthienen, J., Baesens, B., : Active trace clustering for improved process discovery. IEEE Trans. Knowl. Data Eng. **25**(12), 2708–2720 (2013)
9. Choi, S., Kim, E., Oh, S.: Human behavior prediction for smart homes using deep learning, pp. 173–179. IEEE, August 2013
10. Gallicchio, C., Micheli, A.: Experimental analysis of deep echo state networks for ambient assisted living. In: Bandini, S., Cortellessa, G., Palumbo, F. (eds.) Proceedings of the Third Italian Workshop on AI for Ambient Assisted Living. Volume 2061 of CEUR Workshop Proceedings, pp. 44–57. CEUR-WS.org (2017)

11. Wang, A., Chen, G., Shang, C., Zhang, M., Liu, L.: Human activity recognition in a smart home environment with stacked denoising autoencoders. In: Song, S., Tong, Y. (eds.) WAIM 2016. LNCS, vol. 9998, pp. 29–40. Springer, Cham (2016). https://doi.org/10.1007/978-3-319-47121-1_3

12. Wang, J., Chen, Y., Hao, S., Peng, X., Hu, L.: Deep learning for sensor-based activity recognition: a survey. arXiv e-prints (2017)

13. Lee, S., Lin, F.J.: Situation awareness in a smart home environment. In: 2016 IEEE 3rd World Forum on Internet of Things (WF-IoT), pp. 678–683, December 2016

14. Dernbach, S., Das, B., Krishnan, N.C., Thomas, B.L., Cook, D.J.: Simple and complex activity recognition through smart phones. In: Intelligent Environments, pp. 214–221. IEEE (2012)

15. Sztyler, T.: Sensor-based human activity recognition: overcoming issues in a real world setting. Ph.D. thesis, University of Mannheim, Germany (2019)

16. Nguyen, H.D., Tran, K.P., Zeng, X., Koehl, L., Tartare, G.: Wearable sensor data based human activity recognition using machine learning: a new approach. ArXiv abs/1905.03809 (2019)

17. Koschmider, A., Mannhardt, F., Heuser, T.: On the contextualization of event-activity mappings. In: Daniel, F., Sheng, Q.Z., Motahari, H. (eds.) BPM 2018. LNBIP, vol. 342, pp. 445–457. Springer, Cham (2019). https://doi.org/10.1007/978-3-030-11641-5_35

18. Leotta, F., Mecella, M., Mendling, J.: Applying process mining to smart spaces: perspectives and research challenges. In: Persson, A., Stirna, J. (eds.) CAiSE 2015. LNBIP, vol. 215, pp. 298–304. Springer, Cham (2015). https://doi.org/10.1007/978-3-319-19243-7_28

19. van der Aa, H., Leopold, H., Reijers, H.A.: Checking process compliance on the basis of uncertain event-to-activity mappings. In: Dubois, E., Pohl, K. (eds.) CAiSE 2017. LNCS, vol. 10253, pp. 79–93. Springer, Cham (2017). https://doi.org/10.1007/978-3-319-59536-8_6

20. Tax, N., Sidorova, N., Haakma, R., van der Aalst, W.: Mining process model descriptions of daily life through event abstraction. In: Bi, Y., Kapoor, S., Bhatia, R. (eds.) IntelliSys 2016. SCI, vol. 751, pp. 83–104. Springer, Cham (2018). https://doi.org/10.1007/978-3-319-69266-1_5

21. Leotta, F., Mecella, M., Sora, D.: Visual process maps: a visualization tool for discovering habits in smart homes. J. Ambient Intell. Human. Comput. 11(5), 1997–2025 (2019)

22. Mannhardt, F., Tax, N.: Unsupervised event abstraction using pattern abstraction and local process models. In: RADAR+EMISA@CAiSE. Volume 1859 of CEUR Workshop Proceedings, pp. 55–63. CEUR-WS.org (2017)

23. Alharbi, A., Bulpitt, A., Johnson, O.: Towards unsupervised detection of process models in healthcare. Stud. Health Technol. Inf. 247, 381–385 (2018)

24. Tax, N., Alasgarov, E., Sidorova, N., Haakma, R., van der Aalst, W.M.P.: Generating time-based label refinements to discover more precise process models. J. Ambient Intell. Smart Environ. 11(2), 165–182 (2019)

25. Brzychczy, E., Trzcionkowska, A.: Process-oriented approach for analysis of sensor data from longwall monitoring system. In: Burduk, A., Chlebus, E., Nowakowski, T., Tubis, A. (eds.) ISPEM 2018. AISC, vol. 835, pp. 611–621. Springer, Cham (2019). https://doi.org/10.1007/978-3-319-97490-3_58

26. Leemans, S.J.: Robust process mining with guarantees. Ph.D. thesis (2017)

27. van der Aalst, W.M.P., Adriansyah, A., van Dongen, B.F.: Replaying history on process models for conformance checking and performance analysis. WIREs Data Min. Knowl. Discov. 2(2), 182–192 (2012)

28. Munoz-Gama, J., Carmona, J.: A general framework for precision checking. Int. J. Innov. Comput. Inf. Control (IJICIC) **8**(7), 5317–5339 (2012)
29. Augusto, A., et al.: Automated discovery of process models from event logs: review and benchmark. IEEE TKDE **31**(4), 686–705 (2018)
30. Guo, P., Miao, Z.: Multi-person activity recognition through hierarchical and observation decomposed HMM. In: 2010 IEEE International Conference on Multimedia and Expo, pp. 143–148. IEEE, July 2010

Unsupervised Event Abstraction in a Process Mining Context: A Benchmark Study

Greg Van Houdt[1]([⊠]) [iD], Benoît Depaire[1] [iD], and Niels Martin[1,2,3] [iD]

[1] Research Group Business Informatics, UHasselt - Hasselt University, Hasselt, Belgium
{greg.vanhoudt,benoit.depaire}@uhasselt.be
[2] Research Foundation Flanders (FWO), Brussels, Belgium
[3] Data Analytics Laboratory, Vrije Universiteit Brussel, Brussels, Belgium
niels.martin@vub.be

Abstract. Due to the rise of IoT, event data becomes increasingly fine-grained. Faced with such data, process discovery often produces incomprehensible spaghetti-models expressed at a granularity level that doesn't match the mental model of a business user. One approach is to use event abstraction patterns to transform the event log towards a more coarse-grained level and to discover process models from this transformed log. Recent literature has produced various (partial) implementations of this approach, but insights how these techniques compare against each other is still limited.

This paper focuses on the use of Local Process Models and Combination based Behavioural Pattern Mining to discover event abstraction patterns in combination with the approach of Mannhardt et al. [15] to transform the event log. Experiments are conducted to gain insights into the performance of these techniques. Results are very limited with a general decrease in fitness and precision and only a minimal improvement of complexity. Results also show that the combination of the process discovery algorithm and the event abstraction pattern miner matters. In particular, the combination of Local Process Models with Split Miner seems to improve precision.

Keywords: Process mining · Unsupervised learning · Event log · Abstraction

1 Introduction

Process Discovery focuses on the discovery of the process flow from an event log [2], in order to gain insights in the real execution of a business process [1]. However, when event logs are recorded at a fine-grained level, the activities in the discovered process model become increasingly less recognisable to the business users. Furthermore, fine-grained event data often result in incomprehensible spaghetti-models [22].

© Springer Nature Switzerland AG 2021
S. Leemans and H. Leopold (Eds.): ICPM 2020 Workshops, LNBIP 406, pp. 82–93, 2021.
https://doi.org/10.1007/978-3-030-72693-5_7

Against this background, [14] introduced a pattern-based approach to augment low-level events to higher-level activities, resulting in more insightful process models. Their original approach requires domain experts to provide event abstraction patterns which map low-level events to higher-level activities, which are subsequently used to transform the event log.

Mannhardt and Tax [16] studied the use of Local Process Models to learn these event abstraction patterns from data. More recently, [3] introduced a new unsupervised technique to discover event abstraction patterns that are compact and maximal, increasing the options to apply the technique proposed in [16] in an unsupervised manner. This raises the question how well these two options perform with the end goal in mind, i.e. to transform the log such that a process model is discovered which is more comprehensible and remains properly fitting and precise.

This paper describes a benchmark study of Local Process Models (LPMs) and Combination based Behavioral Pattern Mining (COBPAM) in combination with the approach in [16], focused on their capabilities to obtain models of lower complexity without sacrificing fitness and precision. Furthermore, performance differences between these two approaches were explored in order to identify underlying mechanisms at work. This resulted in following contributions:

- In contrast to previous studies, this study also considers the impact of event pattern abstraction on the understandability of the final process models, approximated by a broad set of complexity measures.
- This study provides an empirical comparison between LPM and COBPAM in combination with the method presented in [16], providing initial suggestions which of both event abstraction pattern miners performs best.
- This work provides empirical insights into the interaction between the process discovery algorithm and the event abstraction pattern miner with respect to their conjoint impact on fitness, precision and comprehensibility.

The remainder of this paper is structured as follows. Section 2 gives an outline of related work in the domain. Next, Sect. 3 defines the methodology for the benchmark study, as well as elaborates on the experimental design. Section 4 then gives an overview of the experiment's results before Sect. 5 concludes the paper.

2   Related Work

This paper builds further on the work in [16], which studied the use of LPMs to discover event abstraction abstraction patterns in combination with the approach in [14]. However, their work was slightly different than ours, as they only focused on fitness and precision, whereas we also take process model complexity into account.

LPM [23] can be used to discover event abstraction patterns in an unsupervised manner. It extends frequent pattern mining techniques to more complex patterns and aims to describe frequent behaviour in an event log in local patterns. In [21], LPM is extended with utility functions and constraints to mine more meaningful patterns, while [9] shows how high quality sets of a limited number of LPMs can be constructed. However, both latter approaches require some kind of domain expert interaction, which puts them outside the scope of our study.

Inspired by LPM, Acheli et al. [3] designed the Combination based Behavioural Pattern Mining (COBPAM) approach. It exploits a partial order on potential patterns to discover only those that are compact and maximal, i.e. least redundant.

The recently conducted review study of van Zelst et al. [25] proposes a taxonomy of supervised and unsupervised techniques for event abstraction. Some interesting unsupervised ones are global trace segmentation [11], HLPM-Mine [10], Bose et al. [8], Alharbi et al. [6], RefMod-Miner [18] and the work of Sánchez-Charles et al. [19]. This study focuses on LPM and COBPAM as the setup under consideration employs the approach in [16], which requires a defined process pattern between the low-level events in the event abstraction pattern. Another approach producing compatible patterns for this setup would be the RefMod miner [18]. Unfortunately, as no public implementation was available for this technique, it was not considered in this study.

It is worth noting that the combination of event abstraction pattern miners and the technique in [16] is not the only possibility to discover higher-level process models from low-level event data. For example, [20] presents a framework designed to transform location sensor data to an event log via interaction mining that business users can understand.

3   Methodology and Experimental Setup

This study takes the following algorithmic problem class as a starting point:

> A process model discovered from an event log is too complex to understand because the event log is too fine-grained. What is needed, is a technique which augments the event log to a higher abstraction level such that this transformed event log results in less complex discovered process models which are still correctly representing the underlying process.

This paper considers an algorithmic design to tackle this problem based on the approach in [14] in combination with two unsupervised abstract pattern discovery techniques, i.e. LPM [23] and COPBAM [3].

The quality of the algorithm design is defined on three criteria. Firstly, we want the process model discovered from the transformed log to be more comprehensible. The second and third criteria state that the model discovered from the transformed log should remain fitting and precise with respect to the original data.

3.1   Evaluation Method

To evaluate the comprehensibility of the model discovered from the transformed log, we use complexity as a proxy, which has been shown to be inversely related to the model understandability [17]. In total, ten complexity metrics were used, covering the four complexity dimensions identified in [13][1]. These complexity metrics are computed for the process models discovered from the transformed event logs.

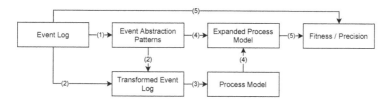

Fig. 1. Method to evaluate fitness and precision: (1) event abstraction patterns are discovered from the original event log, (2) the event log is transformed to higher abstraction level, (3) a process model is discovered, (4) the process model is expanded with the event abstraction patterns to the original granularity level, (5) precision and fitness are computed.

To evaluate the fitness and precision of the model discovered from the transformed event log with respect to the original event log, the same approach as in [16] was used. First, event abstraction patterns are discovered from the original event log. These event abstraction patterns map a local process pattern, defined at the original granularity level, to a higher level activity. Second, these patterns are used to transform the event log. Third, a process model is discovered from this transformed event log. Fourth, the event abstraction patterns are used to expand the process model into an expanded process model which is at the abstraction level of the original event log. This is done by replacing the higher-level activities by its corresponding local pattern. Fifth, the original event log is compared against the expanded process model to calculate fitness and precision values. Fitness is measured by the alignment-based fitness measure [4] and precision is measured with the alignment-based ETC precision measure [5].

3.2   Data

Six publicly available real-life event logs[2] with different characteristics are used in this study. Table 1 illustrates the variation among the logs in terms of number of events, number of activities, number of cases and number of distinct traces.

Regarding the BPI challenge 2019 log, similar to [7], a sample of the event log was taken for performance reasons. We preserved case variants containing at least 50 cases, leaving us with 71% of the events.

---

[1] This is done via the R package `understandBPMN` [13].
[2] The event logs were extracted from the 4TU Centre for Research Data in May 2020.

Table 1. Event log characteristics.

	# events	# activities	# cases	# distinct traces
Road traffic fine management	561.470	11	150.370	231
Hospital billing	451.359	18	100.000	1.020
Sepsis case	15.214	16	1.050	846
BPI 2019	1.135.258	27	224.768	192
BPI 2020 - request for payment	36.796	19	6.886	89
BPI 2020 - domestic declarations	56.437	17	10.500	99

3.3 Experimental Design

Every experiment consists of controlled variables that are of interest to the study. In our setup, the controlled variables are the event abstraction pattern miners and the process discovery algorithms used.

Two event abstraction pattern miners were considered in this study, i.e. LPM and COBPAM. For LPM the ProM implementation [24] was used with default parameter settings, except for the maximum number of transitions (5) and the number of patterns to discover (10). From the 10 patterns discovered, the top 3 according to the model ranking were selected, ignoring patterns subsumed by other better-scoring patterns. For COBPAM, the ProM implementation [24] is used with support threshold, language fit threshold and maximum depth set to respectively 0.7, 0.7 and 2 in accordance to the original work [3]. Patterns are sorted by support value and the top 3 patterns which are not subsumed by other patterns are selected.

Furthermore, two discovery algorithms were used, i.e. the split miner [7] and inductive miner infrequent [12]. Both are configured with their default values. This means there is a conversion step from BPMN to Petri net in the case of split miner[3]. The split miner is implemented as stand-alone Java application, while the inductive miner is accessed via the ProM framework [24].

For all 4 combinations of the 2 event abstraction pattern miners and 2 discovery algorithms, the following experiment was performed. For each event log, an initial process model was discovered and corresponding complexity, fitness and precision values were computed. These values serve as the baseline. Next, event abstraction patterns were mined from the event log and the top three were used to transform the event log to a higher abstraction level using the approach in [14]. The ProM implementation was used [24] and low-level events that were not mapped to higher-level events were kept in the transformed log. All other parameters were set at their default values. Finally, a new process model was discovered from the transformed event log and complexity, precision and fitness

---

[3] Done via the *convert BPMN diagram to Petri Net (Control Flow) plug-in in ProM*.

values were computed as described in Sect. 3.1. These measures can be compared against the baseline values to evaluate the impact of a specific event abstraction pattern miner for a given event log and discovery technique.

4   Empirical Results

This section will explain the results of our experiment per quality dimension. In total, the six event logs, three abstraction levels, two miners and 12 metrics for each model, resulted in 432 metrics. The raw result set is available online[4].

4.1   The Effect of Abstraction on Model Complexity

Complexity is measured by ten metrics in total. These are cognitive weight, token split, connector heterogeneity, control flow complexity, sequentiality, cyclicity, diameter, depth, density and coefficient of network connectivity [13]. From these ten, token split, connector heterogeneity, control flow complexity, sequentiality, cyclicity and the coefficient of network connectivity did not improve on average due to event abstraction for any activity pattern miner, regardless of the process model miner. The results concerning the remaining four are not uniform, however, as is shown in Table 2. The table describes, for each combination of miner and complexity metric, the number of event logs for which an improvement was observed and the average change. Note that a negative delta is considered an improvement, i.e. a reduction in complexity.

Table 2. Abstraction impact on cognitive weight, depth, density and diameter.

		Split miner		Inductive miner	
		# Improvements	Delta	# Improvements	Delta
Cognitive weight	LPM	3/6	−3.37%	2/6	2.54%
	COBPAM	4/6	−2.88%	2/6	3.45%
Depth	LPM	2/6	−7.14%	0/6	33.33%
	COBPAM	4/6	−35.71%	1/6	22.22%
Density	LPM	1/6	12.18%	3/6	−1.52%
	COBPAM	1/6	16.35%	3/6	−0.42%
Diameter	LPM	1/6	4.92%	3/6	−5.41%
	COBPAM	2/6	−2.19%	2/6	2.70%

Cognitive weight, which is the weighted sum of gateways and activities, seems to improve for both LPM and COBPAM when paired with the split miner. Depth, the amount of split minus join gateways, behaves in a similar fashion. The inductive miner has a tendency to generate more (parallel) gateways than

---

[4] https://github.com/gregvanhoudt/UnsupervisedEventAbstraction.

the split miner, and this effect is still present after abstraction. However, it is important to note that the baseline values for the inductive miner are already lower than the split miner's.

Density represents the percentage of sequence flows which are present compared with the theoretical maximum number of sequence flows. It shows the inverse behaviour of cognitive weight and depth, improving when paired with the inductive miner. However, the improvement here is much smaller than the deterioration with the split miner. A small density value indicates that the process is more sequential. The models substantiate this, as the split miner has a tendency to loop back to previous gateways to allow for repetitive behaviour, creating more sequence flows. In that regard, it is possible for depth to improve while density worsens.

The diameter metric only seems to improve, on average, for the combinations COBPAM-split miner and LPM-inductive miner. Also, even if diameter improves on average, the value decreases for the majority of the logs. Given that the results do not seem to correlate with a discovery miner or activity pattern miner, specific conclusions cannot be drawn for this metric. Further experimentation is required to obtain more conclusive results.

When considering all complexity measures simultaneously, we count 15 and 19 improvements for LPM and COBPAM, respectively. Although the difference is small, this seems to indicate COBPAM is slightly better in reducing complexity than LPM, independent of the process discovery algorithm. In general, we can conclude that, in our experimental setting, pattern-based event abstraction does not reduce the complexity of newly learnt models. However, caution is advised as we limited ourselves to only three patterns for each abstraction. Inserting additional patterns might have a positive impact on complexity. Keep in mind this will probably also impact fitness, and potentially precision.

4.2 The Effect of Abstraction on Model Fitness and Precision

The second facet of the study, the accuracy of process models, is measured by fitness and precision. Table 3 summarises the outcomes. A positive delta is now considered an improvement. Note that for precision - LPM - Inductive miner, we were unable to compute two values.

Table 3. Abstraction impact on fitness and precision.

		Split miner		Inductive miner	
		# Improvements	Delta	# Improvements	Delta
Fitness	LPM	0/6	−23.86%	1/6	−13.00%
	COBPAM	0/6	−32.04%	0/6	−22.93%
Precision	LPM	4/6	0.67%	1/4	−34.25%
	COBPAM	1/6	−4.78%	1/6	−26.00%

Regarding fitness, the data shows there is a severe negative effect: only one improvement is observed. Performing a t-test at the 5% significance level, the only insignificant difference was for LPM in combination with the inductive miner. Of course, the numbers have to be nuanced as the split miner generated a baseline of nearly perfect fitness, so an increase will be very difficult to accomplish. However, the magnitude of the decrease makes clear automated pattern-based abstraction negatively affects the fitness of new high-level process models.

One possible explanation is the overlap between activity patterns. Recall that fitness can only be calculated after the high-level model is expanded to again include the low-level event classes. This means one event class can now be present at multiple locations in the model. This can result in the obligation of an event class to be executed multiple times according to the high-level process models, which is not the case on the lower level. Also, the overlaps make it unclear which low-level event belongs to which high-level activity [16], generating potential confusion during the abstraction of the event log.

The precision metric also shows clear evolutions, although not as uniform as fitness. In fact, the average precision metric for LPM in combination with the split miner increased. However, the differences were not significant at the 5% significance level. A potential reason for decreases of precision is that we assume parallel relations between activity patterns. Should patterns overlap, it could be more reasonable to state that two patterns cannot co-exist. If two overlapping high-level patterns are present in the model, this is an introduction of additional behaviour. On the other hand, the goal of event log abstraction is hiding/grouping low-level behaviour, which should have a positive influence on precision.

In general, fitness only increased once for LPM without observing any improvements for COBPAM. The precision metric improved 5 and 2 times for LPM and COBPAM respectively, with the majority of improvements located at LPM-split miner in particular. Results suggest that LPM performs better when interested in fitness and precision of abstracted models.

4.3   Discussion

Overall, the study shows that the use of event abstraction pattern miners to transform an event log with the purpose of discovering less complex process model with good fitness and precision, has limited success. On average, it seems that this approach, using either LPM or COPBAM, results in a decrease of fitness and precision, with only limited effect on complexity.

Fitness typically takes a hit when abstracting the event log, which is not completely unexpected as abstraction patterns hide complex behaviour which can no longer be accounted for by the miner. It is also remarkable that the impact on fitness appears to be correlated to the process discovery algorithm. Before abstraction, the split miner produces the best-fitting models. After abstraction, however, this fitness drops heavily, to the extent that the inductive miner produces better-fitting models at that stage.

Precision also has a tendency to deteriorate, with the exception for the combination of LPM with split miner. For this combination, in the majority of the cases we saw an improvement of precision and the average effect was also positive. It is remarkable that this result is not observed for the combination with COBPAM and that LPM cannot reproduce these effects with inductive miner. This again confirms the pattern that there is some kind of interaction between the process discovery algorithm and the event abstraction pattern miner.

As for complexity, for most of the measures no clear improvement was observed. The only pattern that could be distinguished, which supports the goal of this approach, is the slight improvement of cognitive weight and depth for split miner and the improvement of density for inductive miner. Again, these results hint at an interaction between the discovery miner and abstraction pattern miner.

Overall, we can conclude that this approach combined with LPM and COBPAM has limited results. Future research will be needed to improve these results in order to make them impactfull enough for practical use. Based on our empirical analysis, a potential direction for future research is to delve into the interaction between the discovery algorithm and the abstraction pattern miner. It is clear that there are mechanisms at work and understanding these could open up avenues for improved algorithms. Another path worth investigating is the automatic discovery of how activity patterns interrelate. The approach in [16] has parameters which define which patterns can or cannot co-exist and in what type of interrelation. The current event abstraction pattern miners do not provide this type of information.

Finally, based on these mixed empirical results, one must be careful to draw strong conclusions with respect to the performance of LPM versus COBPAM. One might suggest that both approaches are competitive to each other, with the exception of the combination of split miner with LPM, which actually appears to improve the precision of the models on average. As with respect to reducing complexity, COBPAM seems to have a small edge over LPM, albeit too small to make conclusive statements.

4.4   Limitations

This experiment can be extended in several ways. First of all, a new approach to evaluate LPMs was recently proposed [9], which is not implemented in our work yet. This new evaluation acknowledges the excessive amount of overlapping patterns and disregards confidence and determinism as quality measures. For COBPAM, we only have access to support and language fit scores. A more advanced scoring and selection technique of activity patterns could have improved the experiment, obtaining less overlapping patterns as with the meaningful LPMs [9].

On the other hand, the current study fixes the number of activity patterns that are taken into consideration to three. Mannhardt and Tax [16] concluded that the optimal number of patterns varied per event log. No doubt the same applies to this study.

Next, this experimental setup uses six event logs. Each of them returns six process models: a low-level, a LPM-abstracted and a COBPAM-abstracted model for both the split miner and the inductive miner. Therefore, this study compares 36 process models. To obtain a larger number of observations to draw conclusions from, this number of event logs can easily be increased.

Finally, as discussed in Sect. 2, the RefMod-Miner also satisfies the requirements to take part of this experiment, yet no public implementation is available.

5   Conclusion

In this paper, local process models and the combination based behavioural pattern mining approach are put against each other in unsupervised event log abstraction. The goal was to produce process models at a higher abstraction level with better comprehensibility, while still being well-fitting and properly precise.

However, the experiments show only limited results. While some aspects of complexity show possibilities for improvements, they seem tied to the process model miner. Fitness gets a significant hit overall and precision only tends to improve for the combination of LPM and Split Miner.

Future research is required with respect to the interactions between activity pattern miners and process discovery algorithms. This could allow for more accurate abstraction techniques. Also, discovering meaningful and more precise activity patterns is an interesting research track. But perhaps more importantly, the possibility to discard our assumption about only parallel inter-pattern relations must be explored. Being able to learn this from data could greatly improve the abstraction quality.

References

1. Van der Aalst, W.: Process Mining: Data Science in Action, pp. 3–23. Springer, Heidelberg (2016). https://doi.org/10.1007/978-3-662-49851-4_1
2. van der Aalst, W.M.P., Weijters, T., Maruster, L.: Workflow mining: discovering process models from event logs. IEEE Trans. Knowl. Data Eng. **16**(9), 1128–1142 (2004)
3. Acheli, M., Grigori, D., Weidlich, M.: Efficient discovery of compact maximal behavioral patterns from event logs. In: Giorgini, P., Weber, B. (eds.) CAiSE 2019. LNCS, vol. 11483, pp. 579–594. Springer, Cham (2019). https://doi.org/10.1007/978-3-030-21290-2_36
4. Adriansyah, A., van Dongen, B.F., van der Aalst, W.M.: Conformance checking using cost-based fitness analysis. In: 2011 IEEE 15th International Enterprise Distributed Object Computing Conference, pp. 55–64. IEEE (2011)
5. Adriansyah, A., Munoz-Gama, J., Carmona, J., van Dongen, B.F., van der Aalst, W.M.: Measuring precision of modeled behavior. Inf. Syst. e-bus. Manag. **13**(1), 37–67 (2015)
6. Alharbi, A., Bulpitt, A., Johnson, O.A.: Towards Unsupervised Detection of Process Models in Healthcare. Studies in Health Technology and Informatics, pp. 381–385. IOS Press, Netherlands (2018)

7. Augusto, A., Conforti, R., Dumas, M., La Rosa, M., Polyvyanyy, A.: Split miner: automated discovery of accurate and simple business process models from event logs. Knowl. Inf. Syst. **59**(2), 251–284 (2019)

8. Jagadeesh Chandra Bose, R.P., van der Aalst, W.M.P.: Abstractions in process mining: a taxonomy of patterns. In: Dayal, U., Eder, J., Koehler, J., Reijers, H.A. (eds.) BPM 2009. LNCS, vol. 5701, pp. 159–175. Springer, Heidelberg (2009). https://doi.org/10.1007/978-3-642-03848-8_12

9. Brunings, M., Fahland, D., van Dongen, B.: Defining meaningful local process models. In: Proceedings of the International Workshop on Algorithms and Theories for the Analysis of Event Data 2020. CEUR-WS.org (2020). http://ceur-ws.org/Vol-2625/paper-01.pdf

10. Folino, F., Guarascio, M., Pontieri, L.: Mining multi-variant process models from low-level logs. In: Abramowicz, W. (ed.) BIS 2015. LNBIP, vol. 208, pp. 165–177. Springer, Cham (2015). https://doi.org/10.1007/978-3-319-19027-3_14

11. Günther, C.W., Rozinat, A., van der Aalst, W.M.P.: Activity mining by global trace segmentation. In: Rinderle-Ma, S., Sadiq, S., Leymann, F. (eds.) BPM 2009. LNBIP, vol. 43, pp. 128–139. Springer, Heidelberg (2010). https://doi.org/10.1007/978-3-642-12186-9_13

12. Leemans, S.J.J., Fahland, D., van der Aalst, W.M.P.: Process and deviation exploration with inductive visual miner. In: Limonad, L., Weber, B. (eds.) Business Process Management Demo Sessions (BPMD 2014). CEUR Workshop Proceedings, vol. 1295, pp. 46–50. Eindhoven, The Netherlands. CEUR-WS.org (2014)

13. Lieben, J., Jouck, T., Depaire, B., Jans, M.: An improved way for measuring simplicity during process discovery. In: Pergl, R., Babkin, E., Lock, R., Malyzhenkov, P., Merunka, V. (eds.) EOMAS 2018. LNBIP, vol. 332, pp. 49–62. Springer, Cham (2018). https://doi.org/10.1007/978-3-030-00787-4_4

14. Mannhardt, F., de Leoni, M., Reijers, H.A., van der Aalst, W.M.P., Toussaint, P.J.: From low-level events to activities - a pattern-based approach. In: La Rosa, M., Loos, P., Pastor, O. (eds.) BPM 2016. LNCS, vol. 9850, pp. 125–141. Springer, Cham (2016). https://doi.org/10.1007/978-3-319-45348-4_8

15. Mannhardt, F., de Leoni, M., Reijers, H.A., van der Aalst, W.M., Toussaint, P.J.: Guided process discovery-a pattern-based approach. Inf. Syst. **76**, 1–18 (2018)

16. Mannhardt, F., Tax, N.: Unsupervised event abstraction using pattern abstraction and local process models. In: Gulden, J., et al. (eds.) CEUR Workshop Proceedings, vol. 1859, pp. 55–63. CEUR-WS.org (2017)

17. Mendling, J.: Metrics for Process Models: Empirical Foundations of Verification, Error Prediction, and Guidelines for Correctness, vol. 6. Springer Science & Business Media (2008)

18. Rehse, J.-R., Fettke, P.: Clustering business process activities for identifying reference model components. In: Daniel, F., Sheng, Q.Z., Motahari, H. (eds.) BPM 2018. LNBIP, vol. 342, pp. 5–17. Springer, Cham (2019). https://doi.org/10.1007/978-3-030-11641-5_1

19. Sánchez-Charles, D., Carmona, J., Muntés-Mulero, V., Solé, M.: Reducing event variability in logs by clustering of word embeddings. In: Teniente, E., Weidlich, M. (eds.) BPM 2017. LNBIP, vol. 308, pp. 191–203. Springer, Cham (2018). https://doi.org/10.1007/978-3-319-74030-0_14

20. Senderovich, A., Rogge-Solti, A., Gal, A., Mendling, J., Mandelbaum, A.: The ROAD from sensor data to process instances via interaction mining. In: Nurcan, S., Soffer, P., Bajec, M., Eder, J. (eds.) CAiSE 2016. LNCS, vol. 9694, pp. 257–273. Springer, Cham (2016). https://doi.org/10.1007/978-3-319-39696-5_16

21. Tax, N., Dalmas, B., Sidorova, N., van der Aalst, W.M., Norre, S.: Interest-driven discovery of local process models. Inf. Syst. **77**, 105–117 (2018)
22. Tax, N., Sidorova, N., Haakma, R., van der Aalst, W.M.P.: Event abstraction for process mining using supervised learning techniques. In: Bi, Y., Kapoor, S., Bhatia, R. (eds.) IntelliSys 2016. LNNS, vol. 15, pp. 251–269. Springer, Cham (2018). https://doi.org/10.1007/978-3-319-56994-9_18
23. Tax, N., Sidorova, N., Haakma, R., van der Aalst, W.M.: Mining local process models. J. Innov. Digit. Ecosyst. **3**(2), 183–196 (2016)
24. van Dongen, B.F., de Medeiros, A.K.A., Verbeek, H.M.W., Weijters, A.J.M.M., van der Aalst, W.M.P.: The ProM framework: a new era in process mining tool support. In: Ciardo, G., Darondeau, P. (eds.) ICATPN 2005. LNCS, vol. 3536, pp. 444–454. Springer, Heidelberg (2005). https://doi.org/10.1007/11494744_25
25. van Zelst, S.J., Mannhardt, F., de Leoni, M., Koschmider, A.: Event abstraction in process mining: literature review and taxonomy. Granular Comput. 1–18 (2020)

1st International Workshop on Leveraging Machine Learning in Process Mining (ML4PM)

1st International Workshop in Leveraging Machine Learning for Process Mining (ML4PM 2020)

The field of Machine Learning (ML) continues to grow, with new and promising techniques, and with applications across numerous areas. In the past few years, we have seen strong interest from both industry and academia in leveraging ML techniques in the Process Mining (PM) field. The application of ML to PM is today discussed as the emerging technology that will foster a new paradigm for improving business process management by enabling process task automation and simplification. The intent of the 1st International Workshop on Leveraging Machine Learning for Process Mining was to establish a venue to discuss recent research developments at the intersection of ML and PM by bringing together practitioners and researchers from both communities who are interested in making the ML-generation of PM tools a reality. The open call for contributions solicited submissions in the areas of automated process modeling, predictive process mining, application of deep learning techniques, and online process mining. The workshop attracted fifteen submissions from six different countries (Germany, Italy, Spain, France, Norway, and Egypt), which shows the liveliness of this field. From the received fifteen submissions, seven submissions were passed to the review process and accepted for presentation at the workshop. Each paper was reviewed by three members of the program committee. Papers presented at the workshop were also selected for inclusion in the post-proceedings. These articles are briefly summarized below.

The paper by Baskharon et al. describes a novel ML algorithm to predict the remaining cycle time of running cases. It resorts to survival analysis to learn from incomplete ongoing traces.

The paper by Quishpi et al. investigates how to determine hierarchical/tree patterns through inter-sentence analysis, in order to improve the extraction of process annotations from text.

The paper by Peeperkorn et al. tackles the problem of conformance checking in the supervised setting by training a Recurrent Neural Network classifier.

The paper by Nguyen et al. explores the use of time-aware LSTM (T-LSTM) cells in predictive process monitoring by formulating a cost-sensitive learning approach to account for the common class imbalance in event logs.

The paper by Boltenhagen et al. applies Recurrent Neural Networks and Random Forest classifiers to the problem of classifying traces based on their alignment costs to a reference process model.

The paper of Chiorrini et al. explores the opportunities and issues of applying reinforcement learning to tasks of predictive process monitoring.

The paper of Luettgen et al. illustrates the design of a learning technique that uses word embeddings to encode process cases and evaluates the proposed approach in the context of trace clustering.

In addition to these seven papers, the program of the workshop included the invited talk "Applying AI to BPM: opportunities and pitfalls", given by Ernesto Damiani. Around 120 attendees were present during the workshop presentations, talk, and discussion.

We would like to thank all the authors who submitted papers for publication in this book. We are also grateful to the members of the Program Committee and external referees for their excellent work in reviewing submitted and revised contributions with expertise and patience. Our event received a total of 461 attendees, a fact that demonstrates the interest in the area and paves the way for new editions.

January 2021

Paolo Ceravolo
Sylvio Barbon Jr.
Annalisa Appice

Organization

Workshop Chairs

Paolo Ceravolo ⓘ   Università degli Studi di Milano, Italy
Sylvio Barbon Jr. ⓘ   State University of Londrina, Brazil
Annalisa Appice ⓘ   Università degli Studi di Bari, Italy

Program Committee

María Teresa Gómez-López   Gómez University of Seville
Rafael Accorsi   PwC Digital Services
Niek Tax   Booking.com
Josep Carmona   Universitat Politècnica de Catalunya
Ernesto Damiani   Khalifa University
Chiara Di Francescomarino   Fondazione Bruno Kessler
Antonella Guzzo   Università della Calabria
Mariangela Lazoi   University of Salento
Matthias Ehrendorfer   University of Vienna
Fabrizio Maria Maggi   Free University of Bozen-Bolzano
Paola Mello   Università di Bologna
Gabriel Marques Tavares   Università degli Studi di Milano
Emerson Cabrera Paraiso   Pontifícia Universidade Católica do Paraná
Bruno Bogaz Zarpelão   State University of Londrina
Irene Teinemaa   Booking.com
Michelangelo Ceci   University of Bari Aldo Moro
Natalia Sidorova   Eindhoven University of Technology
Domenico Potena   Università Politechnica delle Marche

Additional Reviewers

Mathilde Boltenhagen
Graziella De Martino
Angelo Impedovo
Vincenzo Pasquadibisceglie

Predicting Remaining Cycle Time from Ongoing Cases: A Survival Analysis-Based Approach

Fadi Baskharon[1], Ahmed Awad[1,2(✉)] (iD), and Chiara Di Francescomarino[3]

[1] Information Technology and Computer Science School, Nile University, Giza, Egypt
{f.zaki,aawad}@nu.edu.eg
[2] University of Tartu, Tartu, Estonia
[3] Fondazione Bruno Kessler, Trento, Italy
dfmchiara@fbk.eu

Abstract. Predicting the remaining cycle time of running cases is one important use case of predictive process monitoring. Different approaches that learn from event logs, e.g., relying on an existing representation of the process or leveraging machine learning approaches, have been proposed in literature to tackle this problem. Machine learning-based techniques have shown superiority over other techniques with respect to the accuracy of the prediction as well as freedom from knowledge about the underlying process models generating the logs. However, all proposed approaches learn from *complete* traces. This might cause delays in starting new training cycles as usually process instances might last over long time periods of hours, days, weeks or even months.

In this paper, we propose a machine learning approach that can learn from *incomplete* ongoing traces. Using a time-aware survival analysis technique, we can train a neural network to predict the remaining cycle time of a running case. Our approach accepts as input both complete and incomplete traces. We have evaluated our approach on different real-life datasets and compared it with a state of the art baseline. Results show that our approach, in many cases, is able to outperform the baseline approach both in accuracy and training time.

Keywords: Predictive process monitoring · Remaining time prediction · Survival analysis · Incomplete traces

1 Introduction

Predictive process monitoring [7] is a sub-field of process mining that is concerned with predicting an outcome of interest while an execution is still running, for instance with the purpose of proactively taking corrective actions before things go wrong. Different types of outcomes can be predicted, as for instance, the remaining time for a case to finish [11,14,15], which activity to be executed next [3,9], or the fulfillment of a certain goal [4,5]. Different techniques have

© Springer Nature Switzerland AG 2021
S. Leemans and H. Leopold (Eds.): ICPM 2020 Workshops, LNBIP 406, pp. 99–111, 2021.
https://doi.org/10.1007/978-3-030-72693-5_8

been used to tackle the prediction challenge, such as machine learning, statistical methods, annotated transition systems and hybrid approaches [7]. However, all these techniques require as a major input a history of *complete* cases, in order to train a model that will be used at run-time to predict the respective outcome for running cases.

Among the techniques for predictive process monitoring, machine learning and deep learning-based approaches have shown superiority with respect to the accuracy of the prediction [14,17]. Yet, as known, training deep learning models requires more resources and a large dataset to obtain better results. Moreover, process instances generally run for long time, days, weeks or even months. A direct threat in this case is the possibility of concept drifts [6] that render the currently-used model for prediction useless and triggers the need to train a new model on a larger set of traces containing newly complete traces. In such case, the retraining has to be delayed until a sufficiently large set of newly completed cases has been collected.

A main limitation of contemporary predictive monitoring techniques is the need for *complete* traces to train their models. This causes delays of retraining cycles until new completed cases are collected. In this paper, we alleviate this limitation by allowing learning from ongoing cases, i.e. *incomplete* traces, by employing survival analysis techniques [2]. Treating incomplete traces as censored data, we are able to train a neural network to predict the remaining time for a running case. Compared to the state-of-the-art, our results show at least comparable accuracy to methods that require complete traces with better results on several data sets; additionally, our training takes much less time to complete.

The rest of this paper is organized as follows. Section 2 summarizes the related work. Section 3 provides the necessary details about survival analysis and the specific techniques we employ. Our contributions are detailed in Sect. 4 for the encoding of (incomplete) traces and the architecture of the neural network and Sect. 5 for the experimental evaluation and comparison to the baseline method. Finally, Sect. 6 concludes the paper.

2   Related Work

In general, literature on predictive process monitoring can be classified based on the type of targeted prediction: next activity, outcome, delays or the remaining cycle time. Due to space limitations, we will discuss the literature related to the prediction of the remaining cycle time, as this is the focus of our work. We refer the user to [7] for a survey on the wider domain of predictive process monitoring.

A recent survey by Verenich et al. [17] has benchmarked the different approaches for predicting the remaining cycle time of a running instance. In general, prediction approaches have been categorized as generative or discriminative. Generative approaches are process-aware, that is, they require a pre-existing representation of the process whose execution generates the traces. Discriminative approaches, instead, are process-agnostic and can learn directly from traces.

Concerning the generative approaches, in [15] the authors discover a transition system from the log and augment it with information about the remaining time of cases; in [11] stochastic Petri nets are leveraged for making predictions; in [18], flow analysis is used to aggregate the remaining cycle time on the case level over its individual activities.

Discriminative methods can rely on different approaches. Some of them leverage non-parametric regression models [12,16], others propose clustering-based techniques, as the work in [1], while others rely on neural networks [14] for the prediction of the remaining cycle time. We will use this latter approach and the work in [14] as baseline for our work.

2.1 Baseline Approach

The baseline approach in [14] predicts the next activity in a process as well as its timestamp. Each event occurring in the trace is transformed into a feature vector $x_1, ..., x_k$ to be fed as input to an LSTM network as follows: (i) **activity type (A)**, i.e., the type of the activity in a one-hot-encoding representation; (ii) **delta t** (fv_{t1}), i.e., the time between the previous and the current event in the trace that allows the network to learn the time dependencies between the process' events; (iii) **two time-based features,** (fv_{t2}) **and** (fv_{t3}), that correspond to the hour of the event within the day in 24-h format and the hour of the event since the start of the week, respectively, so as to learn when the event has happened with respect to a working day or a working week. The LSTM has two outputs: (i) O_a^k that corresponds to a one-hot-encoding representation of the type of the next event with an extra bit representing whether or not this event occurs at the end of the case; and (ii) O_t^k representing the relative time difference between the current event and the next event. The remaining cycle time can be computed by summing O_t^k for all the events from the current event until the last event of the trace. This model needs to be trained with complete traces, as the model needs to learn the process sequence, as well as when the sequence ends. The model also suffers when dealing with sequences containing loops, as loops cause the model to predict overly long sequences.

3 Background

In this section we will briefly introduce the notations we use throughout the rest of the paper and a quick overview on survival analysis as the inspiring method to the contribution of this paper.

3.1 Events, Trace, Logs

The occurrence of an event is the manifestation of the evolution of a running process. Each event contains at least three pieces of information: a reference to the activity, a reference to the process instance, and a timestamp. Events can have more information, e.g., cost, the human resource who executed it, and the lifecycle transition, e.g., *started, completed, etc.* In this paper, we require just the three basic pieces of information.

Let $\mathcal{A}$ be the set of all activities that an event can reference. The set $\mathcal{A}_{end} \in \mathcal{A}$ contains end activities. Additionally, the set $\mathcal{T}$ is the time domain and $\mathcal{C}$ is the universe of case identifiers. Finally, the set $\mathcal{E}$ is the universe of events. Thus, an event $e \in \mathcal{E}$ represents the execution of some activity $a \in \mathcal{A}$ within a case $c \in \mathcal{C}$ that occurred at time $t \in \mathcal{T}$. A shorthand for these notations are $e.a, e.c, e.t$ respectively.

Definition 1 (Trace). *A trace is a finite non-empty sequence of events $\sigma \in \mathcal{E}*$. $|\sigma|$ defines the length of the trace. $\sigma_i \in \mathcal{E}$ is the event at position $i, 1 \leq i \leq |\sigma|$. A trace is called* complete *if and only if $\sigma_{|\sigma|}.a \in \mathcal{A}_{end}$, otherwise it is* incomplete.

A prefix of a trace is defined as a function $pre : \mathcal{E} * \times \mathcal{N} \to \mathcal{E}*$ that returns a sub-sequence of a trace σ up to and including the event at position i in the trace.

A log $L \subset \mathcal{E}*$ is a set of traces where each trace appears at most once.

3.2 Survival Analysis and Censored-Learning

Survival analysis models [2] are key players in statistical studies that focus on analyzing the waiting duration or the remaining time until an event happens, such as a death, failure, churn, or any other event of interest. These kinds of models are capable of answering even more complex questions like "What is the probability that an event does/doesn't happen within an amount of time T?", "What is the probability distribution of the event occurrence over time?" or "What is the proportion of a population which will survive passed a certain time?".

Let T be a random variable denoting the waiting time until the occurrence of an event, survival models provide information about the probability density function $f(t) = Pr(T = t)$ and the survival function $S(t) = Pr(T > t)$, among other functions. The probability density function gives information about the likelihood of the occurrence of the event at time t. The survival function represents the probability of surviving until a certain time t without experiencing the event of interest (See Fig. 1 for the difference between the two functions).

There are three types of survival models: Non-parametric, semi-parametric, and parametric approaches [13]. Only the latter has the ability to extrapolate or predict beyond the data time limit, since it fits the survival curve to a time distribution. There are many suitable distributions to represent a time-based random variable T as stated in [10], such as exponential, log-logistic, log-normal, gamma, Poisson, Geometric, and the Weibull distribution.

A capability of survival models is the ability to learn from unobserved events which are called "censored data". That is, the collected data represent a snapshot in time, where some samples have experienced the event of interest, but most of the samples have not. Yet, these samples contain useful information telling that "at least we have not observed an event until the time t".

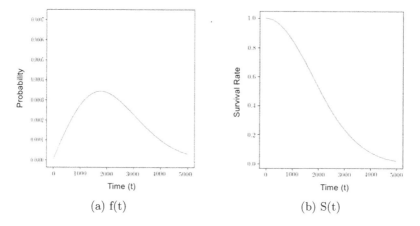

(a) f(t)   (b) S(t)

Fig. 1. Probability density function vs. Survival function

Figure 2 illustrates the random variable T that represents the time to an event of interest. We say that T is observed when we observe the actual waiting time within the period of our study, hence we call it **un-censored (observed) event** $T \in [0,t]$ (e1 in Fig. 2), where the event of interest is exactly observed. We say that T is censored when we partially know about the event occurrence time. We have three types of censored data [13]:

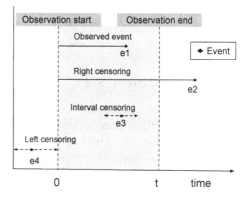

Fig. 2. Observed Vs. Censored events

Right-censored event $T \in [t, \infty)$ refers to cases that have started with the study but the event of interest was not observed during the study time (e2 in Fig. 2). Right-censored events are useful under the assumption of *non-informative censoring*. That is, censoring is independent of the likelihood of the occurrence of the event of interest. In other words, we assume that the cases whose data are censored would have the same distribution of time to event if they were actually observed. **Interval-censored** event $T \in [a,b]$, *where* $0 < a, b < t$, rather than knowing the exact time of the event, all we know is that the event occurred between two known time points (e3 in Fig. 2). **Left-censored** Mathematically, left censored is no different from interval censoring. It indicates that the event occurred at some point prior to the period of study (e4 in Fig. 2).

In the context of predictive process monitoring, only observed events and right-censored events are relevant. That is because, for the set of traces used for learning, we either know the exact end time of the trace, or we don't capture the trace end in our study (in case of incomplete traces).

4   Learning from Incomplete Traces

Traditional survival models are designed to handle records with static features that affect the waiting time. It is not meant to handle time varying features. In the context of predictive process monitoring, cases are time varying as the same activity may have different durations across different cases.

Martisson [8] proposed a model that benefits from the survival analysis interpretation and is able to deal with the time varying features by training a gated recurrent neural network (GRU) that captures the temporal relations between the time steps. The network is trained to predict the parameters of the Weibull distribution by optimizing the log of a special likelihood function to consider both observed and censored events, as explained in Sect. 4.2. The Weibull distribution turns out to be a suitable choice since it is controlled by two parameters α and β, that makes it flexible to interpret complex outputs, and because of its ability to fit both discrete and continuous problems.

In our work, we adapt the network architecture of the baseline method [14] by changing the encoding of the input traces to account for incomplete traces (Definition 1) and adapt the likelihood function from [8] to train the network to predict the parameters of the Weibull distribution that fits the time to the end event of a case. We discuss these two steps in detail in the following subsections.

4.1   Neural Network Setup

The problem of measuring the remaining time till a process ends can be tackled using a similar approach like [8]. However, the original work was designed to predict the waiting time to recurrent events, e.g. the time to the next failure of a machine. In our case, we are interested in the time until a process instance ends.

To adapt the approach to the prediction of the remaining time to an end event, we kept the same loss function and Weibull parameters as in [8]. Then, we adapted the network design from many-to-many to many-to-one to account for non-recurrent end events.

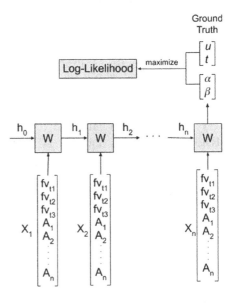

Fig. 3. Network architecture

Considering a log $\mathcal{L}$, we use N_{max} to denote the length of the longest trace $\sigma_l \in \mathcal{L}$. We train a model for each possible prefix $pre(\sigma, p)$ (Definition 1), where $\sigma \in \mathcal{L}$ and $p \in \{2, 3, 4, \ldots, N_{max} - 1\}$.

As shown in Fig. 3, the model consists of an input layer $[X_1, X_2, \ldots X_n]$, which is the vector representation of the prefix, as explained in Sect. 2.1, connected to two GRU layers; and a dense layer of 2 neurons for the output representing the α and the β of the distribution of the random variable T for a given trace. Having the Weibull distribution, we need to find the most likely value in the curve, which turns out to be the mode of the distribution.

4.2 Optimization Function

Let T be a random variable for the waiting time having some parameters θ, and t the observed cycle time, we are interested in the negative likelihood as a loss function. In other words, we aim at maximizing the likelihood of T being around the true observation t for complete traces or at pushing it to the right beyond the censored point t in case of incomplete traces:

$$\mathcal{L}(t, \theta) \propto \begin{cases} P(T = t|\theta) & \text{Observed events (complete trace)} \\ P(T > t|\theta) & \text{Censored events (incomplete trace)} \end{cases} \tag{1}$$

This can be expressed mathematically as follows (detailed proof in [8]):

$$\mathcal{L}(t, \theta) \propto log\left[f_T(t)^u S_T(t)^{1-u}\right]$$
$$= u.\left[\beta.log(\frac{t}{\alpha}) + log(\beta)\right] - \left(\frac{t}{\alpha}\right)^\beta \tag{2}$$

where u is the event indicator, meaning that $u = 1$ in case of observed events and $u = 0$ in case of censored events. This is equivalent to optimizing θ to maximize the PDF $f_T(t)$ around t for the observed cases, and maximize the survival function $S_T(t)$ beyond t for the censored cases. Figure 4 illustrates what the objective function aims to do.

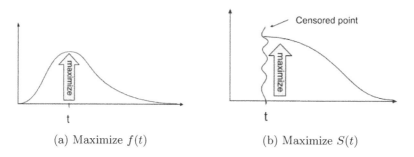

(a) Maximize $f(t)$ (b) Maximize $S(t)$

Fig. 4. Illustration of the optimization function

In order to train the neural network, we need a (t, u) pair for each observation. For complete traces, t is the actual time till the end of the trace and u = 1 means we observed the end of the trace. For incomplete traces, t is the time till the last event observed (not the end event) and u = 0 means that we have not observed the trace end till this time.

5   Evaluation

In this section we report on the implementation, used datasets, procedure and results of the experiments performed for the evaluation of the proposed approach and its comparison with the baseline.

Implementation
We have used Tensorflow 2.0.1 to build and train our network. All experiments were run on an Intel Core i7-8650U CPU @ 1.90 GHz 2.11 GHz. The code for the network and the experiments can be found on our Git hub repo.

Datasets
We have evaluated our approach on four datasets (real life logs) that are described next.

a. **Helpdesk dataset:** This log contains events from a help desk ticketing system of an Italian software company[1]. The process consists of 9 activities, and all cases start with the ticket creation into the system. Each case ends when the issue is resolved and the ticket is closed.
b. **BPI'12 subprocess W dataset:**[2]. The log contains data from the application procedure for financial products at a large financial institution. This process consists of three sub-processes. Two of them have events corresponding to automatic activities, whereas the third sub-process (items sub-process) contains events for manual (human executed) tasks. We only used the items sub-process.
c. **BPI'12 subprocess W dataset with no repetition:** This is the same dataset as "b" but without loops.
d. **Environmental permit dataset:** This is a log of an environmental permitting process at a Dutch municipality[3]. Each case refers to one permit application.

The four datasets have very different characteristics in terms of trace length and the number of unique activities, as shown in Fig. 5 and Fig. 6 respectively. Datasets a and c are the simplest with few unique activities and short traces, while dataset b contains loops in the process which affects the performance of the baseline method. Finally, dataset d is considered to be the most complex with very long traces and different activities. Moreover, it also presents a large variation of the distribution of the log duration.

---

[1] https://doi.org/10.17632/39bp3vv62t.1
[2] https://doi.org/10.4121/uuid:3926db30-f712-4394-aebc-75976070e91f
[3] https://doi.org/10.4121/uuid:26aba40d-8b2d-435b-b5af-6d4bfbd7a270

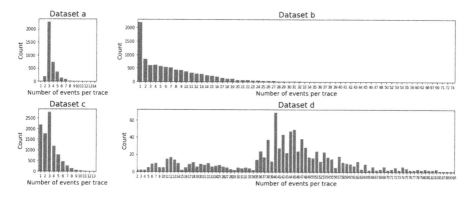

Fig. 5. Length of traces in each dataset

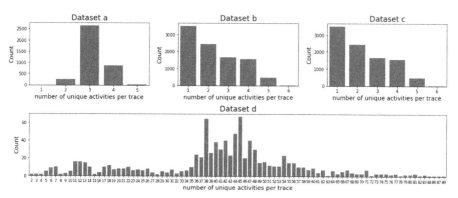

Fig. 6. Unique activities per trace in each dataset

Experimental Procedure

Each dataset is pre-processed in order to (i) remove the traces with length below $p + 1$, for each prefix $p \in \{2, ..., N_{max} - 1\}$, with N_{max} the longest trace of the log; and (ii) build the needed features using the first p events and the time till the last event. The dataset is then split into three equal parts: training set (TS), validation set and test set.

In order to evaluate the effect of censored data, we built a special training set (TS_C) composed of 50% of complete traces and 50% of incomplete traces[4]. To this aim, we split the training set TS further into two sets of traces: the first representing observed cases and the second representing censored cases, which are simulated by randomly cutting the traces. If n is the trace length and p the length of the prefix used for the features, the time till the last observed event is computed looking at the event at position n, if the trace belongs to the first set, or looking at the event in position j, where j is randomly chosen between

---

[4] The choice of 50% as a fixed ratio for complete/incomplete traces is to reduce the variables in the experiments for better comparison.

$p + 1$ to n, if the trace belongs to the second set. For censored data, traces need to have at least $p + 2$ events.

The following three experiments have been conducted:

- **Experiment 1**: training on TS using the time to event approach (Sect. 4).
- **Experiment 2**: training on TS using the time to event approach and transformation of the output to a new less biased space. For the output transformation we used a root cubic transformation where $\phi(x) = \sqrt[3]{x}$, and inverse transformation where $f(x) = x^3$, which turns out to give the best results given the output distribution (See Fig. 7).
- **Experiment 3**: training on TS_C using the time to event approach and transformation of the output to a new less biased space.

We used the same hyperparameters for simplicity, and preserve the same validation and test sets per dataset and prefix throughout the 3 models. The validation set is used to avoid over-fitting, and the test set to compute the performance.

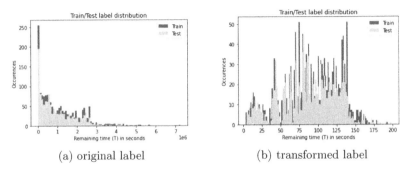

(a) original label   (b) transformed label

Fig. 7. Label distribution before and after transformation

Results

We have executed the three experiments described above and compared the results with the ones of the baseline [14] in terms of the mean absolute error (MAE) expressed in days. Figure 8 summarizes the results.

In general, our approach better captures long term dependencies for longer traces (datasets b and d). The accuracy is further enhanced when transforming the label (Experiment 2). With 50% censored data, we find that the accuracy of the model is not harmfully impacted. Instead, it sometimes outperforms the baseline due to its ability to further remove any bias. The number on top of each prefix represents the number of traces used for testing. Obviously, it decreases as the prefix increases since we have less traces matching the length criteria. Yet, our model is able to learn from the smallest set and outperforms the baseline.

There is a little variation in performance with datasets a and c due to their short traces and limited number of possible activities. However, we can see that training with 50% of right-censored traces improves the model performance in

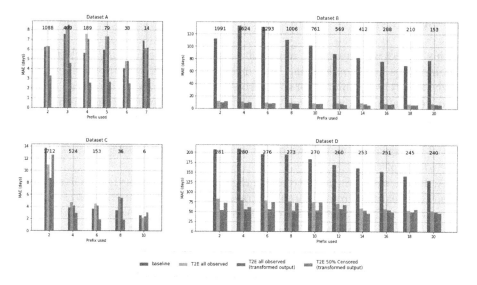

Fig. 8. MAE for experiments 1–3 and the baseline method for the four datasets

almost all the prefixes. This is due to the balance achieved from having both observed and censored traces. In other words, when the majority of the traces are very short, the network tends to predict zero remaining time producing a large MAE for the large traces, affecting the overall performance. The existence of right-censored traces reduces this tendency and forces the network to compromise between short and long traces. Datasets b and d are very challenging due to their very long traces, loops, and random behavior especially in dataset d (Fig. 6). This is obviously affecting the performance of the baseline method since it tries to predict all the remaining events.

Our three experiments have close performance, and surprisingly having 50% of right-censored traces did not harm the performance. This empirically proves that incomplete traces are quite insightful and the network did learn from them.

Figure 9 reports the training time of our approach compared to the baseline. We run the training using the setup mentioned in Sect. 4.1. This huge difference in training time is expected because the network focuses only on predicting the remaining time instead of learning the actual events sequence till the end of the trace.

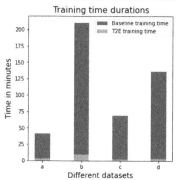

Fig. 9. Training duration

6   Conclusion

In this paper, we present an approach to predict the remaining cycle time of ongoing cases based on learning from incomplete traces. The approach employs survival analysis techniques for this purpose. Our results show, in general, besides a reduced training time, a lower MAE compared to the baseline approach. Using incomplete traces can be useful in several cases, especially when concept drifts occur. Waiting until collecting complete traces, indeed, might compound the impact of degrading model performance as process instances usually take long time to complete.

As future work, we will investigate the applicability of survival analysis techniques and learning from incomplete traces to predict the outcome of a running case. Additionally, we intend to experiment with more data sets and with different percentages of censored traces.

References

1. Folino, F., Guarascio, M., Pontieri, L.: Discovering context-aware models for predicting business process performances. In: Meersman, R., et al. (eds.) OTM 2012. LNCS, vol. 7565, pp. 287–304. Springer, Heidelberg (2012). https://doi.org/10.1007/978-3-642-33606-5_18
2. Klein, J.P., Moeschberger, M.L.: Survival Analysis: Techniques for Censored and Truncated Data. Springer, New York (2006). https://doi.org/10.1007/978-1-4757-2728-9
3. Lakshmanan, G.T., Shamsi, D., Doganata, Y.N., Unuvar, M., Khalaf, R.: A Markov prediction model for data-driven semi-structured business processes. Knowl. Inf. Syst. **42**(1), 97–126 (2013). https://doi.org/10.1007/s10115-013-0697-8
4. Leontjeva, A., Conforti, R., Di Francescomarino, C., Dumas, M., Maggi, F.M.: Complex symbolic sequence encodings for predictive monitoring of business processes. In: Motahari-Nezhad, H.R., Recker, J., Weidlich, M. (eds.) BPM 2015. LNCS, vol. 9253, pp. 297–313. Springer, Cham (2015). https://doi.org/10.1007/978-3-319-23063-4_21
5. Maggi, F.M., Di Francescomarino, C., Dumas, M., Ghidini, C.: Predictive monitoring of business processes. In: Jarke, M., et al. (eds.) CAiSE 2014. LNCS, vol. 8484, pp. 457–472. Springer, Cham (2014). https://doi.org/10.1007/978-3-319-07881-6_31
6. Maisenbacher, M., Weidlich, M.: Handling concept drift in predictive process monitoring. In: SCC, pp. 1–8. IEEE Computer Society (2017)
7. Márquez-Chamorro, A.E., Resinas, M., Ruiz-Cortés, A.: Predictive monitoring of business processes: a survey. IEEE Trans. Serv. Comput. **11**(6), 962–977 (2018)
8. Martinsson, E.: WTTE-RNN: Weibull time to event recurrent neural network. Ph.D. thesis, Chalmers University of Technology & University of Gothenburg (2016)
9. Polato, M., Sperduti, A., Burattin, A., Leoni, M.: Time and activity sequence prediction of business process instances. Computing **100**(9), 1005–1031 (2018). https://doi.org/10.1007/s00607-018-0593-x
10. Rodrıguez, G.: Parametric survival models. Princeton University, Rapport technique, Princeton (2010)

11. Rogge-Solti, A., Weske, M.: Prediction of remaining service execution time using stochastic petri nets with arbitrary firing delays. In: Basu, S., Pautasso, C., Zhang, L., Fu, X. (eds.) ICSOC 2013. LNCS, vol. 8274, pp. 389–403. Springer, Heidelberg (2013). https://doi.org/10.1007/978-3-642-45005-1_27

12. Senderovich, A., Di Francescomarino, C., Maggi, F.M.: From knowledge-driven to data-driven inter-case feature encoding in predictive process monitoring. Inf. Syst. **84**, 255–264 (2019)

13. StataCorp LLC: Stata survival analysis reference manual (2017)

14. Tax, N., Verenich, I., La Rosa, M., Dumas, M.: Predictive business process monitoring with LSTM neural networks. In: Dubois, E., Pohl, K. (eds.) CAiSE 2017. LNCS, vol. 10253, pp. 477–492. Springer, Cham (2017). https://doi.org/10.1007/978-3-319-59536-8_30

15. van der Aalst, W.M.P., Schonenberg, M.H., Song, M.: Time prediction based on process mining. Inf. Syst. **36**(2), 450–475 (2011)

16. van Dongen, B.F., Crooy, R.A., van der Aalst, W.M.P.: Cycle time prediction: when will this case finally be finished? In: Meersman, R., Tari, Z. (eds.) OTM 2008. LNCS, vol. 5331, pp. 319–336. Springer, Heidelberg (2008). https://doi.org/10.1007/978-3-540-88871-0_22

17. Verenich, I., Dumas, M., La Rosa, M., Maggi, F.M., Teinemaa, I.: Survey and cross-benchmark comparison of remaining time prediction methods in business process monitoring. ACM Trans. TIST **10**(4), 1–34 (2019)

18. Verenich, I., Dumas, M., La Rosa, M., Nguyen, H.: Predicting process performance: a white-box approach based on process models. J. Softw. Evol. Process. **31**(6) (2019)

Time Matters: Time-Aware LSTMs for Predictive Business Process Monitoring

An Nguyen[1]($\boxtimes$), Srijeet Chatterjee[1], SvenWeinzierl[2], Leo Schwinn[1],
Martin Matzner[2], and Bjoern Eskofier[1]

[1] Department of Computer Science, Friedrich-Alexander-University
Erlangen-Nürnberg (FAU), Erlangen, Germany
{an.nguyen,srijeet.chatterjee,leo.schwinn,bjoern.eskofier}@fau.de
[2] Institute of Information Systems, Friedrich-Alexander-University
Erlangen-Nürnberg (FAU), Nürnberg, Germany
{sven.weinzierl,martin.matzner}@fau.de

Abstract. Predictive business process monitoring (PBPM) aims to predict future process behavior during ongoing process executions based on event log data. Especially, techniques for the next activity and timestamp prediction can help to improve the performance of operational business processes. Recently, many PBPM solutions based on deep learning were proposed by researchers. Due to the sequential nature of event log data, a common choice is to apply recurrent neural networks with long short-term memory (LSTM) cells. We argue, that the elapsed time between events is informative. However, current PBPM techniques mainly use "vanilla" LSTM cells and hand-crafted time-related control flow features. To better model the time dependencies between events, we propose a new PBPM technique based on time-aware LSTM (T-LSTM) cells. T-LSTM cells incorporate the elapsed time between consecutive events inherently to adjust the cell memory. Furthermore, we introduce cost-sensitive learning to account for the common class imbalance in event logs. Our experiments on publicly available benchmark event logs indicate the effectiveness of the introduced techniques.

Keywords: Predictive business process monitoring · Deep learning ·
Recurrent neural network · LSTM · Time-Awareness

1 Introduction

In the last years, a variety of predictive business process monitoring (PBPM) techniques that base on machine learning (ML) were proposed by researchers [6] to improve the performance of operational business processes [4]. PBPM is a class of techniques aiming at predicting future process characteristics in running process instances [12], like next activities, next timestamps or process-related performance indicators. Such PBPM techniques produce predictions through

---

S. Chatterjee—Equal contribution with An Nguyen.

© Springer Nature Switzerland AG 2021
S. Leemans and H. Leopold (Eds.): ICPM 2020 Workshops, LNBIP 406, pp. 112–123, 2021.
https://doi.org/10.1007/978-3-030-72693-5_9

predictive models. These models are in turn constructed based on historical event log data.

A current trend in PBPM is to apply deep neural networks (DNNs) to learn more accurate predictive models from event log data than with "traditional" ML algorithms like probabilistic automata [7]. DNNs belong to the ML-sub-field deep learning (DL) and achieve that by identifying the intricate structures in high-dimensional data through multi-representation learning [11].

Existing DL-based PBPM techniques often rely on DNN architectures consisting of out-of-the-box constructs like layers with a "vanilla" long short-term memory (LSTM) cell [9] or state-of-the-art loss functions for parameter learning.

Event logs can be seen as sequences of events in continuous time with irregular intervals (i.e., elapsed time between consecutive events). We argue that these time intervals are informative in the case of event logs in PBPM. Intuitively, these time intervals describe human behavior of executing business processes. Thus, a time-aware PBPM technique considering information on time intervals could potentially achieve a higher predictive quality. Time information is currently only exploited via hand-crafted control-flow features as inputs to "vanilla" LSTM cells [15]. To better account for the time information in event log data, we propose a new PBPM techniques using time-aware LSTM (T-LSTM). T-LSTM extends the "vanilla" LSTM cells by incorporating the elapsed time between consecutive events in order to adjust the memory state and is inspired by work from Baytas et al. [2].

Furthermore, the problem of next activity prediction is commonly modeled as a supervised multi-class classification problem. The distribution of activities in event logs are commonly skewed. Therefore, we additionally introduce cost-sensitive learning to address the inherent class-imbalances.

The main contributions of this work are summarized below:

- We introduce a time-aware LSTM model for the tasks of predicting next activities and timestamps in PBPM
- We tackle the problem of skewed class distributions via cost-sensitive learning

We evaluate the effectiveness of our proposed techniques by conducting experiments for the next activity and timestamp prediction on publicly available benchmark event logs commonly used for PBPM.

The remainder of the paper is structured as follows: Sect. 2 presents related work on DL-based next activity and timestamp prediction. Section 3 introduces preliminaries and the concept of a LSTM. Sections 4 and 5 describes the architecture of T-LSTM and our experimental setup respectively. Then, in Sects. 6 and 7, we present and discuss our results. Section 8 concludes our paper and discusses future research directions.

2   Related Work

Inspired by the field of natural language processing (NLP), Evermann et al. [7] applied recurrent neural network-based and LSTM-based DNN architectures for the next activity and next sequence of activity prediction in PBPM. They made use of word embeddings to encode activities of event log's process instances.

Navarin et al. [14] used a "vanilla" LSTM-based DNN architecture for predicting the completion time of running process instances. They one-hot encoded the activity attributes, computed temporal control-flow attributes, and considered additional real-valued or categorical context attributes.

Tax et al.[15] proposed a multitask learning approach using "vanilla" LSTM cells for next activity and timestamp prediction respectively. Like in [14], they one-hot encoded the activity and computed temporal control-flow features. However, the authors did not consider additional data attributes in their approach. This work acts as a baseline for a variety of other techniques such as [18].

Khan et al. [10] introduced memory augmented neural networks (MANNs) in PBPM. MANNs reduce the number of trainable parameters. In general, the network's architecture consists of an externalized state memory and two "vanilla" LSTM cells manipulating the memory. One LSTM cell works as encoder and the other one as decoder. Concerning the predictive quality, their approach is comparable to the one presented in [15].

Camagro et al. [5] extended the implementation of [15] and fed the resource attribute into the DNN model. Additionally, instead of one-hot encoding, they applied embeddings, as proposed by Evermann et al. [7].

Taymouri et al. [16] introduced generative adversarial networks (GANs) for the next activity and timestamp prediction. The network's architecture comprises two "vanilla" LSTM cells. One for the generator and the other one for the discriminator.

To date, several studies have investigated DNN-based PBPM techniques. None of the related works proposes a DL-architecture that explicitly models the elapsed time between two successive events. We address this gap by adapting time-aware LSTM cells [2]. Further, Mehdijev et al. [13] tackle the class imbalance problem in the context of the DNN-based prediction of next activities through a second neural network, namely radial basis function neural network, which generates semi-artificial data of the minority class in the pre-processing phase. In contrast, we adapt cost-sensitive learning to investigate the class-imbalance problem for DL-architectures comprising T-LSTM cells.

3   Background

3.1   Preliminaries

Definition 1 (Event, Trace, Event Log). *An event is a tuple (c, a, ts) where c is the case id, a is the activity (label) and ts is the timestamp. A trace is a non-empty sequence $\sigma = \langle e_1, \ldots, e_{|\sigma|} \rangle$ of events such that $\forall i, j \in \{1, \ldots, |\sigma|\}$ $e_i.c = e_j.c$ and $e_i.ts \leq e_j.ts$, for $1 \leq i < j \leq |\sigma|$. An event log L is a set $\{\sigma_1, \ldots, \sigma_{|L|}\}$ of traces. A trace can also be considered as a sequence of vectors which contain derived control flow information or features. Formally, $\sigma = \langle \mathbf{x}^{(1)}, \mathbf{x}^{(2)}, \ldots, \mathbf{x}^{(|\sigma|)} \rangle$, where $\mathbf{x}^{(t)} \in \mathbb{R}^{n \times 1}$ is a vector, and the superscript* *indicates the time-order upon which the events happened. n is the number of features derived for each event.*

Definition 2 (Prefix and Label). *Given a trace* $\sigma = \langle e_1, \ldots, e_k, \ldots, e_{|\sigma|} \rangle$, *a prefix of length k, that is a non-empty sequence, is defined as* $f_p^{(k)}(\sigma) = \langle e_1, \ldots, e_k \rangle$, *with* $0 < k < |\sigma_c|$. *A next activity label for a prefix of length k is defined as* $f_{l,a}^{(k)}(\sigma) = e_{k+1}.a$, *whereas a next timestamp label for a prefix of length k is defined as* $f_{l,ts}^{(k)}(\sigma) = e_{k+1}.ts$. *The above definition also holds for an input trace representing a sequence of vectors. For example, the tuple of all possible prefixes, the tuple of all possible next activity labels and the tuple of all possible next timestamp labels for* $\sigma = \langle \mathbf{x}^{(1)}, \mathbf{x}^{(2)}, \mathbf{x}^{(3)} \rangle$ *are* $\langle\langle \mathbf{x}^{(1)} \rangle, \langle \mathbf{x}^{(1)}, \mathbf{x}^{(2)} \rangle\rangle$, $\langle e_2.a, e_3.a \rangle$, *and* $\langle e_2.ts, e_3.ts \rangle$.

3.2 Long Short-Term Memory Cells

Most of the DNN architectures proposed for the next activity and timestamp prediction in PBPM [17] use "vanilla" LSTM cells [9]. LSTMs belong to the class of recurrent neural networks [11] and are designed to handle temporal dependencies in sequential prediction problems [3].

Given a sequence of inputs $\sigma = \langle \mathbf{x}^{(1)}, \mathbf{x}^{(2)}, \mathbf{x}^{(3)}, ..., \mathbf{x}^{(k)} \rangle$, a LSTM computes sequences of outputs $\langle \mathbf{h}^{(1)}, \mathbf{h}^{(2)}, \mathbf{h}^{(3)}, ..., \mathbf{h}^{(k)} \rangle$ via the following recurrent equations:

$$\mathbf{f}_g^{(t)} = sigmoid(\mathbf{U}_f \mathbf{h}^{(t-1)} + \mathbf{W}_f \mathbf{x}^{(t)} + \mathbf{b}_f) \qquad \text{(forget gate)},$$
$$\mathbf{i}_g^{(t)} = sigmoid(\mathbf{U}_i \mathbf{h}^{(t-1)} + \mathbf{W}_i \mathbf{x}^{(t)} + \mathbf{b}_i) \qquad \text{(input gate)},$$
$$\tilde{\mathbf{c}}^{(t)} = tanh(\mathbf{U}_g \mathbf{h}^{(t-1)} + \mathbf{W}_g \mathbf{x}^{(t)} + \mathbf{b}_g) \qquad \text{(candidate memory)},$$
$$\mathbf{c}^{(t)} = \mathbf{f}_g^{(t)} \circ \mathbf{c}^{(t-1)} + \mathbf{i}_g^{(t)} \circ \tilde{\mathbf{c}}^{(t)} \qquad \text{(current memory)}, \qquad (1)$$
$$\mathbf{o}_g^{(t)} = sigmoid(\mathbf{U}_o \mathbf{h}^{(t-1)} + \mathbf{W}_o \mathbf{x}^{(t)} + \mathbf{b}_o) \qquad \text{(output gate)},$$
$$\mathbf{h}^{(t)} = \mathbf{o}_g^{(t)} \circ tanh(\mathbf{c}^{(t)}) \qquad \text{(current hidden state)},$$
$$\forall t \in \{1, 2, \ldots, k\}.$$

$\{\mathbf{U}_{f,i,g,o}, \mathbf{W}_{f,i,g,o}, \mathbf{b}_{f,i,g,o}\}$ are trainable parameters, $\circ$ denotes the Hadamard product (element-wise product), $\mathbf{h}^{(t)}$ and $\mathbf{c}^{(t)}$ are the hidden state and cell memory of a LSTM cell. Additionally, a LSTM cell uses four gates to manage its states over time to avoid the problem of exploding/vanishing gradients in the case of longer sequences [3]. $\mathbf{f}_g^{(t)}$ (forget gate) determines how much of the previous memory is kept, $\mathbf{i}_g^{(t)}$ (input gate) controls the amount new information is stored into memory, $\tilde{\mathbf{c}}^{(t)}$ (candidate memory) defines how much information is stored into memory and $\mathbf{o}_g^{(t)}$ (output gate) determines how much information is read out of the memory. The hidden state $\mathbf{h}^{(t)}$ is commonly forwarded to a successive layer.

4   Methodology

4.1   Time-Aware Long Short-Term Memory Cells

"Vanilla" LSTM cells, as described in Sect. 3.2, assume a uniform distribution of the elapsed time between events ($\boldsymbol{\Delta}^{(t)} := x_{ts}^{(t)} - x_{ts}^{(t-1)}$). This assumption does not hold for most event logs analyzed in PBPM though (see Fig. 4). The elapsed time between consecutive events might have an impact on the next activity and timestamp prediction. Hence, a LSTM cell should be able to take irregular elapsed times into account when processing event logs.

Time-aware long short-term memory (T-LSTM) cells are an extension of the LSTM. Figure 1 depicts the T-LSTM cell and highlights its differences with regard to the LSTM cell.

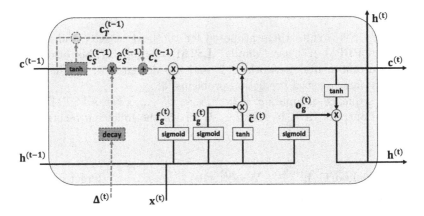

Fig. 1. Illustration of a T-LSTM cell with its computational components at time step t. The dashed and blue components indicate the extensions to the "vanilla" LSTM cell. The previous cell memory $\mathbf{c}_S^{(t-1)}$ is adjusted to $\mathbf{c}_*^{(t-1)}$ (see Eq. (2)) and is then processed together with $\mathbf{h}^{(t-1)}$ and $\mathbf{x}^{(t)}$ via the LSTM computations, as formalized in Eq. (1).

The main idea behind T-LSTM is to perform a subspace decomposition of the previous cell memory $\mathbf{c}^{(t-1)}$. First, a short term memory component $\mathbf{c}_S^{(t-1)}$ is extracted via a network. Next, the short term memory is discounted via a decay function of the elapsed time and yields $\hat{\mathbf{c}}_s^{(t-1)}$. Then, the long term memory ($\mathbf{c}_T^{(t-1)} = \mathbf{c}^{(t-1)} - \mathbf{c}_S^{(t-1)}$) is calculated. Finally, the previous cell memory is adjusted $\mathbf{c}_*^{(t-1)} = \mathbf{c}_T^{(t-1)} + \hat{\mathbf{c}}_s^{(t-1)}$). The adjusted previous memory $\mathbf{c}_*^{(t-1)}$ is then, together with $\mathbf{h}^{(t-1)}$ and $\mathbf{x}^{(t)}$, further processed as in LSTM cells by substituting $\mathbf{c}^{(t-1)}$ with $\mathbf{c}_*^{(t-1)}$ in Eq. (1). The following equations summarize the T-LSTM specific computations for the subspace decomposition and adjustment of the previous memory.

$$\mathbf{c}_S^{(t-1)} = tanh(\mathbf{W}_d\mathbf{c}^{(t-1)} + \mathbf{b}_d) \qquad \text{(short term memory)},$$

$$\hat{\mathbf{c}}_s^{(t-1)} = \mathbf{c}_S^{(t-1)} * decay(\mathbf{\Delta}^{(t)}) \qquad \text{(discounted short term memory)},$$

$$\mathbf{c}_T^{(t-1)} = \mathbf{c}^{(t-1)} - \mathbf{c}_S^{(t-1)} \qquad \text{(long term memory)}, \qquad (2)$$

$$\mathbf{c}_*^{(t-1)} = \mathbf{c}_T^{(t-1)} + \hat{\mathbf{c}}_s^{(t-1)} \qquad \text{(adjusted previous memory)},$$

$$... \qquad \text{(LSTM computations as in Eq. (1))},$$

$$\forall t \in \{1, 2, \ldots, k\}.$$

Note, that we only add $\{\mathbf{W}_d, \mathbf{b}_d\}$ as trainable parameters compared to the LSTM cell. As recommended in Baytas et al. [2], we chose $decay(\mathbf{\Delta}^{(t)}) = 1/log(e + \mathbf{\Delta}^{(t)})$ since we input the elapsed times in seconds and therefore have large values for $\mathbf{\Delta}^t$. Any other monotonic decreasing function and scale for $\mathbf{\Delta}^t$ would be valid as well, but our initial choice proved to be effective. The intuition behind the subspace decomposition is that the short term memory should be discounted if the elapsed time is very large, while the long term memory should be maintained in the adjusted previous cell memory $\mathbf{c}_*^{(t-1)}$. Similar as for LSTMs, the hidden state $\mathbf{h}^{(t)}$ is forwarded to successive layer for further processing. Hence, it is straightforward to substitute LSTM with T-LSTM cells in a given DNN architecture.

4.2   Network Architecture

We adapted the multitask architecture proposed by Tax et al. [15] as a baseline (see Fig. 2). The predicted next activity $\hat{e}_{k+1}.a$ is the output of a softmax activation after the last dense layer, where the output dimension is equal to the number of unique activity labels. $\hat{e}_{k+1}.a$ is evaluated against the one-hot encoded ground truth label $e_{k+1}.a$ by using the Cross-Entropy (CE) loss. The predicted next timestamp $\hat{e}_{k+1}.ts$ is a scalar output of a dense layer. We do not apply any additional activation after the time specific dense layer to be consistent with the implementation[1] of Tax et al. [15]. $\hat{e}_{k+1}.ts$ is compared with the ground truth timestamp $e_{k+1}.ts$ using the Mean Absolute Error (MAE). The total loss is the sum of both losses, as implemented in Tax et al. [15]. Further, they applied one-hot encoding for the activities and compute time-related control-flow features, which we also used in our experiments. We refer to the baseline architecture as "**Tax**". We performed an ablation study and made three modifications to the baseline DNN architecture:

- We weighted the CE loss function based on the distribution of activity labels in the training set. Hence, the classification of under-represented event classes had larger influence during training. We refer to this model as "**Tax+CS**".
- We replaced all LSTM layers with T-LSTM layers and refer to this model as "**Tax+T-LSTM**".
- We added cost-sensitive learning and replaced all LSTM layers with T-LSTM layers. We call this model "**Tax+CS+T-LSTM**".

---

[1] https://github.com/verenich/ProcessSequencePrediction.

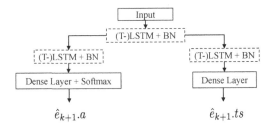

Fig. 2. Network architecture for this work based on the multitask learning approach proposed by Tax et al. [15]. The dashed components are either LSTM or T-LSTM layers. The input is of the network is a sequence of vectors representing a prefix $\langle e_1, \ldots, e_k \rangle$ as in Tax et al. [15]. For the baseline architecture we applied one-hot encoding and LSTM layers as in [15]. The outputs of the model are the predicted next activity ($\hat{e}_{k+1}.a$) and timestamp ($\hat{e}_{k+1}.ts$). Each of the LSTM layers is followed by a batch normalization layer (BN) to speed up training, as used in Tax et al. [15].

5   Experiments

5.1   Datasets

We performed our experiments on the same publicly available datasets as Tax et al. [15] to validate the effectiveness of our proposed techniques. Figure 3 shows the distribution of the activities (labels) for the different datasets. It is evident that the distributions of activities are skewed for both event logs. Table 1 presents descriptive statistics of the datasets used in this work.

Helpdesk[2]: This event log originates from a ticket management process of an Italian software company.

BPI'12 W Subprocess[3] **(BPI12W):** The Business Process Intelligence (BPI) 2012 challenge provided this event log from a German financial institution. The data come from a loan application process. The 'W' indicates state of the work item for the application.

5.2   Preprocessing

We used the cleaned and prepared datasets as in Tax et al. [15]. The datasets can be found on the corresponding GitHub repository[4]. The preprocessing steps include splitting the data into training and test set, calculating time divisors, and ASCII encoding activities and sequence generation. Datasets were split into 2/3rd and 1/3rd for training and testing respectively and preserve the temporal order of cases. We additionally used the last 20% of the training data as a validation set in order to tune the hyperparameters. We adapted the sequence

---

[2] https://doi.org/10.4121/uuid:0c60edf1-6f83-4e75-9367-4c63b3e9d5bb.

[3] https://doi.org/10.4121/uuid:3926db30-f712-4394-aebc-75976070e91f.

[4] https://github.com/verenich/ProcessSequencePrediction/tree/master/data.

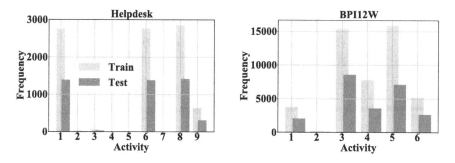

Fig. 3. Activity distribution in training and test set for Helpdesk and BPI12W datasets. It is evident that the distributions of the activity labels are skewed.

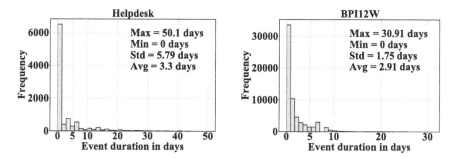

Fig. 4. Event duration distribution for the complete Helpdesk and BPI12W datasets. It can be observed that the majority of the events are completed within one day. However, there are many events with longer duration. Note that we input the elapsed time between events (Δ^t) in seconds for T-LSTM.

and feature generation methods by Tax et al. [15]. The features include the activity of the event, position of the event in the case, time since the last event, time from the starting event of the case, time from midnight, and day of the week. We create one-hot encoded versions of the ground truth labels $e_{k+1}.a$ for the next activity prediction in order to compare them with the predicted next activity labels $\hat{e}_{k+1}.a$.

5.3   Training Setup

For hyperparameter tuning, we performed a grid search on the training set and chose the model with the lowest validation loss. The validation loss is the sum of activity-related validation loss and time-related validation loss. The number of LSTM or T-LSTM units were set to 64 or 100. For the dropout rate (for both dense layers), we tried the values 0.0 and 0.2. We choose Nadam as an optimization algorithm, as used in [15]. Nesterov accelerated gradient (NAG) calculates the step using the 'lookahead' algorithm, which approximates the next parameters. Adam optimizer estimates learning rates based on ini-

Table 1. Descriptive statistics of the datasets used in this study.

Characteristic	Helpdesk	BPI12W
Number of instances	3,804	9,658
Case variants	154	2,263
Unique activities	9	6
Events	13,710	72,413
Max # events per case	14	74
Min # events per case	1	1
Avg # events per case	3.604	7.497

tial moments of the gradients. Nadam is a combination of both and is robust in noisy datasets. Furthermore, we tested a range of different learning rates $\{0.0001, 0.0002, 0.001, 0.002, 0.01\}$ since this is known to have a large impact on LSTMs [8]. We trained each model for 150 epochs, with a batch size of 64 and apply early stopping with patience 25 for regularization.

5.4   Evaluation

We applied the same evaluation metrics as in [15]. We used the *Accuracy* metric to evaluate the next activity prediction. For the next timestamp prediction, we used the *Mean Absolute Error (MAE)* to evaluate our models.

5.5   Implementation

We conducted all experiments on a workstation with 24 CPU cores, 748 GB RAM and a singe GPU NVIDEA QUADRO RTX6000. We implemented the experiments in Python 3.7. We used the DL framework TensorFlow 2.1[5]. The source code is available on GitHub[6].

6   Results

6.1   Next Activity Prediction

Table 2 shows the results for the next activity prediction in terms of Accuracy. For Helpdesk and BPI12W, the approach Tax+CS+T-LSTM achieved the highest Accuracy (0.724 and 0.778) among all approaches. The approach's improvement compared to the baseline is 0.012 and 0.018. While the two approaches, Tax+CS and Tax+T-LSTM, outperformed the baseline for Helpdesk, these approaches achieved a lower Accuracy for BPI12W than the baseline.

---

[5] https://www.tensorflow.org.
[6] https://github.com/annguy/time-aware-pbpm.

Table 2. Results for the next activity prediction in terms of Accuracy. The best result for each dataset is highlighted (larger is better).

Approach	Helpdesk	BPI12W
Tax (baseline)	0.712	0.760
Tax+CS	0.713	0.757
Tax+T-LSTM	0.718	0.693
Tax+CS+T-LSTM	**0.724**	**0.778**

6.2   Next Timestamp Prediction

Table 3 shows the results for the next timestamp prediction task in terms of MAE in days. All approaches with a T-LSTM cell, clearly outperformed the baseline for both event logs. Thereby, the approach Tax+CS achieved the lowest MAE of 2.87 days and 0.88 days for Helpdesk and BPI12W respectively. Compared to the baseline, this approach reduced the MAE by 0.88 days (Helpdesk) and 0.68 days (BPI12W). The other two approaches, Tax+T-LSTM and Tax+CS+T-LSTM, achieved a slightly worse MAE values compared to Tax+CS for both event logs. It is worth noticing that for Helpdesk Tax+CS+T-LSTM and for BPI12W Tax+T-LSTM yielded the second best results with MAE close to Tax+CS.

Table 3. Results for next step time prediction in terms of MAE in days. The best result for each dataset is highlighted (lower is better).

Approach	Helpdesk	BPI12W
Tax (baseline)	3.75	1.56
Tax+CS	**2.87**	**0.88**
Tax+T-LSTM	3.01	0.88
Tax+CS+T-LSTM	2.94	0.90

7   Discussion

In this paper, we argued that the elapsed time between consecutive events carries valuable information on human behavior in running business processes. Therefore, we introduced T-LSTM cells for PBPM which inherently model the elapsed time between consecutive events. Further, we introduced of cost-sensitive learning to better cope with the problem of imbalanced data.

   The obtained results indicate that the elapsed time between consecutive events is informative and that a DNN architecture relying on T-LSTM cells cab yield more accurate models for PBPM. Especially, with the approach Tax+CS+T-LSTM, we could outperform the baseline (Tax) for both datasets (i.e., Helpdesk and BPI12W) and both prediction tasks (i.e., next activity prediction and next timestamp prediction). Thereby, we could observe that cost-sensitive learning plays a crucial role for the predictive quality of a DNN architecture using T-LSTM cells instead of "vanilla" LSTM cells. Interestingly, the effectiveness of the introduced techniques is more evident for the next timestamp prediction compared to the next activity prediction

Even though our presented results on DNN architectures using T-LSTMs seem promising, there are a few limitations to our work. First, we need to verify our findings by performing experiments on more datasets. Second, a better hyperparameter tuning approach like Bayesian optimization [1] could be applied for all configurations to get a better estimate of their effectiveness. Further, several runs with random initialization should be performed to estimate the stability of the models.

8 Conclusion and Future Work

We propose T-LSTM as an alternative to the commonly used "vanilla" LSTM cell to better exploit information on the elapsed time between consecutive events. Furthermore, we introduced the concept of cost-sensitive learning to account for the common class-imbalance in event log data. Our results indicate the effectiveness of the introduced techniques for the next activity and timestamp prediction. This suggests that integrating specific mechanisms into neural network layers to incorporate event log specific characteristics might be an interesting direction for future research. Here, we mainly demonstrated the benefit of replacing a normal LSTM with a time-aware LSTM cell for a given baseline approach [15].

An avenue for future research is to investigate if T-LSTM cells might also improve other LSTM-based PBPM approaches such as Camargo et al. [5] involving resource attributes or Taymouri et al. [16] generating fake event logs. Another direction for future research is to further customize an LSTM cell in terms specifically for PBPM. For example, a process-aware LSTM cell could not only deal with time information but also with resource information.

Author contributions. Srijeet Chatterjee: Equal contribution with An Nguyen

References

1. Akiba, T., Sano, S., Yanase, T., Ohta, T., Koyama, M.: Optuna: a next-generation hyperparameter optimization framework. In: Proceedings of the 25rd International Conference on Knowledge Discovery and Data Mining (KDD) (2019)
2. Baytas, I.M., Xiao, C., Zhang, X., Wang, F., Jain, A.K., Zhou, J.: Patient subtyping via time-aware LSTM networks. In: Proceedings of the 23rd International Conference on Knowledge Discovery and Data Mining (KDD), pp. 65–74 (2017)
3. Bengio, Y., Simard, P., Frasconi, P., et al.: Learning long-term dependencies with gradient descent is difficult. Trans. Neural Networks 5(2), 157–166 (1994)
4. Breuker, D., Matzner, M., Delfmann, P., Becker, J.: Comprehensible predictive models for business processes. MIS Q. 40(4), 1009–1034 (2016)

5. Camargo, M., Dumas, M., González-Rojas, O.: Learning accurate LSTM models of business processes. In: Hildebrandt, T., van Dongen, B.F., Röglinger, M., Mendling, J. (eds.) BPM 2019. LNCS, vol. 11675, pp. 286–302. Springer, Cham (2019). https://doi.org/10.1007/978-3-030-26619-6_19

6. Di Francescomarino, C., Ghidini, C., Maggi, F.M., Milani, F.: Predictive process monitoring methods: which one suits me best? In: Weske, M., Montali, M., Weber, I., vom Brocke, J. (eds.) BPM 2018. LNCS, vol. 11080, pp. 462–479. Springer, Cham (2018). https://doi.org/10.1007/978-3-319-98648-7_27

7. Evermann, J., Rehse, J.R., Fettke, P.: Predicting process behaviour using deep learning. Decis. Support Syst. **100**, 129–140 (2017). https://www.evermann2017predicting

8. Greff, K., Srivastava, R.K., Koutník, J., Steunebrink, B.R., Schmidhuber, J.: LSTM: a search space odyssey. IEEE Trans. Neural Networks Learn. Syst. **28**(10), 2222–2232 (2017)

9. Hochreiter, S., Schmidhuber, J.: Long short-term memory. Neural Comput. **9**(8), 1735–1780 (1997)

10. Khan, A., et al.: Memory-augmented neural networks for predictive process analytics. arXiv preprint arXiv:1802.00938 (2018)

11. LeCun, Y., Bengio, Y., Hinton, G.: Deep learning. Nature **521**(7553), 436 (2015)

12. Maggi, F.M., Di Francescomarino, C., Dumas, M., Ghidini, C.: Predictive monitoring of business processes. In: Jarke, M., et al. (eds.) CAiSE 2014. LNCS, vol. 8484, pp. 457–472. Springer, Cham (2014). https://doi.org/10.1007/978-3-319-07881-6_31

13. Mehdiyev, N., Evermann, J., Fettke, P.: A novel business process prediction model using a deep learning method. Bus. Inf. Syst. Eng. **62**(2), 143–157 (2018). https://doi.org/10.1007/s12599-018-0551-3

14. Navarin, N., Vincenzi, B., Polato, M., Sperduti, A.: LSTM networks for data-aware remaining time prediction of business process instances. In: IEEE Symposium Series on Computational Intelligence (SSCI), pp. 1–7. IEEE (2017)

15. Tax, N., Verenich, I., La Rosa, M., Dumas, M.: Predictive business process monitoring with LSTM neural networks. In: Dubois, E., Pohl, K. (eds.) CAiSE 2017. LNCS, vol. 10253, pp. 477–492. Springer, Cham (2017). https://doi.org/10.1007/978-3-319-59536-8_30

16. Taymouri, F., La Rosa, M., Erfani, S., Bozorgi, Z.D., Verenich, I.: Predictive business process monitoring via generative adversarial nets: the case of next event prediction. In: Proceedings of the 18th International Conference on Business Process Management (BPM) (2020)

17. Weinzierl, S., et al.: An empirical comparison of deep-neural-network architectures for next activity prediction using context-enriched process event logs. arXiv:2005.01194 (2020)

18. Weinzierl, S., Dunzer, S., Zilker, S., Matzner, M.: Prescriptive business process monitoring for recommending next best actions. In: Fahland, D., Ghidini, C., Becker, J., Dumas, M. (eds.) BPM 2020. LNBIP, vol. 392, pp. 193–209. Springer, Cham (2020). https://doi.org/10.1007/978-3-030-58638-6_12

A Preliminary Study on the Application of Reinforcement Learning for Predictive Process Monitoring

Andrea Chiorrini[✉], Claudia Diamantini, Alex Mircoli, and Domenico Potena

Department of Information Engineering, Polytechnic University of Marche, Ancona, Italy
a.chiorrini@pm.univpm.it, {c.diamantini,a.mircoli,d.potena}@univpm.it

Abstract. The present paper explores the opportunity of applying reinforcement learning to various typical tasks in the field of predictive process monitoring. The tasks considered are the prediction of both next event activity and time completion as well as the prediction of the whole progression of running cases. Experiments have been conducted on the popular benchmark dataset, BPI' 2012, on which we compare the proposed learning system with state of the art methods adopting LSTM networks trained through supervised learning. Results enlighten promising features of the approach and interesting research issues and challenges, as well as proving the applicability of reinforcement learning to predictive process monitoring.

Keywords: Predictive process monitoring · Reinforcement learning · Outcome and time prediction · Process mining

1 Introduction

Recently the field of predictive process monitoring is receiving increasing attention [7,8]. It aims at improving process monitoring through the introduction of predictive capabilities, allowing both process improvement and proactive problem handling. Predictive process monitoring relies on models, obtained from historical process logs, able to forecast the evolution of a process instance based only on the first part of it. In detail, predictive process monitoring tries to handle several different problems, such as the one step ahead prediction of the next activity that will be performed and the estimation of its execution time, or the prediction of all the activities to be performed until the end of the trace, i.e. the trace suffix, and the total execution time of the trace, i.e. the trace time. Recently, different machine learning techniques have been adopted to deal with predictive process monitoring tasks, with a particular focus on Long Short-Term Memory (LSTM) networks. In this paper we investigate the use of reinforcement learning to predict, both suffix and one step ahead, activities and execution times. Reinforcement Learning (RL) is a particular kind of machine learning paradigm that

S. Leemans and H. Leopold (Eds.): ICPM 2020 Workshops, LNBIP 406, pp. 124–135, 2021.
https://doi.org/10.1007/978-3-030-72693-5_10

trains models to directly maximize a reward signal, without assigning any label or necessarily trying to find some hidden structure in the data. Reinforcement learning has been gaining increasing attention since 2015, when [9] trained an agent that bested many human professional players over various Atari games. This has led the scientific community to further investigate the techniques, leading to various interesting results (e.g., [13,14]), up until the latest astonishing artificial agent [17] that managed to beat professional human player in Starcraft II, an extremely complex real time strategy game. An interesting feature of this family of algorithms is that learning is guided by an objective function that takes into account all the chain of future decisions and its effects, instead of focusing only on the decision at hand. This could be an interesting feature in the predictive process mining field, where events to be predicted are conditioned by the process workflow.

Motivated by past success of RL and this observation, we set as our goal to study if it is possible to apply reinforcement learning to the field of predictive process monitoring. At the best of our knowledge this is the first study of this kind. The only other study that applied reinforcement learning in the process mining field is [4], where the problem of efficient resource allocation is considered. Our results enlighten promising features of the approach and interesting research issues.

The rest of the paper is organized as follows: Sect. 2 is devoted to introduce related work. In Sect. 3 background knowledge about reinforcement learning is provided, while Sect. 4 explains our proposed approach. Section 5 presents the performed experiments and discusses achieved results. Finally, Sect. 6 concludes the paper and outlines some directions for future works.

2 Related Work

Recent efforts in predictive process monitoring exploits Deep Neural Networks, specifically LSTM and CNN [1,3,6,10,11,16].

In particular, [16] trained an LSTM for the one step ahead event prediction, in particular the activity associated to the next event and its completion time, as well as the suffix of the trace, iteratively using the one step ahead event prediction. They encoded each event into a feature vector that is a combination of the one-hot encode of the associated activity and three temporal features related to the event's timestamp such the time of the day, the time since the previous event, and the accumulated duration since the start of a process case. LSTM has also been used in [3] to predict the activity associated with the next event of a case, but this approach, differently from [16], uses the embedded dimension of LSTM to both reduce the input's size and include extra attributes like the resource associated to each event. Their experiments show that the proposed approach sometimes outperforms [16] for the prediction of the next event. However, [3] only focuses on predicting event activity types and cannot predict the next event's timestamp as it cannot handle numerical variables.

In [1] a combination of approaches in [3] and [16] is proposed for the one step ahead event prediction. The approach considers and predicts the next activity, the timestamp and type of resource of next event. To do so they introduce a notion of abstract class of resource, i.e. group of resources that usually performs similar activities, this way they manage to avoid the main limit of [3] which is the inability of handling numerical variables and therefore predicting next event's timestamp.

Another LSTM model has been proposed by Lin et al. [6] for the prediction of the next activity and all the other categorical attributes of the next event (e.g. the associated resource) of the past events, using an approach similar to an attention mechanism for weighting the event attributes on the basis of their relevance in the prediction of future events. Again this approach suffers from the inability of handling numerical values and therefore predicting timestamps.

Even more recently, the usage of CNN has been investigated. The basic idea is to convert the temporal data enclosed in an event log into spatial data so as to treat them as images [10] This idea has been further extended in [11], where an RGB encoding of process instances is used to train a 2-D CNN based on two inception blocks. Both papers only tackle the one step ahead activity prediction task.

3   Background

In this section we provide general background knowledge on RL.

Reinforcement learning is learning what to do—how to map situations to actions—so as to maximize a numerical reward signal. The learner is not told which actions to take, but instead must discover which actions yield the most reward by trying them. A reinforcement learning problem is formalized using ideas from dynamical systems theory, specifically, as the optimal control of incompletely-known Markov decision process [15].

The learner (agent) interacts with an environment during a sequence of timesteps composing the learning episode. In the domain of process mining we can think the learning episode as the evolution of the trace, and the occurrence of an event in it as a timestep. At each timestep, an interaction between agent and environment occurs through observations (x) of the environment, actions (a) and rewards (r). The observation is the trace event, the action is the prediction for the next event, and the reward is derived from the comparison between the next event information and the predicted one. The agent's goal is to maximize its cumulative future reward performing its actions, with respect to the state of the environment. The state s_i of the environment is defined as the full sequence of observations and actions performed until timestep i, formally: $s_i = x_1, a_1, x_2, ..., a_{i-1}, x_i$. However, it is complex to use a state composed of a variable number of observations and actions as input. Hence, it is usually preferred to use a constant fixed number of observations and actions. In this paper we refer to the number of past timesteps considered to define the state as *window size*. Having fixed the window size to a generic k, the state is written as $s_i = x_{i-k}, a_{i-k}, x_{i-k+1}, ..., a_{i-1}, x_i$.

The objective function of the agent at timestep i can be expressed as:

$$R_i = \sum_{t=i}^{T} \gamma^{t-i} r_t, \tag{1}$$

where $\gamma < 1$ is the discount factor of future rewards, used to prioritize more recent rewards, and T is the number of timesteps in the whole learning episode. In particular, the methodology adopted in this paper considers Q learning agents. This type of agents tries to approximate the optimal action-value function $Q^*(s, a)$, learning it from the transitions from a state s_i to a next state s_{i+1} on the basis of the performed action a_i and the received reward r_i. The optimal action-value function may be expressed as:

$$Q^*(s, a) = \max_{\pi} \mathbf{E}\left[R_i | s = s_i, a = a_i, \pi\right] \tag{2}$$

which is the maximum expected reward achievable after seeing sequence s and taking action a, by following any behaviour policy π for mapping sequences to actions.

This reinforcement learning algorithm is based on the fact that knowing $Q^*(s, a)$ an agent can choose the best sequence of actions at any state, maximizing its reward.

Obviously, perfectly knowing $Q^*(s, a)$ it is not always possible, especially in complex environment. Still it is possible to discover $Q(s, a, \theta)$, through a machine learning model, where θ are the parameters of the trained model so that $Q(s, a, \theta) \approx Q^*(s, a)$. In the case of deep Q network (DQN) agents, the model adopted is a deep neural network, and θ are its weights, used to approximate the optimal action-value function. In our study we used as underlying model to approximate the Q-function an LSTM based neural network. The network weights may be adjusted through training using as loss function, that varies at each timestep, the mean squared error defined as follows:

$$L_i(\theta_i) = \mathbf{E}_{(s,a,r,s')}\left[r + \gamma \max_{a'} \tilde{Q}(s', a', \theta_i^-) - Q(s, a, \theta_i)\right]^2 \tag{3}$$

in which γ is a discount factor determining a penalty for more future reward, θ_i are the parameters of the Q-network at iteration i and θ_i^- are the network parameters used to compute the target at iteration i used in place of the optimal and unknown $max_{a'}Q^*(s', a')$.

Therefore, in contrast to supervised learning where targets are fixed before learning begins, the targets depends on the network weights. Though, since θ_i^- is kept fixed at the ith optimization of $L_i(\theta_i)$, all the optimization problems at each iteration are well defined. In our case we used a so called soft update of θ which updates the parameters at each iteration on the basis of a coefficient β accordingly to the formula: $\theta_{i+1}^- = (1 - \beta)\theta_i^- + \beta\theta_i$, where $\beta \in (0, 1)$.

It is worth noting that this algorithm is model-free as it solves the reinforcement learning task directly, without estimating the system transition dynamics. Also it is an off-policy algorithm, since it learns a greedy policy where

$a = argmax_{a'}Q(s, a', \theta)$ but it still ensure, through its behaviour policy, an adequate exploration of the state space through a random action. This allows to discover if there are better actions to perform with respect to the recommended one. In our particular case, for training, we used a Boltzmann Q Policy, which builds a probability law on q values and returns an action selected randomly according to this law while, for prediction purposes, a GreedyQPolicy is adopted which selects the action with the highest reward.

For further details, the full description of the algorithm can be found in [9], which originally proposed it.

4 Methodology

This section is devoted to describe the proposed methodology, which uses two agents trained through reinforcement learning to predict activity and execution time of both the one step ahead event as well as the activities suffix and trace time. First, we give some preliminary definitions of event, trace, and event log.

Let ε be the event universe, i.e., the set of all possible event identifiers. An event $e_i \in \varepsilon$ is characterized by a set of properties. In the context of the present paper, we assume the availability of the following properties: the activity associated to the event, denoted by a_i and the complete timestamp t_i. A trace is a finite non-empty sequence of events $\sigma = <e_1, e_2, \ldots, e_k>, e_i \in \varepsilon, e_i \neq e_j$ for $i \neq j$. We assume the events are ordered with respect to their timestamp, i.e. $t_i < t_j$ for $i < j$. An event log L is a set of traces such that each event appears at most in one trace.

In the following, we describe the pre-processing performed to make event log data suitable for being fed to our system, and the details on the architecture adopted.

4.1 Pre-processing

As it will be clear in the following we use an LSTM as agent's model. Here we describe how log data are processed to generate the input sequences to the model. An event e_i in the sequence is logically represented by 4 components, namely the activity a_i, and three temporal features. Each activity is expressed by a binary vector built using the one-hot encoding of the activity type. One-hot encoding has been chosen as it is an effective and popular way of representing categorical data. Its main advantage is that one-hot encoding transformation does not introduce any order or similarity among the representation of categorical data.

Regarding the three added temporal features, the first is the time passed between Sunday midnight and the event e_i (t_{w_i} in Eq. 4) which is useful to express the seasonality of the process. The second is the time passed between the completion of an event e_i and the completion of the previous event e_{i-1} (t_{e_i} in Eq. 5), thus substantially corresponding to the event duration (plus possible idle time between the two events). The last temporal feature is the time passed between the start of the trace and the event e_i (t_{t_i} in Eq. 6), which gives information about the progression of the trace. This last one is particularly relevant since there may be a strong correlation between the performed actions and the "age" of the process case.

Formally:

$$t_{w_i} = \frac{t_i - t_{w_0}}{\Delta t_w} \tag{4}$$

$$\begin{cases} t_{e_i} = 0 & \text{if } i = 1, \\ t_{e_i} = \frac{t_i - t_{i-1}}{\Delta_{\max_e}} & \text{otherwise} \end{cases} \tag{5}$$

$$t_{t_i} = \frac{t_i - t_0}{\Delta_{\max_t}} \tag{6}$$

where t_i is the timestamp of the event at index i, t_{w_0} is the timestamp of the last passed Sunday midnight, and t_0 is the start timestamp of the process. Δt_w $\Delta_{\max_e}$ and $\Delta_{\max_t}$ are normalization factors to make features varying in the range $[0, 1]$, as it improves the performance of the network. Δt_w is the amount of time in a week, while $\Delta_{\max_e}$ and $\Delta_{\max_t}$ are, respectively, the maximum event duration and the maximum trace duration. It is also worth noting that given t_{e_i} and both $t_{w_{i-1}}$ and $t_{t_{i-1}}$, it is possible to derive the value of both t_{w_i} and t_{t_i}.

4.2   Learning Architecture

The overall architecture is shown in Fig. 1. In the Figure, dashed lines enlighten the learning phase, while solid lines refer to the prediction phase. In the system, we have two different agents. Both take as input a sequence of events, in which every event is defined by the three temporal features and the one-hot encoding of the activity as explained before. One agent predicts the one step ahead activity, the next one that will be performed, while the other is devoted to predict its completion time. As said, every DQN agent has an underlying neural network that models the reward function. For each of our two agents, we used an LSTM based neural network to learn and approximate the optimal $Q^*(s, a)$, instead of training them using ground-truth labels, typical of supervised learning. This is done, through the agents' interaction with their respective environment, thanks to which they receive their reward. The LSTM architecture have been chosen because of its widespread adoption in predictive process mining. We hasten to note that DQN agents only work with a discrete action space and therefore they are unable to produce continuous outputs. To address this issue we divided the output time in bins, each representing the range in which the estimated time falls, and we designed the time agent so as to produce bin indexes as outputs.

As explained in Sect. 3, the learning process of our RL agents is based on the notion of transition from a state of the environment to another on the basis of the performed action and its associated reward at each timestep.

We set the reward functions in each environment as binary reward functions: in the time environment, the reward gives a plus one when the predicted bin included the true time, and zero otherwise; similarly in the activity environment the reward gives a plus one when the prediction is correct and a zero otherwise.

For the one step ahead prediction of the next activity and time the two agents work in isolation exploiting their underlying LSTM network model to perform their prediction. For suffix prediction the situation is more complex, as each agent has access only to the information of the first part of the trace. In particular, it reads only the first k events where k is the window size. Hence, each agent needs to rely both on its own prediction and on the other agent's prediction to have all the required inputs for predicting more than one step ahead, as the true information is not available. In a way, the two agents cooperates exchanging messages to inform the other of their prediction, at each timestep. This way the whole sequence may be predicted using the predicted information when the true one is not available. All this is iterated until the end of the trace is predicted.

Formally, at the first iteration we consider the sequence $\sigma_k = <e_1, e_2, \ldots, e_k>$ of events of length k (window size), where e_j be the j-th event of a trace, which is characterized by the tuple $<a_j, t_{w_j}, t_{e_j}, t_{t_j}>$. The time predictor agent α_t and the action predictor agent α_a are defined as follows:

$$\alpha_t : \sigma_k \mapsto t'_{e_{k+1}},$$

$$\alpha_a : \sigma_k \mapsto a'_{k+1},$$

where apex denotes the predicted value. Each agent will inform the other of its prediction and therefore the predicted next event e'_{k+1} will be characterized by the tuple $<a'_{k+1}, t'_{w_{k+1}}, t'_{e_{k+1}}, t'_{t_{k+1}}>$, where $t'_{w_{k+1}}$ and $t'_{t_{k+1}}$ are derived from $t'_{e_{k+1}}$, t_{t_k} and t_{w_k}. Then a new prediction will be performed by each agent using as input $\sigma_{k+1} = <e_2, \ldots, e_k, e'_{k+1}>$. Iterating at the i-th step, the sequence σ_i will be formed by $k-i$ real events and i predicted ones. The algorithm is iterated until the end event of the process is predicted.

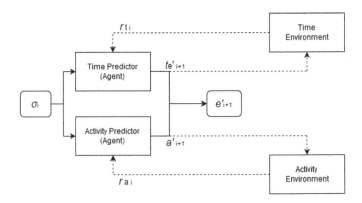

Fig. 1. Overall architecture.

5 Evaluation

In this section we empirically evaluate the performance of the proposed approach. Results are compared with those of other approaches using LSTM networks for uniformity reasons, so to remark the contribution of RL paradigm.

The following subsections describe the experimental setup, reference metrics and results.

5.1 Experimental Setup

Dataset. The experimental dataset is a subset of an event log from the Business Process Intelligence Challenge (BPI'12) [2] which contains data from the application procedure for financial products at a large financial institution. This process consists of three subprocesses: one that tracks the state of the application, one that tracks the states of work items associated with the application, and a third one that tracks the state of the offer. Since our goal is to predict the coming events and their timestamps, events that are performed automatically aren't considered relevant. Therefore, we limit our evaluation to the work items subprocess (BPI'2012 W): the one containing events that are manually executed. As done in [16], to perform our experiments we used chronologically ordered first 2/3 of the traces as training data, and evaluate the activity and time predictions on the remaining 1/3 of the traces.

The dataset has been pre-processed as explained in Sect. 4. For what concerns the setting of bins defining the output values of the time agent, we analyzed the whole distribution of events duration in the dataset. This allowed to set the various ranges so as to both balance the number of elements in a bin and to maintain a reasonable similarity between elements in the same bin.

The resulting bin endpoints are [0, 1, 10, 60, 120, 240, 480, 1440, 2880, 4320, 7200, 10080, 14400, 20160, 30240, 40320, 50400] expressed in minutes. Also note that the chosen endpoints correspond to meaningful time frames such as hours, days or weeks. Figure 2 shows the distribution of events duration in the dataset. The x-axis is in logarithmic scale for visualization purposes.

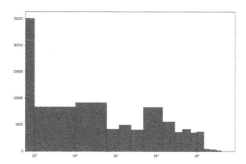

Fig. 2. t_e distribution in bins

Agents. We performed the experiments using Keras-rl [12], running on a machine with two NVIDIA GeForce GTX 1080, a i7 8700K CPU @3.70 GHz and 32 GB RAM. Each agent was trained for 600000 steps and is characterized by the use of a sequential memory of dimension 500000, a BoltzmannQPolicy clipped in range $(-15, 15)$ as behaviour train policy, and a GreedyQPolicy as test policy; the target function was updated through soft update using $\beta = 10^{-2}$ as coefficient. The underlying neural network has two hidden LSTM layers with 200 neurons each and ReLU activation; during training we used an Adam optimizer with a learning rate of 10^{-3}. This configuration was kept for all the tested window size, as it had the best performance for approximating the Q function, between those tested.

5.2 Metrics

In order to properly compare our results with previous work, we adopted the same evaluation metrics.

One-Step Ahead Prediction. We evaluate our results in terms of accuracy, for the next activity prediction, and in terms of mean absolute error (MAE) in days, for the predicted time. For the purpose of comparison with the baselines we use the MAE but it is important to remember that our time agent predicts ranges of time. Therefore, since we need a continuous value for the time in order to compute the MAE, we choose for this the inferior endpoint of the bin predicted as value. For example, if the predicted bin is the third one, which corresponds to range $[10, 60)$, the time used for computing the MAE will be 10 min.

Suffix Prediction. For the suffix completion time prediction we consider the absolute trace duration error (TDE)

$$TDE = |t'_{t_f} - t_{t_f}| \tag{7}$$

where, with some abuse of notation, f refers to the final event in the true and estimate trace, hence t_{t_f} (t'_{t_f}) represents the total duration of the true (estimated) trace. The TDE is then averaged over all traces.

For evaluating the accuracy of the activity suffix prediction the most well known and used distance is the Damerau-Levenshtein distance, which is defined as the minimum number of deletion, insertion, substitution and transposition operations needed to transform the first string to the second. In particular, this distance can be normalized dividing its value for the length of the longer string. What we adopted for comparison purposes is the Damerau-Levenshtein similarity expressed as one minus the normalized Damerau-Levenshtein distance.

5.3 Results

In Table 1 we present the performances achieved for the one step ahead prediction tasks. For next completion time prediction (Table 1(a)) we compare our results with the best reported by Tax et al. [16] for different window sizes. Table 1(b) reports the accuracy of the next activity prediction of our method, and the ones reported by Tax et al. [16] and Camargo et al. [1]. In [1] the next completion time task is not addressed. In Table 1(a) the row "All" reports the average performance over all the tested window sizes. In [16] these correspond to all the values in the range [2, 20], whereas in our case we considered the set {2, 3, 4, 5, 6, 7, 10, 20}.

It can be seen that our performance in the next completion time prediction are clearly better than the baseline, whilst our accuracy is worse. In particular, the relative improvement in the case of completion time prediction is about 27%, and the relative accuracy degradation is only about 8% with respect to best result provided by [1]. These results may be justified as follows. DQN agents optimize a cumulative reward function that takes into account rewards on future actions, in a sense trying to simulate the future. Completion times show a form of dependency on the total trace duration. For instance, overestimating the duration of early events will lead to an excessively long overall trace duration estimate. This may guide the learner through states with a better generalization ability. On the contrary, a similar relation does not exist for activities in the considered setting, where only the structural perspective of the process (i.e. the workflow) is taken into account. Thus enriching the log with other perspectives and in particular with data regarding case-specific and event-specific properties may likely highlight dependencies among activities and thus lead to improved results. We plan to verify such hypothesis in future work.

Table 1. Comparison of performances for the one step ahead prediction tasks. (a) Next completion time. (b) Next activity.

Window size	MAE (days)	
	Ours	Tax et al.
2	1.34	1.69
10	1.05	1.45
20	0.62	0.98
All	1.17	1.59

Accuracy		
Ours	Tax et al.	Camargo et al.
71.3%	76%	77.8%

(a)   (b)

We also show in Table 2 the performance achieved in suffix prediction tasks. The results confirm a better behavior of the proposed RL architecture on the completion time prediction than on the activity prediction task. For the former, the relative improvement is about 21%, which is in line with the one step ahead performance. For the latter, we observe a much worse performance degradation of about the 66% with respect to [1], and about 50% with respect to [16]. This is in part due to an expected error propagation effect, since errors committed at the early suffix prediction stages progressively compromise all the subsequent ones. As another issue reducing our systems performance, we observed that the event agent struggle to predict the end of the trace, leading to excessively long traces. To verify this, we calculated the DL-similarity truncating the predicted traces to the length of the true traces, discovering that performances improves up to a DL-similarity of 0.2974, which is comparable with the accuracy obtained by [16]. For what concerns computational complexity, clearly the time required to train an RL agent is much higher than the LSTM alone. We experimented an increase factor of about 20× of the required training time. This is a well known characteristics of RL training although alternative techniques with better computational performance have been proposed [5]. We plan to investigate them in future work.

Table 2. Comparison of performances for the suffix prediction tasks. (a) Completion time. (b) Next activities.

Window size	mean TDE (days)		
	Ours	Tax et al.	Camargo et al.
2	12.66	≈ 14	≈ 11
10	6.17	≈ 9	≈ 9
20	4.63	≈ 6	≈ 9

(a)

DL-similarity		
Ours	Tax et al.	Camargo et al.
0.174	0.3533	0.525

(b)

6   Conclusions and Future Works

The main contribution of this paper is to provide a preliminary study of the applicability of reinforcement learning techniques to predictive process monitoring tasks. In particular we used Deep Q Networks agents to address both the one step ahead activity and completion time prediction, and the trace suffix outcome prediction. Through our experiments on the BPI'2012 popular benchmark dataset, we showed that DQN agents can fully exploit time information, achieving results that significantly outperforms state of the art approaches based on LSTM architectures, while the plain workflow information seems to be insufficient to train an RL agent for the activity prediction task. The present paper also highlights several interesting research directions. First of all, as already noticed the proposed approach may be further refined through the use of case-specific data, and event-specific data. Second, more complex reward functions may be used in order to weight the activity and/or case importance, for instance something like the amount of money involved in a loan procedure or the cost to perform a specific activity. Third, alternative RL techniques can be considered, to investigate both their efficiency and accuracy performances.

References

1. Camargo, M., Dumas, M., González-Rojas, O.: Learning accurate LSTM models of business processes. In: Hildebrandt, T., van Dongen, B.F., Röglinger, M., Mendling, J. (eds.) BPM 2019. LNCS, vol. 11675, pp. 286–302. Springer, Cham (2019). https://doi.org/10.1007/978-3-030-26619-6_19

2. van Dongen, B.: BPI Challenge 2012 event log (2012). https://doi.org/10.4121/uuid:3926db30-f712-4394-aebc-75976070e91f

3. Evermann, J., Rehse, J.R., Fettke, P.: Predicting process behaviour using deep learning. Decis. Support Syst. **100**, 129–140 (2017). Smart Business Process Management

4. Huang, Z., van der Aalst, W., Lu, X., Duan, H.: Reinforcement learning based resource allocation in business process management. Data Knowl. Eng. **70**(1), 127–145 (2011)

5. Lange, S., Gabel, T., Riedmiller, M.: Batch reinforcement learning. In: Wiering, M., van Otterlo, M. (eds.) Reinforcement Learning. Adaptation, Learning, and Optimization, vol. 12. Springer, Heidelberg (2012). https://doi.org/10.1007/978-3-642-27645-3_2

6. Lin, L., Wen, L., Wang, J.: MM-Pred: a deep predictive model for multi-attribute event sequence. In: Proceedings, Society for Industrial and Applied Mathematics, pp. 118–126 (2019)

7. Maggi, F.M., Di Francescomarino, C., Dumas, M., Ghidini, C.: Predictive monitoring of business processes. In: Jarke, M.M., et al. (eds.) CAiSE 2014. LNCS, vol. 8484, pp. 457–472. Springer, Cham (2014). https://doi.org/10.1007/978-3-319-07881-6_31

8. Marquez-Chamorro, A., Resinas, M., Ruiz-Cortes, A.: Predictive monitoring of business processes: a survey. IEEE Trans. Serv. Comput. **11**(6), 962–977 (2018)

9. Mnih, V., et al.: Human-level control through deep reinforcement learning. Nature **518**(7540), 529–533 (2015)

10. Pasquadibisceglie, V., Appice, A., Castellano, G., Malerba, D.: Using convolutional neural networks for predictive process analytics. In: 2019 International Conference on Process Mining (ICPM 2019), pp. 129–136 (2019)

11. Pasquadibisceglie, V., Appice, A., Castellano, G., Malerba, D.: Predictive process mining meets computer vision. In: Fahland, D., Ghidini, C., Becker, J., Dumas, M. (eds.) BPM 2020. LNBIP, vol. 392, pp. 176–192. Springer, Cham (2020). https://doi.org/10.1007/978-3-030-58638-6_11

12. Plappert, M.: keras-rl (2016). https://github.com/keras-rl/keras-rl

13. Silver, D., et al.: A general reinforcement learning algorithm that masters chess, shogi, and Go through self-play. Science **362**(6419), 1140–1144 (2018)

14. Silver, D., Huang, A., et al.: Mastering the game of go with deep neural networks and tree search. Nature **529**(7587), 484–489 (2016)

15. Sutton, R.S., Barto, A.G.: Reinforcement Learning: An Introduction, 2nd edn. The MIT Press, Cambridge (2018)

16. Tax, N., Verenich, I., La Rosa, M., Dumas, M.: Predictive business process monitoring with LSTM neural networks. In: Dubois, E., Pohl, K. (eds.) CAiSE 2017. LNCS, vol. 10253, pp. 477–492. Springer, Cham (2017). https://doi.org/10.1007/978-3-319-59536-8_30

17. Vinyals, O., Babuschkin, I., et al.: Grandmaster level in starcraft "II" using multi-agent reinforcement learning. Nature **575**(7782), 350–354 (2019)

An Alignment Cost-Based Classification of Log Traces Using Machine-Learning

Mathilde Boltenhagen[1(✉)], Benjamin Chetioui[2], and Laurine Huber[3]

[1] Université Paris-Saclay, ENS Paris-Saclay, CNRS, LSV, Gif-sur-Yvette, France
mathilde.boltenhagen@lsv.fr
[2] University of Bergen, Department of Informatics, Bergen, Hordaland, Norway
benjamin.chetioui@uib.no
[3] Université de Lorraine, LORIA (UMR 7503), Nancy, France
laurine.huber@loria.fr

Abstract. Conformance checking is an important aspect of process mining that identifies the differences between the behaviors recorded in a log and those exhibited by an associated process model. Machine learning and deep learning methods perform extremely well in sequence analysis. We successfully apply both a Recurrent Neural Network and a Random Forest classifiers to the problem of evaluating whether the alignment cost of a log trace to a process model is below an arbitrary threshold, and provide a lower bound for the fitness of the process model based on the classification.

1 Introduction

With the cost of computer memory becoming negligible, organizations have become able to store extremely complex event logs from their systems. Process Mining (PM) is a field of study that attempts to make sense of these logs by producing corresponding *process models*. As decision makers increasingly rely on such models, it is crucial to ensure that they model the targeted systems reliably. *Conformance checking* is an entire subfield of PM that aims at defining the key criteria of a good process model [1]. As of today, the four main criteria that are considered are *fitness, precision, generalization*, and *simplicity*. Because of the complexity of the involved data and of the resulting process models, the fitness criterion is the only one unanimously accepted in the community. Computing the fitness requires alignments of the event logs with the process model, which often is costly [2,3] and for which a trade-off is possible between higher result quality and lower computational complexity. The need for such a compromise begs the question: is it possible to extract high-quality conformance checking information through a less complex process?

To motivate such research, the 2016 *Process Discovery Contest* invited scientists to study model compliance from a classification-oriented perspective [4]. The event logs were classified in two classes—*compliant* and *deviant*—using pure data mining techniques. By encoding event logs into sequences of activities called

© Springer Nature Switzerland AG 2021
S. Leemans and H. Leopold (Eds.): ICPM 2020 Workshops, LNBIP 406, pp. 136–148, 2021.
https://doi.org/10.1007/978-3-030-72693-5_11

log traces, it is possible to perform such a classification using Recurrent Neural Networks (RNNs). RNNs are at the core of significant progress in other fields of Computer Science such as Natural Language Processing, or Bioinformatics [5]. The PM community has recently shown significant interest in RNNs, but principally on the topic of *Predictive Business Process Monitoring* [6–11].

In this paper, we focus on the efficiency of Deep Learning (DL) and classical Machine Learning (ML) methods in conformance checking scenarios. Our core contribution is an application of a RNN and a Random Forest (RF) classifier to the problem of classifying traces based on their alignment costs to a reference process model. We provide some theoretical properties of the fitness along with reproducible experiments.

2   Related Work

The classification of log traces has been studied in the context of system deviation analysis. Such works generally consider two classes of processes (*normal* and *deviant*) and aim at explaining why discrepancies occur and deviant processes arise. Nguyen et al. defined trace classes from data attributes and investigated the problem of classification using decision trees, the k-Nearest Neighbors algorithm and neural networks [12]; Sun et al. and Bose et al. investigated labeled traces and association rules mining methods that can be used to extract human readable results from them [13,14]. Similarly, Bellodi et al. provided a method to classify log traces using Markov Logic formulas [15]. One glaring difference between these works and ours is that we have an oracle at our disposal to classify our traces, i.e. a process model.

The application of Long Short-Term Memory (LSTM) networks to the problem of predicting the next event in a business process was previously investigated in several works [6–9]. In lieu of RNNs, Pasquadibisceglie et al. investigated Convolutional Neural Networks for the same purpose [10]. Building on top of all these approaches, Taymouri et al. tackled the problem by implementing a Generative Adversarial Network, with promising results [11].

The present paper is probably most similar to the work of Nolle et al. [16], whose results, which are based on RNN-based alignments, are extremely promising, though they perform anomaly detection instead of log trace classification.

3   Preliminaries

In this section, we provide some background and notation for both PM and ML.

3.1   Log Traces, Process Model, Fitness and Alignments

We represent event data as log traces.

Definition 1 (Log traces). *Let $\mathcal{A}$ be a set of activities. We define a* log L *as a finite multiset of sequences $\sigma \in \mathcal{A}^*$, which we refer to as* log traces.

$\langle open, read, wait, wait, close \rangle$
$\langle open, read, close \rangle$
$\langle write, wait, close \rangle$
$\langle open, wait, write, close \rangle$

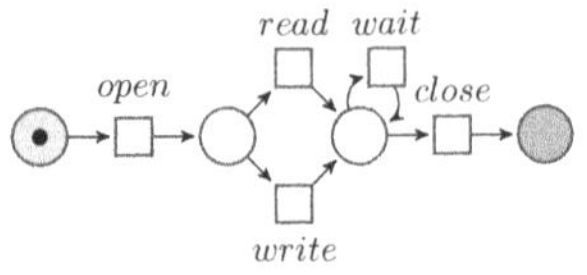

Fig. 1. A log L and an associated process model M

Process models can be generated from an event log; these models extrapolate a set of possible *runs* from the recorded log traces exhibited in the aforementioned event log. An example of a log and associated process model is provided in Fig. 1.

Definition 2 (Runs of a process model). *Let M be a process model defined over a set of activities $\mathcal{A}$. We write $Runs(M) \subseteq \mathcal{A}^*$ the set of sequences generated by M.*

This paper does not discuss the structure of process models; for a given model M, we consider the set $Runs(M)$ to be a sufficient description of M. How well M models a log is measured by the *fitness* criterion and can be computed based on $Runs(M)$ as the minimal cost of aligning each log trace to a run of M.

Definition 3 (Alignment Cost, Optimal Alignment). *Given a log trace $\sigma = \langle \sigma_1, \ldots, \sigma_m \rangle \in L$, and a process model M, we define the alignments of σ with M as sequences of pairs (moves) $\langle (\sigma'_1, u'_1), \ldots, (\sigma'_p, u'_p) \rangle$ with $p \leq m + n$ such that, for a given index i and a given run $u = \langle u_1, \ldots, u_n \rangle \in Runs(M)$:*

- *each move (σ'_i, u'_i) is either: a synchronous move (σ_j, u_k) with $\sigma_j = u_k$, a log move $(\sigma_j, \gg)$, which represents the deletion of σ_j in σ, or a model move $(\gg, u_k)$, which represents the insertion of u_k in σ, where $j \in \{1, \ldots, m\}$ and $k \in \{1, \ldots, n\}$;*
- *the left projection $\langle \sigma'_1, \ldots, \sigma'_p \rangle$ of the alignment to $\mathcal{A}^*$ (which drops the occurrences of $\gg$), yields σ;*
- *the right projection $\langle u'_1, \ldots, u'_p \rangle$ of the alignment to $\mathcal{A}^*$ (which drops the occurrences of $\gg$), yields u.*

We call alignment cost *the count of non-synchronous moves in the alignment. An optimal alignment is an alignment in which the alignment cost is the minimum possible given σ and M.*

The table below describes an optimal alignment of the log trace $\langle open, wait, write, close \rangle$ with the process model drawn in Fig. 1. Since the alignment contains one non-synchronous move, its cost is 1.

trace	open	wait	write	close
run	open	$\gg$	write	close

We compute the fitness of a process model with regards to a trace as follows:

$$\text{fitness}(\sigma, M) = 1 - \frac{\text{minCost}(\sigma, \text{select}(\sigma, M))}{|\sigma| + \min_{u' \in Runs(M)} |u'|} \tag{1}$$

where $\text{select}(\sigma, M)$ returns a run $u \in Runs(M)$ such that the set of alignments of σ with M using u contains an optimal alignment, and $\text{minCost}(\sigma, u)$ returns the minimum cost of aligning σ with M using a run u.

A trace is said to be *fitting* when its fitness is 1, i.e. when its optimal alignment has a cost of 0. We define the fitness of a process model M with regards to a log L to be the average of the fitness of M with regards to each log trace of L.

3.2   Supervised Learning from Sequences

There are several approaches towards training classification models from sequential data in a supervised way. They have in common that they must encode sequences of variable lengths as fixed-size vectors; these vectors are subsequently used as training examples for the classifier, which learns a classification model from them. The quality of the model is then assessed using several metrics and methods, based on its ability to accurately classify new inputs.

Building a Model. One can construct the vectors referenced above in different ways, e.g. by ignoring the order of the sequences (Bag-of-words) in the hope that knowledge about the frequency of each word in the sequences is sufficient to train a classifier (e.g. a RF classifier), or by training Deep Neural Networks able to encode the ordering of the sequences in the vectors (e.g. a LSTM network).

Long Short-Term Memory networks are RNNs able to learn and remember over long sequences of inputs [5]. They achieve that by integrating neurons specifically designed to determine whether a piece of information should be remembered or forgotten, depending on whether it is relevant for classification. Figure 2 gives the structure and relevant equations of an LSTM cell.

When one uses LSTM networks for sequence classification, the sequences (represented as sequences of integers) are usually first passed through an embedding layer before being passed through the LSTM layer; the prediction is then the output of a dense layer. One may add dropout layers to the network, in order to randomly ignore a percentage of units during training to avoid overfitting. The specificity of this architecture is that the whole sequence is fed as input to the network and that the embedding is learned through the training process; this permits learning a representation of the sequence that somehow embeds its sequential properties.

Definition 4 (Bag-of-words (BoW) encoding). *For an alphabet $\mathcal{A}$ and a sequence $\sigma \in \mathcal{A}^*$, a Bag-of-words encoding canonically maps σ to a multiset of words of $\mathcal{A}$.*

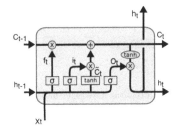

$$f_t = \sigma(W_f \cdot [h_{t-1}, x_t] + b_f)$$
$$i_t = \sigma(W_i \cdot [h_{t-1}, x_t] + b_i)$$
$$\tilde{C}_t = \tanh(W_c \cdot [h_{t-1}, x_t] + b_c)$$
$$o_t = \sigma(W_o \cdot [h_{t-1}, x_t] + b_o)$$
$$C_t = f_t * C_{t-1} + i_t * \tilde{C}_t$$
$$h_t = o_t * \tanh(C_t)$$

where b_g is the bias added at gate g, W_g is the weight vector for gate g, x_t is the current input, C_{t-1} the memory of last hidden unit, and h_{t-1} the output of last hidden unit.

Fig. 2. LSTM cell (adapted from [17])

In its simplest version, the multiset is encoded as a vector of integers $\mathcal{X}_\sigma$, and the element at index i in $\mathcal{X}_\sigma$ gives the count of the word at index i in $O_{\mathcal{A}}$ in the sequence σ, where $O_{\mathcal{A}}$ is a vector containing exactly all the elements of $\mathcal{A}$ in some arbitrary order. For instance, the respective BoW encodings of the log traces in Fig. 1 are $\langle 1, 1, 2, 1, 0 \rangle$, $\langle 1, 1, 0, 1, 0 \rangle$, $\langle 0, 0, 1, 1, 1 \rangle$, and $\langle 1, 0, 1, 1, 1 \rangle$ for $O_{\mathcal{A}} = \langle open, read, wait, close, write \rangle$.

Random Forests (RFs) are an ensemble learning method for classification. A RF constructs a bootstrapped collection of decision trees, i.e. a collection of decision trees that are sampled with replacement. Each decision tree consists of inner dichotomous nodes representing tests on random subsets of features, and of leaf nodes representing the possible output classes. The class of a given input can be predicted by taking the majority vote of the classification trees [18]. These decision trees can help to understand which features are important for classification, since every output can be represented as a list of decisions taken at the dichotomous nodes.

Validating a Model. We recall some metrics used to evaluate classification models, as well as one famous validation technique, namely the *K-fold cross-validation*.

Definition 5. *In the following, given a classification model C and a given input i, we write $y_{C,i}$ the actual class of the input and $\hat{y}_{C,i}$ its predicted class by C.*

Definition 6 (Accuracy). *For a given classification model C and an input i, we say that the classification is* accurate *when $y_{C,i} = \hat{y}_{C,i}$. For a set of inputs S, we define $E_{S,C} = \{i : i \in S, \ \hat{y}_{C,i} = y_{C,i}\}$. The accuracy $acc_C(S)$ of the classification of S by C is given as $acc_C(S) = \frac{|E_{S,C}|}{|S|}$.*

Definition 7 (Cross-Entropy Loss). *For a given binary classification model C and a given set of inputs S, there exists an error function called* cross-entropy loss $loss_C(S)$ *defined by $loss_C(S) = \frac{1}{|S|} \sum_{i \in S} -\log(P(\hat{y}_{C_i} = y_{C_i}))$.*

K-fold Cross-validation. K-fold cross-validation is a model validation technique used to lower the biases that may emerge when one only selects one training set and one testing set. Given $K \in \mathbb{N}^*$, the dataset D is split into K i-indexed subsets D_i. For each subset, one trains a model using $D \setminus D_i$ as the training set, and subsequently evaluates it using D_i as the testing set. The performance of the model is then summarized using the mean and variance of the evaluation scores.

4   Classifying Traces and Bounding the Fitness of a Model

The fitness of a log trace to a process model represents important information in conformance checking. Computing the fitness requires computing alignments of the trace with the model, which is a costly process. In this section, we present a binary classification of log traces based on their closeness to a process model: the *Alignment Cost Threshold-based Classification* (ACTC). This classification provides means of extracting relevant information at a much lower cost than alignments, while still guaranteeing a lower bound for the fitness of a process model to a log.

Definition 8 (Alignment Cost Threshold-based Classification). *Let M be a process model and L be a log. For a given alignment cost threshold $t_{AC} \in \mathbb{N}$, the ACTC maps each log trace $\sigma \in L$ to one of two classes depending on its minimal alignment cost $c_{\sigma,M}$:*

- *the* positive *class L_{pos} if $c_{\sigma,M} \leq t_{AC}$;*
- *the* negative *class L_{neg} otherwise.*

The t_{AC} parameter allows us to have more flexibility—in that we can now work with arbitrarily close traces instead of only fitting ones—and to control the balance of our two classes.

Theorem 1. *Given the ACTC of a log L for a model M and a cost threshold t_{AC}, the following holds:*

$$fitness(L, M) \geq \frac{\displaystyle\sum_{\sigma \in L_{pos}} 1 - \frac{t_{AC}}{|\sigma| + \min_{u \in Runs(M)} |u|}}{|L|} \qquad (2)$$

i.e. $fitness(L, M)$ is bounded from below.

Proof. The fitness of a process model M with regards to a log L is defined as the average of the fitness of M with regards to each log trace of L, i.e.

$$fitness(L, M) = 1 - \frac{\displaystyle\sum_{\sigma \in L} \frac{minCost(\sigma, select(\sigma, M))}{|\sigma| + \min_{u' \in Runs(M)} |u'|}}{|L|}. \qquad (3)$$

Let there be an ACTC of cost threshold t_{AC}. For every $\sigma \in L$, we have

$$\text{fitness}(\sigma, M) \geq \begin{cases} 0 & \text{if } \sigma \in L_{\text{neg}} \\ 1 - \dfrac{t_{AC}}{|\sigma| + \min\limits_{u' \in Runs(M)} |u'|} & \text{if } \sigma \in L_{\text{pos}} \end{cases} \tag{4}$$

since t_{AC} is the highest allowed alignment cost for a trace to be classified into L_{pos}. It follows trivially that:

$$\text{fitness}(L, M) \geq \frac{\sum\limits_{\sigma \in L_{\text{pos}}} 1 - \dfrac{t_{AC}}{|\sigma| + \min\limits_{u \in Runs(M)} |u|}}{|L|}. \tag{5}$$

In the following, we write $B = \text{fitness}(\sigma, M)$ for any $\sigma \in L_{\text{pos}}$.   □

Taking a small value for t_{AC} results in a large B, but a potentially smaller cardinality for L_{pos}; on the other hand, a large t_{AC} will induce a larger cardinality for L_{pos} but a smaller B. The aim of the following is to compute B from predictions, i.e. in a case where L_{pos} is built using a predictive approach. In this case, there is a risk that traces will be classified erroneously. We show in the next sections that classification models are good enough in practice to guarantee a lower bound of their fitness that is very close to the one outlined above.

5   Experiments

In this section, we present our datasets; we follow by describing how we parameterize our classification models; finally, we present our experimental results.

5.1   Alignment Datasets

The ACTC requires a training set of alignments; for that purpose, we have created alignments datasets that contain the trace variants of each dataset (i.e. the unique sequences in the log) and their minimal alignment costs for several process models[1]; that way, we rid our results of the noise induced by duplicate traces.

We ran our experiments on the three largest logs from the Business Process Intelligence Challenges available at the time of writing, using models from the work of Augusto et al. [19]. The models were discovered using the preprocessing method of Conforti et al. [20], and then either the Inductive Miner (IM) [21], the Split Miner (SM) [22], or the Heuristic Miner (SHM) [23]. Table 1 summarizes the relevant pieces of information pertaining to the datasets.

For each log, we also generate a set of 1000 random mock traces of lengths varying between 1 and the length of the longest trace in the log. These traces have, in most cases, a very high alignment cost with regards to the process models.

---

[1] https://github.com/BoltMaud/An-Alignment-Cost-Based-Classification-of-Log-Traces-Using-ML.

Table 1. Event log description and alignment costs

Log	Number of trace variants	Method of model discovery	Average alignment cost	Median alignment cost	Dataset name
BPIC_2012	4 366	Noise filter + IM	2.14	2.00	A_{2012}^{im}
		Noise filter + SM	3.02	3.00	A_{2012}^{sm}
		Noise filter + SHM	7.60	6.00	A_{2012}^{shm}
BPIC_2017	15 930	Noise filter + IM	14.90	13.00	A_{2017}^{im}
		Noise filter + SM	15.03	13.00	A_{2017}^{sm}
		Noise filter + SHM	16.31	14.00	A_{2017}^{shm}
BPIC_2019	11 973	Noise filter + IM	24.38	6.00	A_{2019}^{im}

5.2 Learning Methods

We train two classifiers, namely a RF on BoW-encoded sequences, and a LSTM network on sequences whose encoding embeds the sequential properties of the activities. The general overview of the training process is shown in Fig. 3.

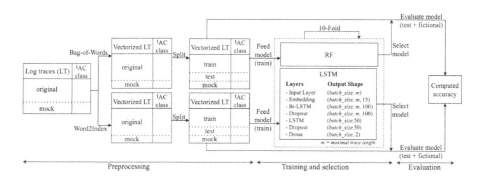

Fig. 3. Overview of the experimental setup

LSTM Network. This model takes constant-length vectors of integers as inputs, in which a given integer corresponds to exactly one activity. Traces that are shorter than the expected length of the vectors are padded as needed.

The architecture of the model we train is given in Fig. 3. The *input* layer takes a vector of size m (corresponding to the length of the longest trace in the log) containing elements belonging to the set of all the actions taken in the log traces. The vector is encoded into a vector of 15 elements using an *embedding* layer. The resulting vector is then fed to a *bi-LSTM* layer—ensuring that the left and right contexts of the actions in the input traces are remembered—and then to another simpler LSTM layer. *Dropout* layers with a frequency rate of 0.5 are added to prevent overfitting. The *dense* layer uses the softmax activation function to output the predicted classes, thus ensuring that they are mutually exclusive. We train the model for 10 epochs and with a batch size of 50 instances[2].

RF Classifier. The RF classifier does not take into account the order of the events, as it takes as input vectors that represent an ensemble of features, in our case activities. The classifier is thus trained with vectors resulting from a BoW encoding of the traces.

The target values, i.e. the prediction classes, are 0 (negative) or 1 (positive) depending on the minimal alignment cost of the sequence.

We set up 3 verification steps: first, we split the dataset into a training set (67%) and a testing set (33%) using a 10-fold cross-validation on the training sets to find the best predictive model in terms of accuracy. Second, we predict the classes of the sequences in the testing sets, and compare the accuracy during training to the accuracy during testing; they should be similar. Finally, we feed randomly generated traces with a high alignment cost to the predictive model; they should always be classified negatively.

5.3 Results and Interpretation

We built two distinct classifiers—one RNN and one RF—for each pair (d, m), with d one of the 7 datasets in Table 1, and m one of the possible alignment costs for the model; each pair represents an ACTC problem.

Table 2 summarizes the results of the experiments, where t_{AC} is the median of the alignment costs given in Table 1. The table contains the accuracies and losses for our testing data, and we compare our running times with the ones of ProM[3] for computing the alignments.

Both learning models exhibit good accuracy and low losses, thus confirming the potential of predictive approaches for the problem of alignment. The predicted lower bound of the fitness is computed from the traces classified as positive and is very close to the exact fitness lower bound. However, we note a significant difference between the actual fitness and these lower bounds. This is because the fitness function we use is coarse-grained, in that it gives a purely binary score denoting whether a log trace is classified as negative or positive. Despite this weakness, it remains somewhat useful as a heuristic to decide which of two models better fits a trace. It is also worth noting that our binary classification is straightforward to understand, whereas understanding alignments tends to require more expertise; such a classification is therefore likely to be very valuable to decision makers.

---

[2] The size of the embedding layer, the number of epochs, the batch size, and the stack of LSTM layers were chosen after several initial experiments, as they were the parameters that yielded the best results.

[3] https://www.promtools.org.

Table 2. Alignment cost threshold-based classification by using a RNN and a Random Forest classifier, with t_{AC} the median of the alignment costs.

Alignments	Fitness	t_{AC}	% of positive	Fitness lower bound	RNN				Random Forest				ProM avg. run-time (ms)
					Acc	Loss	Predicted fitness lower bound	Avg. run-time (ms)	Acc	Loss	Predicted fitness lower bound	Avg. run-time (ms)	
A^{im}_{2012}	0.950	2	73	0.695	0.999	0.011	0.695	12.00	0.988	0.057	0.700	0.06	42.28
A^{sm}_{2012}	0.932	3	73	0.670	0.829	0.377	0.745	19.72	0.820	0.472	0.713	0.08	52.85
A^{shm}_{2012}	0.837	6	56	0.476	0.969	0.104	0.491	23.75	0.972	0.136	0.479	0.06	99.89
A^{im}_{2017}	0.874	13	53	0.463	0.984	0.047	0.473	10.01	0.979	0.056	0.467	0.03	5.12
A^{sm}_{2017}	0.819	13	52	0.415	0.985	0.049	0.420	2.70	0.985	0.053	0.421	0.03	7.72
A^{shm}_{2017}	0.794	14	52	0.400	0.981	0.055	0.410	4.05	0.984	0.055	0.405	0.03	33.23
A^{im}_{2019}	0.561	6	53	0.328	0.973	0.078	0.338	15.11	0.958	0.103	0.344	0.03	1.09

Once the model has been trained, predicting the class of a trace is, in most cases, significantly faster than computing its exact alignment, as summarized in Table 2. One glaring exception is in the case of A^{im}_{2019}, in which computing exact alignments remains roughly 14 times more efficient than performing a prediction using the RNN. This is because the model is very simple (made of only 13 transitions, without loops); this is not surprising and should not matter in practice, as predictive approaches are tools designed to outperform exact approaches in complex cases with big or even intractable search spaces. One noteworthy caveat of using predictive approaches, however, is the fact that the models must be trained before they become able to output predictions. In our experiments, training a model took from 3.18 s to 8.97 s for our RF classifier, and from 2675.87 s to 34837.31 s for our LSTM network—both of which involved a 10-fold cross validation.

To better assess the impact of t_{AC} on our results, we perform a comparison of the predictions with varying t_{AC} values in Table 3. We summarize the accuracy, loss, and distribution into the two output classes for the testing data, as well as for randomly generated mock data. We notice that the accuracy drops very fast as t_{AC} grows larger for the mock data; this is induced by an equally quick drop in the percentage of log traces classified as negative. Given actual log traces however, both classifiers are reasonably accurate in each one of the considered cases. As was the case in Table 2, we note that the predicted lower bound of the fitness is close to the one given by our exact formula. This is also a nice result, although the actual fitness of the process model with regards to the log is pretty far off at 0.837.

Table 3. Comparison of the prediction results for different t_{AC} values for the testing set of A_{2012}^{shm}. The exact fitness for the used sublog is 0.837.

t_{AC}	Class	%	Fitness lower bound	RNN			Random Forest		
				Acc	Loss	Predicted fitness lower bound	Acc	Loss	Predicted fitness lower bound
2	all	100	0.071	0.992	0.029	0.076	0.998	0.009	0.073
	pos	8		0.982	0.214		1.000	0.043	
	neg	92		0.992	0.013		0.998	0.006	
	mock	100	/	0.961	0.108	/	0.904	0.207	/
4	all	100	0.169	0.991	0.042	0.166	0.999	0.016	0.170
	pos	20		0.968	0.151		1.000	0.021	
	neg	80		0.997	0.016		0.998	0.015	
	mock	100	/	0.937	0.303	/	0.876	0.317	/
6	all	100	0.476	0.971	0.104	0.491	0.972	0.150	0.479
	pos	56		0.990	0.066		0.978	0.063	
	neg	44		0.944	0.156		0.962	0.268	
	mock	100	/	0.871	0.543	/	0.837	0.548	/
8	all	100	0.500	0.976	0.092	0.498	0.984	0.077	0.501
	pos	65		0.980	0.079		0.989	0.031	
	neg	35		0.970	0.116		0.974	0.161	
	mock	100	/	0.818	1.189	/	0.782	0.911	/
10	all	100	0.524	0.937	0.165	0.508	0.971	0.103	0.522
	pos	73		0.943	0.100		0.979	0.055	
	neg	27		0.921	0.336		0.949	0.233	
	mock	100	/	0.364	3.759	/	0.620	1.650	/

6 Conclusion and Opening

We presented a compelling use of ML for conformance checking by constructing binary oracles—using a RF classifier and a LSTM network—that are able to predict with high accuracy whether the minimal alignment cost of a log trace with regards to a process model is below an arbitrary threshold. The method we proposed is more flexible, cheaper, and easier to understand for humans than the one usually used for exact alignments. We furthermore proved the existence of a lower bound for the fitness of a process model. Our work shows that there is a lot of value to be gained in exploring the use of ML methods in conformance checking. Future investigations may include whether the exact minimal alignment cost of a trace with a process model can be predicted from a regression model; another interesting project could build on the work of Nolle et al. [16] to predict optimal alignments of a log trace to a process model.

References

1. Carmona, J., van Dongen, B., Solti, A., Weidlich, M.: Conformance Checking. Springer, Cham (2018). https://doi.org/10.1007/978-3-319-99414-7
2. Adriansyah, A.: Aligning observed and modeled behavior (2014)
3. van Dongen, B., Carmona, J., Chatain, T., Taymouri, F.: Aligning modeled and observed behavior: a compromise between computation complexity and quality. In: Dubois, E., Pohl, K. (eds.) CAiSE 2017. LNCS, vol. 10253, pp. 94–109. Springer, Cham (2017). https://doi.org/10.1007/978-3-319-59536-8_7

4. Carmona, J., de Leoni, M., Depaire, B., Jouck, T.: Summary of the process discovery contest 2016. In: Proceedings of the Business Process Management Workshops, Springer (2016)
5. Hochreiter, S., Schmidhuber, J.: Long short-term memory. Neural Comput. **9**, 1735–1780 (1997)
6. Evermann, J., Rehse, J.R., Fettke, P.: Predicting process behaviour using deep learning. Decis. Support Syst. **100**, 129–140 (2017)
7. Tax, N., Verenich, I., La Rosa, M., Dumas, M.: Predictive business process monitoring with LSTM neural networks. In: Dubois, E., Pohl, K. (eds.) CAiSE 2017. LNCS, vol. 10253, pp. 477–492. Springer, Cham (2017). https://doi.org/10.1007/978-3-319-59536-8_30
8. Camargo, M., Dumas, M., González-Rojas, O.: Learning accurate LSTM models of business processes. In: Hildebrandt, T., van Dongen, B.F., Röglinger, M., Mendling, J. (eds.) BPM 2019. LNCS, vol. 11675, pp. 286–302. Springer, Cham (2019). https://doi.org/10.1007/978-3-030-26619-6_19
9. Lin, L., Wen, L., Wang, J.: Mm-Pred: a deep predictive model for multi-attribute event sequence. In: International Conference on Data Mining, SIAM (2019)
10. Pasquadibisceglie, V., Appice, A., Castellano, G., Malerba, D.: Using convolutional neural networks for predictive process analytics. In: International Conference on Process Mining, ICPM, IEEE (2019)
11. Taymouri, F., Rosa, M.L., Erfani, S., Bozorgi, Z.D., Verenich, I.: Predictive business process monitoring via generative adversarial nets: the case of next event prediction. In: Fahland, D., Ghidini, C., Becker, J., Dumas, M. (eds.) BPM 2020. LNCS, vol. 12168, pp. 237–256. Springer, Cham (2020). https://doi.org/10.1007/978-3-030-58666-9_14
12. Nguyen, H., Dumas, M., La Rosa, M., Maggi, F.M., Suriadi, S.: Mining business process deviance: a quest for accuracy. In: Meersman, R., et al. (eds.) OTM 2014. LNCS, vol. 8841, pp. 436–445. Springer, Heidelberg (2014). https://doi.org/10.1007/978-3-662-45563-0_25
13. Sun, C., Du, J., Chen, N., Khoo, S.C., Yang, Y.: Mining explicit rules for software process evaluation. In: Proceedings of the 2013 International Conference on Software and System Process (2013)
14. Bose, R.J.C., van der Aalst, W.M.: Discovering signature patterns from event logs. In: 2013 IEEE Symposium on Computational Intelligence and Data Mining (CIDM). IEEE (2013)
15. Bellodi, E., Riguzzi, F., Lamma, E.: Probabilistic declarative process mining. In: Bi, Y., Williams, M.-A. (eds.) KSEM 2010. LNCS (LNAI), vol. 6291, pp. 292–303. Springer, Heidelberg (2010). https://doi.org/10.1007/978-3-642-15280-1_28
16. Nolle, T., Seeliger, A., Thoma, N., Mühlhäuser, M.: DeepAlign: alignment-based process anomaly correction using recurrent neural networks. In: Dustdar, S., Yu, E., Salinesi, C., Rieu, D., Pant, V. (eds.) CAiSE 2020. LNCS, vol. 12127, pp. 319–333. Springer, Cham (2020). https://doi.org/10.1007/978-3-030-49435-3_20
17. Olah, C.: Understanding LSTM networks, August 2015. Accepted 02 Sept 2020
18. Breiman, L.: Random forests. Mach. Learn. **45**, 5–32 (2001)
19. Augusto, A., Armas-Cervantes, A., Conforti, R., Dumas, M., La Rosa, M., Reissner, D.: Abstract-and-compare: a family of scalable precision measures for automated process discovery. In: Weske, M., Montali, M., Weber, I., vom Brocke, J. (eds.) BPM 2018. LNCS, vol. 11080, pp. 158–175. Springer, Cham (2018). https://doi.org/10.1007/978-3-319-98648-7_10

20. Conforti, R., La Rosa, M., ter Hofstede, A.H.: Filtering out infrequent behavior from business process event logs. IEEE Trans. Knowl. Data Eng. **29**, 300–314 (2016)
21. Leemans, S.J.J., Fahland, D., van der Aalst, W.M.P.: Discovering block-structured process models from event logs containing infrequent behaviour. In: Lohmann, N., Song, M., Wohed, P. (eds.) BPM 2013. LNBIP, vol. 171, pp. 66–78. Springer, Cham (2014). https://doi.org/10.1007/978-3-319-06257-0_6
22. Augusto, A., Conforti, R., Dumas, M., La Rosa, M., Polyvyanyy, A.: Split miner: automated discovery of accurate and simple business process models from event logs. Knowl. Inf. Syst. **59**, 251–284 (2019)
23. Augusto, A., Conforti, R., Dumas, M., La Rosa, M., Bruno, G.: Automated discovery of structured process models from event logs: the discover-and-structure approach. Data Knowl. Eng. **117**, 373–392 (2018)

Improving the Extraction of Process Annotations from Text with Inter-sentence Analysis

Luis Quishpi[✉], Josep Carmona[✉], and Lluís Padró[✉]

Computer Science Department, Universitat Politècnica de Catalunya,
Barcelona, Spain
{quishpi,jcarmona,padro}@cs.upc.edu

Abstract. The automatic extraction of formal process information from textual descriptions of processes is a challenging problem, but worth exploring, since it enables organizations to align complementary information that talks about processes. In this paper we continue our previous work on this area, based on defining hierarchical/tree patterns on the dependency trees that arise from the linguistic analysis. We now incorporate a new abstraction layer on these patterns, that consider relationships between nearby sentences. The aim of this extension is to capture inter-sentence relationships that typically arise in textual descriptions of processes. The experiments done on publicly available benchmarks corroborate this intuition, showing a significant rise in the ability to capture all the important control-flow relationships defined in the text.

1 Introduction

As it has been recently acknowledged, there are quite important challenges on applying Natural Language Processing (NLP) techniques in the field of Business Process Management (BPM) [12]. Among the important ones, the extraction of process models from textual process descriptions is a very attractive use case, since the creation of process models consumes up to 60% of the time spent on process management projects. This paper focuses on this challenging task.

Although different approaches have been considered in the last years (see Sect. 2), a number of open challenges remain for reaching a maturity level enabling its widespread adoption. For instance, techniques must be able to identify sentences that provide contextual information, rather than describe process steps. Furthermore, the inherent ambiguity of natural language can lead to different interpretations regarding the process that is described [14].

In this paper we significantly expand the techniques and results recently presented in [9], where we described robust tree-based patterns to be queried over the dependency trees arising from the NLP analysis of the textual descriptions. Patterns in [9] where only applicable in the context of a single sentence, which made our approach unable to extract inter-sentence relationships. The contribution of this paper is therefore the extension of our previous contribution with a

© Springer Nature Switzerland AG 2021
S. Leemans and H. Leopold (Eds.): ICPM 2020 Workshops, LNBIP 406, pp. 149–161, 2021.
https://doi.org/10.1007/978-3-030-72693-5_12

more general set of patterns, resulting in a significant boost in the recall of the original framework (see experiment results in Sect. 5).

The paper is organized as follows: next section shortly describes the work related to this contribution. Section 3 overviews the main components of our proposal, presented in Sect. 4. Experiments and tool support are reported in Sect. 5, whilst Sect. 6 concludes the paper and outlines future work.

2   Related Work

For the sake of space, we only report here the related work that focuses on the extraction of process knowledge from textual descriptions [1,4,13], or the work that considers textual annotations in the scope of BPM [7,10].

For the former, the work by Gonçalves et al. [1] adopts important steps to extract the different BPMN elements and the work by Friedrich et al. [4] is acknowledged as the state-of-the-art for extracting process representations from textual descriptions, so we focus our comparison on this approach. As we will see in the evaluation section, our approach is significantly more accurate with respect to the state-of-the-art in the extraction of the main process elements. Likewise, we have incorporated as well the patterns from [13], and a similar outcome is reported in the experiments. The main reason is that approach relies on a deep NLP analysis and patterns on the syntactic structure of the sentence, instead of a shallow analysis and flat patterns.

For the later type of techniques [7,10], we see these frameworks as the principal application for our techniques. In particular, we have already demonstrated in the platform https://modeljudge.cs.upc.edu an application of the use of annotations in the scope of teaching and learning process modeling[1].

3   Preliminaries

The core of our proposal is the use of deep NLP analyzers to convert a textual description of a process into a syntax-semantic structure. Then, this structure is mined using tree-shaped patterns to obtain a conceptual representation of the process. Although other tools could be used, we resort to FreeLing as a NLP analyzer, TRregex as a tree-oriented pattern matching tool, and ATDP as a conceptual representation support. We describe each of them below.

3.1   Natural Language Processing

Linguistic analysis tools can be used as a means to structure information contained in texts for its later processing in applications less related to language itself. This is our case: we use NLP analyzers to convert a textual description of a process model into a structured representation.

---

[1] The reader can see a tutorial for annotating process modeling exercises in the ModelJudge platform at https://modeljudge.cs.upc.edu/modeljudge_tutorial/.

The NLP processing software used in this work is FreeLing[2] [8], an open–source library of language analyzers providing a variety of analysis modules for a wide range of languages. More specifically, the natural language processing layers used in this work are: tokenization & sentence splitting, morphological analysis, PoS-Tagging, Named Entity Recognition, Word sense disambiguation, Dependency parsing, Semantic role labeling and Coreference resolution. The three last steps are of special relevance since they allow the top-level predicate construction, and the identification of actors throughout the whole text: dependency parsing identifies syntactic subjects and objects (which may vary depending, e.g., on whether the sentence is active or passive), while semantic role labeling identifies semantic relations (the *agent* of an action is the same regardless of whether the sentence is active or passive). Coreference resolution links several mentions of an actor as referring to the same entity.

3.2   Annotated Textual Descriptions of Processes (ATDP)

ATDP is a formalism proposed in [10], aiming to represent process models on top of textual descriptions. This formalism naturally enables the representation of a wide range of behaviors, ranging from procedural to completely declarative, but also hybrid ones. Different from classical conceptual modeling principles, this highlight ambiguities that can arise from a textual description of a process, so that a specification can have more than one possible interpretation[3].

ATDP specifications can be translated into linear temporal logic over finite traces [2,5], opening the door to formal reasoning, automatic construction of formal models (e.g. in BPMN) from text, and other interesting applications such as simulation: to generate end-to-end executions (i.e., an *event log* [15]) that correspond to the process described in the text, which would allow the application of *process mining* algorithms.

ATDP models are defined over an input text, which is marked with *typed text fragments*, which may correspond to *entities*, or *activities*. Marked fragments can be related among them via a set of *fragment relations*.

Entity Fragments. The types of entity fragments defined in ATDP are:

- *Role.* The role fragment type is used to represent types of autonomous actors involved in the process, and consequently responsible for the execution of activities contained therein.
- *Business Object.* This type is used to mark all the relevant elements of the process that do not take an active part in it, but that are used/manipulated by process activities.

---

[2] http://nlp.cs.upc.edu/freeling.
[3] In this work we consider a flattened version of the ATDP language, i.e., without the notion of *scopes*.

Activity Fragments. ATDP distinguishes two types of activity fragments:

- *Condition.* It is considered discourse markers that mark conditional statements, like: *if, whether* and *either.* Each discourse marker needs to be tailored
  to a specific grammatical structure.
- *Task and Event.* Those fragment types are used to represent the atomic units
  of work within the business process described by the text. Usually, these
  fragments are associated with verbs. Event fragments are used to annotate
  other occurrences in the process that are relevant from the point of view of
  the control flow, but are exogenous to the organization responsible for the
  execution of the process.

Fragment Relations. Text fragments can be related to each other by means of
different relations, used to express properties of the process emerging from the
text:

- *Agent.* Indicates the role responsible for the execution of an activity.
- *Patient.* Indicates the role or business object on which an activity is performed.
- *Coreference.* Induces a coreference graph where each connected component
  denotes a distinct process entity.
- *Sequential.* Indicates the sequential execution of two activity fragments A1 and
  A2 in a sentence. We consider two important relations from [10]: Precedence
  and Response. Moreover, to cover situations where ambiguities in the text
  prevent selecting any of the two aforementioned relations, we also incorporate a less restrictive constraint WeakOrder, that only applies in case both
  activities occur in a trace.
- *Conflicting.* A conflict relation between two condition activity fragments
  $\langle C1, C2 \rangle$ in a sentence indicates that one and only one of them can
  be executed, thus capturing a choice. This corresponds to the relation
  NonCoOccurrence from [10].

3.3   TRegex

In this paper, we use Tregex[4] [6], a query language that allows the definition of
regular-expression-like patterns over tree structures. Tregex is designed to match
patterns involving the content of tree nodes and the hierarchical relations among
them. In our case we will be using Tregex to find substructures within syntactic
dependency trees. Applying Tregex patterns on a dependency tree allows us to
search for complex labeled tree dominance relations involving different types of
information in the nodes. The nodes can contain symbols or a string of characters
(e.g. lemmas, word forms, PoS tags) and Tregex patterns may combine those
tags with the available dominance operators to specify conditions on the tree.
Additionally, as in any regular expression library, subpatterns of interest may

---

[4] https://nlp.stanford.edu/software/tregex.html.

be specified and the matching subtree can be retrieved for later use. This is achieved in Tregex using unification variables as shown in pattern (2) in Fig. 1.

Figure 1 describes the main Tregex operators used in this research to specify pattern queries.

Operator	Meaning
X << Y	X dominates Y
X >> Y	X is dominated by Y
X !>> Y	X is not dominated by Y
X < Y	X immediately dominates Y
X > Y	X is immediately dominated by Y
X >, Y	X is the first child of Y
X >- Y	X is the last child of Y
X >: Y	X is the only child of Y
X $-- Y	X is a right sibling of Y
X $. Y	X is the immediate left sibling of Y

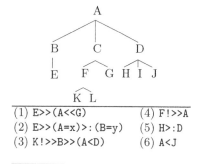

(1) E>>(A<<G)   (4) F!>>A
(2) E>>(A=x)>:(B=y)   (5) H>:D
(3) K!>>B>>(A<D)   (6) A<J

Fig. 1. Some operators provided by Tregex (left). The tree on the right would match patterns (1), (2), (3), and would not match patterns (4), (5), (6). Note that unless parenthesized, all operators refer to the first element in the pattern. Pattern (2) uses operator = to capture nodes A and B into variables x and y respectively.

4   Generalized Approach

4.1   Basic Approach: Intra-sentence Analysis

In this paper we describe an extension to the approach presented in [9]. This subsection summarizes the basic original approach, and following subsections provide details on the added extensions, which mainly consist of the extraction of relations between actions or conditions in different sentences, as well as an extended evaluation covering not only entities and actions, but also relations.

In [9] we presented a proposal to extract Business Process elements (entities, actions, conditions, events, and relations) from a process textual description.

The approach consists of: (a) Use a full-fledged NLP analysis pipeline [8] to analyze the text and extract verbal predicates, involved actors and objects, syntactic trees of all sentences, and coreferences between different mentions of the same actor/object, and (b) apply a cascade of TRegEx patterns on the output of the NLP preprocess to extract and elaborate the relevant process information. These patterns perform the following tasks:

1. Select the appropriate description for an entity or object. For instance, in the sentence "*The process starts when the female patient is examined by an outpatient physician, who decides whether she is healthy or needs to undertake an additional examination*" the results of the NLP semantic role labeling

step for *Agent* would return the whole subtree headed by *physician* (i.e. *an outpatient physician, who decides... examination*).

The used Tregex patterns will strip down such a long description removing the determiner and the relative clause, while keeping the core actor/object and its main modifiers, thus extracting respectively `outpatient physician` as a role, and `female patient` as a business object.

2. Next step is identifying relevant activities. The NLP preprocess detects all predicates in the text (mainly, all verbs are considered a predicate, plus some deverbal nouns such as "reception", "meeting", etc.). However, although many verbs in a process description may be predicates from a linguistic perspective, they do not correspond to actual process activities. Thus, we use a set of patterns that discard predicates unlikely to be describing a relevant process task, or relabel them as *condition* or *event* fragments:

   (a) More specifically, we use a set of predicates that check for syntactic structures involving conditional clauses (*if, whether, either, ...*) and the appropriate nodes in the tree are marked as *condition* fragments. In this step, we determine, for instance, that `she is healthy` and `needs an additional examination` are conditions in the sentence "*... who decides whether she is healthy or needs an additional examination.*".

   (b) Another set of patterns deal with syntactic structures involving keywords like *when, once, as soon, whenever*, etc., and mark the related predicates as *event* fragments. These patterns allow us to identify the fragment `confirm (payment)` as an event fragment in the sentence "*Once the payment is confirmed, the ZooClub department can print the card...*"

   (c) A third batch of patterns takes care of discarding activities that are not relevant to the process. To this end, we use two different strategies: one is removing all activities related to auxiliary, control, or subjective verbs (*be, have, start, want, think, believe*, etc.) which are unlikely to describe an actual process task. The second strategy relies on removing actions described in a subordinate clause. For instance, in the sentence "*..., the examination is prepared based on the information provided by the outpatient section*", the verbs *base* and *provide* would be removed as activities, since the main action described by this sentence is just *prepare (examination)*, and the subordinate clause just gives additional details on the object or procedure, but not on the actual process activity.

3. The last set of patterns deal with relations between activities. In our original work we tackled only relations between two activities in the same sentence. We considered different types of relations:

   (a) Precedence: We use patterns to detect sentences relating one *event* and one *activity* in a precedence relation. E.g. in the sentence "*An intaker keeps this registration with him at times when visiting the patient*", it would extract the sequential relation from `visit (patient)` to `keep (registration)`.

(b) Response: This relation is identified between *condition* and *activity* fragments, which typically occur in conditional sentences such as *"If the patient signs an informed consent, a delegate of the physician arranges an appointment with one of the wards and updates the HIS selecting the first available slot"*. From this sentence, we would extract the relation that `arrange (appointment)` *responds to* `sign (consent)`.

(c) Weak Order: There are many pairs of activities appearing in the same sentence where some kind of sequential order can be deduced, but it is not possible for an automatic system to determine their exact kind of relation. In these cases, we take a conservative approach and extract the least restrictive constraint, WeakOrder. For instance, in the sentence *"The Payment Office of SSP generates a payment report and then pays the vendor"*, we could extract that `generate` and `pay` are in WeakOrder.

(d) Conflict: Conflict relations can be determined between condition fragments, provided they are in the right syntactic structure. In this way, we can extract the constraint that the sample can not be safely used and contaminated at the same time from the sentence *"... decides whether the sample can be used for analysis or whether it is contaminated"*, or that conditional fragments `approve` and `deny` from the sentence *"The next step is for the IT department to analyse the request and either approve or deny it."* are considered in conflict.

4.2   Inter-sentence Analysis

Patterns used in [9] for relation extraction summarized in Sect. 4.1 aimed to capture relations between two activities/events mentioned in the same sentence. The main contribution of this paper is the extension of these patterns to capture also relations between activities or events located in different sentences.

To achieve this goal, since TRegEx is able to handle a single tree at a time, we first need to join together the syntactic trees for all sentences in the text in a single tree. For this, we add two kinds of artificial parent nodes: A `<PARAGRAPH>` node that has as children the root nodes for each of the sentences in the same paragraph, and a `<DOCUMENT>` node that has as children all the `<PARAGRAPH>` nodes. With that, we obtain a unique tree for all the document, and we can apply TRegEx patterns that span over more than one sentence. Figure 2 shows an example of a tree representing a short document. We apply patterns on the document tree to extract conflict and sequence relations between activities, events, or conditions detected in previous steps (see Sect. 4.1).

Conflicts. Conflicts between activities in the same sentence are detected using patterns described in [9]. The following patterns deal with conflicts between activities in different sentences. Their goal is to instantiate in variables `originRef` and `destinationRef` verbs that head sentences which may contain nodes marked as `<ACTIVITY>` or `<CONDITION>` on which the relation will be extracted.

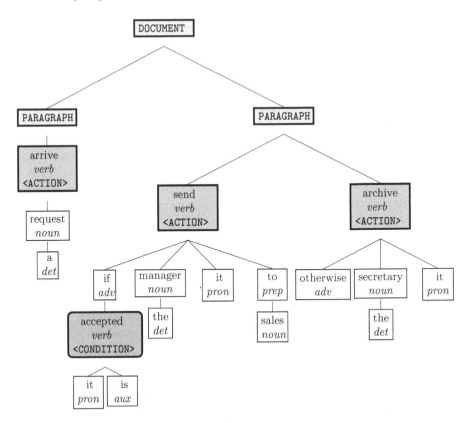

Fig. 2. Document tree for a text with two paragraphs: The first one with the sentence "*A request arrives*", the second with two sentences: "*If it is accepted, the manager sends it to sales. Otherwise, the secretary archives it*". Nodes in the syntactic dependency trees have been marked as <ACTION> or <CONDITION> in previous steps.

```
PC1  /verb/=originRef > /<PARAGRAPH>/
                   << /<CONDITION>/
                   $. (/verb/=destinationRef << /<CONDITION>/)
PC2  /verb/=originRef > /<PARAGRAPH>/
                   << /<CONDITION>/
                   $. (/verb/ !<< /<CONDITION>/
                           $. (/verb/=destinationRef << /<CONDITION>/))
```

Pattern PC1 checks for a verb directly under a <PARAGRAPH> (i.e. main sentence verb) that has a condition as a child, and that its right sibling (i.e., main verb in the following sentence) also has a condition. This would extract a conflict between *proceed* and *repeat* in the pair of sentences "*If sample is ok, proceed with examination. If contamination is detected, repeat sampling.*" Pattern PC2 captures the same kind of structure, when there is an additional sentence without a condition in between (e.g. "*If sample is ok, proceed with examination. Fill out treatment request form. If contamination is detected, repeat sampling.*")

Sequences. A second batch of patterns takes care of extracting sequence relations between activities in contiguous sentences. As in the case of conflicts, the patterns instantiate the variables `originRef` and `destinationRef` to candidate subtrees that are then searched for `<ACTIVITY>` or `<CONDITION>` nodes. Some patterns directly instantiate the variable `destination`, the actual target of the extracted relation.

```
PS1  /verb/=originRef > /<PARAGRAPH>/
                      $. (/verb/=destinationRef
                             < /afterwards|then|immediately/)
PS2  /verb/=originRef > /<PARAGRAPH>/
                      $. (/verb/ << (/<CONDITION>/=destination << /or/))
PS3  /verb/=originRef > /<PARAGRAPH>/
                      $. (/verb/ << (/or/ << /<CONDITION>/=destination))
PS4  /verb/=originRef > /<PARAGRAPH>/
                      $. (/verb/ << /<CONDITION>/=destination
                             < /otherwise|else/)
PS5  /verb/=originRef > /<PARAGRAPH>/
                      $. (/verb/ << /<CONDITION>/
                             < (/otherwise|else/=destination)
```

Pattern PS1 extracts a sequence relation between the main verb of a sequence and the main verb of the next one provided the latter has a modifier such as *afterwards, then, immediately*, etc. Patterns PS2 and PS3 establish sequence relations between an activity and or-ed conditions in the following sentence (e.g. extract sequences *send→fill* and *send→reject* in sentences *"Send form to customer. The customer can fill the form or reject to do it."*) Patterns PS4 and PS5 check a similar case, but where the second sentence has an "if-else" structure. They would extract the sequence relations *send→accept* and *send→cancel* in the sentence *"A budget is sent to the customer. If he accepts it, the bill is issued, otherwise the operation is cancelled."*

5   Tool Support and Experiments

This section presents experiments evaluating the performance gain obtained when including patterns to capture relations between activities or events located in different sentences. We report two different results: First, we report relations extraction performance using a baseline based on [9] where we extract relations only for pairs in the same sentence. Second, we report results applying patterns to extract both intra- and inter-sentence relations.

The evaluation is performed comparing the relations extracted against gold standard manual annotations. For both Table 1 and Table 2, the test data set used in our experiments are the same as that used in the original proposal [9], which consists of 18 text-model pairs, each example includes a textual process description paired with the corresponding BPMN models created by a human.

The first 13 models stem from material in the appendix of [3], and the last 5 from our academic dataset[5] used in [11].

As a gold reference for evaluation, we manually created one ATDP for each example following the activities and relations in those BPMN models, i.e. marking as activity fragments only the text pieces that had a corresponding element in the BPMN model and connecting only the activities fragments that had a corresponding relation in the BPMN model.

Intra-sentence. Results for the first scenario (only intra-sentence patterns) are shown in Table 1, and correspond to results obtained using the patterns described in [9], which rely on extracting relations just within sentences. Precision is the percentage of right relations over predicted relations ($P = \#ok/\#pred$). Recall is the percentage of expected relations extracted ($R = \#ok/\#gold$). F_1 score is the harmonic mean of precision and recall ($F_1 = 2PR/(P + R)$). We only count extracted relations as right if they match the gold annotations in type

Table 1. Evaluation of relation extraction using only intra-sentence patterns. Sequence relations are evaluated on the transitive clausure of both the sets of gold annotations and annotations produced by the system.

Source	Conflict						Sequence					
	#gold	#pred	#ok	P	R	F_1	#gold	#pred	#ok	P	R	F_1
1-1_bicycle_manufacturing	2	1	1	100	50	67	59	8	6	75	10	18
1-2_computer_repair	1	0	0	0	0	0	59	9	7	78	12	21
2-1_sla_violation	5	2	0	0	0	0	372	46	14	30	4	7
3-1_2009-1_mc_finalice_sct	0	0	0	0	0	0	52	4	3	75	6	11
3-2_2009-2_conduct	1	0	0	0	0	0	20	8	4	50	20	29
3-6_2010-1_claims_notification	2	0	0	0	0	0	63	9	7	78	11	19
4-1_intaker_workflow	0	0	0	0	0	0	596	17	5	29	1	2
5-1_active_vos_tutorial	3	0	0	0	0	0	15	4	3	75	20	32
6-1_acme-	1	0	0	0	0	0	340	22	11	50	3	6
7-1_calling_leads	1	0	0	0	0	0	13	2	1	50	8	13
8-1_hr_process_simple	0	0	0	0	0	0	15	7	6	86	40	55
9-2_exercise_2	3	0	0	0	0	0	11	6	6	100	55	71
10-2_process_b3	3	0	0	0	0	0	114	7	3	43	3	5
1081511532_rev3	1	0	0	0	0	0	47	9	5	56	11	18
1120589054_rev4-	0	0	0	0	0	0	66	9	6	67	9	16
1364308140_rev4	1	0	0	0	0	0	21	8	4	50	19	28
20818304_rev1	3	2	2	100	67	80	36	11	6	55	17	26
784358570_rev2	2	0	0	0	0	0	126	11	6	55	5	9
TOTAL	29	5	3	**60**	**10**	**18**	2025	197	103	**52**	**5**	**9**

---

[5] https://github.com/setzer22/alignment_model_text/tree/master/datasets/
NewDataset.

(<SEQUENCE>, <CONFLICT>). In both experiments, sequence relations are evaluated over the transitive closure of the sequence annotations.

Inter-sentence. In the second evaluation scenario, in addition to patterns created in [9], we use inter-sentence patterns described in Sect. 4.2.

Obtained results presented in Table 2 show that our new contribution extracts more relations, thus obtaining a large boost in recall (from 0.05 to 0.70 overall) with a very mild loss of precision (from 0.52 to 0.50 overall). Recall is boosted both for conflict and sequence relations, while precision is increased for conflicts, but slightly decreased for sequences.

Table 2. Evaluation of relation extraction using both intra- and inter- sentence patterns. Sequence relations are evaluated on the transitive clausure of both the sets of gold annotations and annotations produced by the system.

Source	Conflict						Sequence					
	#gold	#pred	#ok	P	R	F_1	#gold	#pred	#ok	P	R	F_1
1-1_bicycle_manufacturing	2	2	2	100	100	100	59	90	54	60	92	72
1-2_computer_repair	1	0	0	0	0	0	59	65	33	51	56	53
2-1_sla_violation	5	4	2	50	40	44	372	572	152	27	41	32
3-1_2009-1_mc_finalice	0	0	0	0	0	0	52	57	42	74	81	77
3-2_2009-2_conduct	1	0	0	0	0	0	20	33	19	58	95	72
3-6_2010-1_claims	2	1	1	100	50	67	63	76	53	70	84	76
4-1_intaker_workflow	0	0	0	0	0	0	596	906	455	50	76	61
5-1_active_vos_tutorial	3	3	3	100	100	100	15	16	12	75	80	77
6-1_acme-	1	0	0	0	0	0	340	561	287	48	84	61
7-1_calling_leads	1	1	1	100	100	100	13	41	13	32	100	48
8-1_hr_process_simple	0	0	0	0	0	0	15	21	15	71	100	83
9-2_exercise_2	3	6	3	50	100	67	11	10	9	90	82	86
10-2_process_b3	3	1	1	100	33	50	114	138	83	60	73	66
1081511532_rev3	1	1	1	100	100	100	47	41	30	73	64	68
1120589054_rev4	0	0	0	0	0	0	66	78	66	85	100	92
1364308140_rev4	1	0	0	0	0	0	21	26	10	38	48	43
20818304_rev1	3	3	3	100	100	100	36	29	19	66	53	58
784358570_rev2	2	3	2	67	100	80	126	118	85	72	67	70
TOTAL	29	25	19	**76**	**66**	**70**	2025	2878	1437	**49**	**71**	**59**

6 Conclusions and Future Work

We have presented an extension of our work in [9], consisting in adding syntax-tree based patterns to capture relations between activities or events located in different sentences. Results show that crossing the sentence boundaries is a highly productive strategy, since many more relations can be extracted. Also, the fact of using syntax-aware patterns, and not just flat regular expressions allows this extension to be done with almost no loss of precision.

Acknowledgments. This work has been supported by MINECO and FEDER funds under grant TIN2017-86727-C2-1-R, and by the Ecuadorian National Secretary of Higher Education, Science and Technology (SENESCYT).

References

1. Gonçalves, J.C.A.R., Santoro, F.M., Baião, F.A.: Business process mining from group stories. In: Proceedings of the 13th International Conference on Computers Supported Cooperative Work in Design, CSCWD 2009, Santiago, Chile, 22–24 April 2009, pp. 161–166. IEEE (2009)
2. De Giacomo, G., De Masellis, R., Montali, M.: Reasoning on LTL on finite traces: insensitivity to infiniteness. In: Proceedings of the 28th AAAI Conference on Artificial Intelligence, pp. 1027–1033. AAAI Press (2014)
3. Friedrich, F.: Automated generation of business process models from natural language input. School of Business and Economics, Humboldt-Universität (2010)
4. Friedrich, F., Mendling, J., Puhlmann, F.: Process model generation from natural language text. In: Mouratidis, H., Rolland, C. (eds.) CAiSE 2011. LNCS, vol. 6741, pp. 482–496. Springer, Heidelberg (2011). https://doi.org/10.1007/978-3-642-21640-4_36
5. De Giacomo, G., Vardi, M.Y.: Linear temporal logic and linear dynamic logic on finite traces. In: IJCAI (2013)
6. Levy, R., Andrew, G.: Tregex and Tsurgeon: tools for querying and manipulating tree data structures. In: LREC, pp. 2231–2234. Citeseer (2006)
7. López, H.A., Debois, S., Hildebrandt, T.T., Marquard, M.: The process highlighter: from texts to declarative processes and back. In: Proceedings of the Dissertation Award, Demonstration, and Industrial Track at BPM 2018 Co-Located with 16th International Conference on Business Process Management (BPM 2018), Sydney, Australia, 9–14 September 2018, pp. 66–70 (2018)
8. Padró, L., Stanilovsky, E.: Freeling 3.0: towards wider multilinguality. In: Proceedings of the Eighth International Conference on Language Resources and Evaluation (LREC), pp. 2473–2479 (2012)
9. Quishpi, L., Carmona, J., Padró, L.: Extracting annotations from textual descriptions of processes. In: Fahland, D., Ghidini, C., Becker, J., Dumas, M. (eds.) BPM 2020. LNCS, vol. 12168, pp. 184–201. Springer, Cham (2020). https://doi.org/10.1007/978-3-030-58666-9_11
10. Sànchez-Ferreres, J., Burattin, A., Carmona, J., Montali, M., Padró, L.: Formal reasoning on natural language descriptions of processes. In: Hildebrandt, T., van Dongen, B.F., Röglinger, M., Mendling, J. (eds.) BPM 2019. LNCS, vol. 11675, pp. 86–101. Springer, Cham (2019). https://doi.org/10.1007/978-3-030-26619-6_8
11. Sànchez-Ferreres, J., van der Aa, H., Carmona, J., Padró, L.: Aligning textual and model-based process descriptions. Data Knowl. Eng. **118**, 25–40 (2018)
12. van der Aa, H., Carmona, J., Leopold, H., Mendling, J., Padró, L.: Challenges and opportunities of applying natural language processing in business process management. In: Proceedings of the 27th International Conference on Computational Linguistics (COLING), pp. 2791–2801 (2018)

13. van der Aa, H., Di Ciccio, C., Leopold, H., Reijers, H.A.: Extracting declarative process models from natural language. In: Giorgini, P., Weber, B. (eds.) CAiSE 2019. LNCS, vol. 11483, pp. 365–382. Springer, Cham (2019). https://doi.org/10.1007/978-3-030-21290-2_23

14. van der Aa, H., Leopold, H., Reijers, H.A.: Dealing with behavioral ambiguity in textual process descriptions. In: La Rosa, M., Loos, P., Pastor, O. (eds.) BPM 2016. LNCS, vol. 9850, pp. 271–288. Springer, Cham (2016). https://doi.org/10.1007/978-3-319-45348-4_16

15. van der Aalst, W.M.P.: Process Mining, 2nd edn. Springer, Heidelberg (2016). https://doi.org/10.1007/978-3-662-49851-4

Case2vec: Advances in Representation Learning for Business Processes

Stefan Luettgen[ORCID], Alexander Seeliger[✉][ORCID], Timo Nolle[ORCID],
and Max Mühlhäuser[ORCID]

Telecooperation Lab, Technical University of Darmstadt, Darmstadt, Germany
{luettgen,seeliger,nolle,max}@tk.tu-darmstadt.de

Abstract. The execution of a business process is often determined by the surrounding context, e.g., department, product, or other attributes an event provides. Process discovery mainly focuses on the executed activities, although the context of a case may be needed to accurately represent a process instance, e.g., for clustering, prediction, or anomaly detection. Hence, in this paper, we present a representation learning technique (Case2vec) using word embeddings for business process data to better encode process instances. Our work extends Trace2vec and incorporates an additional semantic level by using not only the activity name but also the attributes and thereby incorporating the context. We evaluate our approach in the context of trace clustering. Additionally, we show that Case2vec can be used to abstract events which are semantically similar but syntactically different. We also show that word embeddings allow for interpretability when employing vector space arithmetic.

Keywords: Representation learning · Word embeddings · Process context

1 Introduction

In recent years, process mining has become an important technology for organizations analyzing their business processes. Event logs recorded by process-aware information systems can be analyzed with process mining to obtain valuable insights about how a business process is executed in reality. However, process mining techniques primarily focus on the control-flow of a process without considering the context a case is executed in, e.g., department, product, customer, or other attributes an event provides. This additional process context may help to further reveal patterns within the event log, which are not visible in the control-flow perspective, to enhance process mining techniques. Our goal is to learn vector representations of process cases that include this context information that can be used in various process mining techniques.

Vector representations of cases are required by many techniques in process mining such as trace clustering [4,11,12], prediction [3], and anomaly detection [10,13]. Trace clustering aims to improve the discovery of process models

© Springer Nature Switzerland AG 2021
S. Leemans and H. Leopold (Eds.): ICPM 2020 Workshops, LNBIP 406, pp. 162–174, 2021.
https://doi.org/10.1007/978-3-030-72693-5_13

by grouping similar cases. Clusters of cases that are executed in similar contexts can be generated, allowing the user to compare process models of different contexts. Improved prediction models can be learned that also consider the process contexts. Furthermore, anomaly detection methods based on extended vector representations can provide more reliable results. These are just a few examples for potential use cases of context-including vector representations.

Our work is based on a technique proposed in the area of natural language processing (NLP) for learning vector representations of words and sentences. Similar to a sentence with words, a case of a business process consists of a sequence of activities. Activities are also not executed in random order, but according to a predefined grammar, the underlying process model. The core idea is to model similarities and intentionally avoid comparing by words only, because we know that different words or sentences can bear the same meaning.

A previous work, Trace2vec [4], showed that the representation learning approach Word2vec [8], which constructs a vector space of the words of a corpus to capture similarities, can also be used on process data. To model such similarities, a large event log is crawled to order activities which occur together within this vector space. However, Trace2vec also showed some difficulties in the experiments with the BPIC15 event log: First, the vocabulary of event logs is much smaller compared to the vocabulary of natural language. Second, the context of a case is not taken into account, which can provide further details about the dependencies between activities and attributes. For instance, if the BPIC15 event log is clustered into the municipalities without considering the process context, it is assumed that the control-flow alone clearly determines the municipality. In highly standardized processes like governmental processes, the control-flow is the very part that does not separate one trace from another, but rather its context, e.g., an officer working exclusively in one or a few municipalities.

In this paper, we present an extended approach based on Trace2vec that can indeed lead to sensible results when evaluating these representations for a trace clustering task. We name this extension Case2vec, because it uses event and case attributes to capture the process context. Our extension increases the vocabulary that allows to better exploit case relationships. Besides our extension, we examine a proper hyperparameter strategy that can better deal with the sparse vocabulary in business process data. We revisit the original approach using the BPIC15 event log and show how parameter tuning and especially incorporating attributes improves results. We also show a wider range of results on the BPIC19 event log, which holds not only more traces, but also more attributes.

As additional tasks we investigate two useful applications of the neural network architecture presented: (1) *Event abstraction* allows to show that syntactically different activities are semantically similar, given enough traces in a similar context. (2) Arithmetic operations within the vector space keep semantic meaning which we show in an *interpretability task*. This is done on an artificial paper writing process to show the task more clearly because we know how the activities in this process depend on each other.

2   Related Work

Process case representations are used by various process mining techniques such as trace clustering, anomaly detection, and prediction. Different representations have been proposed in the related work. A simple representation technique is the bag-of-words model which is used to compute the similarity of sentences based on the co-occurrences. Song et al. [12] encode sequences of activities as one hot vectors, in which each component corresponds to an activity. Transitions between activities are used instead by Bose et al. [1] to compare cases.

Besides manually defined case representations, automatically generated representation vectors can be learned. For instance, a word embedding is a feature learning technique in which words are mapped to a vector space. Words appearing together frequently within a text corpus will be mapped close together within a vector space to capture their semantic relationship. Word embeddings do not rely on syntactical features and, therefore, can compute a similarity value of two sentences, even if none of the words of each sentence is the same. De Koninck et al. [4] transferred the idea of Word2vec [8] and Doc2vec [7] to process data. An LSTM and CBOW-based approach was introduced by Bui et al. [2]. A supervised representation learning approach based on conditional random fields for event abstraction was introduced by Tax et al. [14].

Representation learning has been used for different analysis methods. Trace2vec representations were used to cluster traces into similar groups in [4]. Tavares et al. [13] use the same representations to identify anomalous cases.

A drawback of most related work in this field is the limitation to the pure control-flow, namely the sequence of activities to learn case representations. Thus, the process context of the cases is not considered.

3   Case2vec

In NLP, word embeddings use the context of the words in a document to exploit semantic similarities of words by mapping them to a vector space. The closer these words appear together in the document the closer they are mapped together in the vector space. Thereby, semantic similarity of different statements can be confirmed as long as they are mapped close together.

As already mentioned, a popular technique for modeling word embeddings is Word2vec [8]. The task is to model what is in the neighborhood of a word. This can be done using two different approaches. We can predict a word given its surrounding words (continuous bag of words, short CBOW), or the other way round, predict the surrounding words given one word (skip-gram). For example, in the sentence *I like ... process mining*, continuous bag of words would insert words of similar representation to fill the gap, e.g., *business*. Vice versa, skip-gram would take the word *business* and amend it with preceding and succeeding words given by example sentences in the training data.

Word2vec learns a CBOW or skip-gram model using a neural network and implicitly constructs an abstract representation of the vocabulary and its

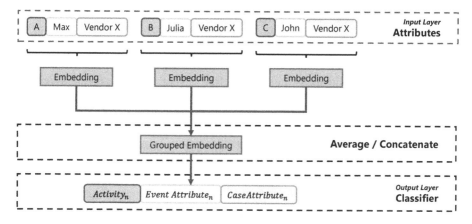

Fig. 1. Architectural overview of Case2vec where each trace's activity names, event, and case attributes are concatenated as single words.

relationships between each other. Similarly, activities within a sequence of a business process are also dependent on the preceding and succeeding activities which form the context. We can employ the idea of word embeddings and map activities to a vector space such that activities in similar regions are related to each other according to their function in the underlying process. Doc2vec serves as a representation of a collection of words, namely a document. Analogous to the Word2vec model, in Doc2vec a word is a document and we want to predict the surrounding documents. The structure of a trace from an event log is of similar form when considering activity names as words in a trace sequence. The resulting embedding space is a representation where activities and traces, given enough sample traces, are projected according to their role in the overall process model.

The embedding on the control-flow level is constructed by using the activity name as a single word. The set of different activity names forms the vocabulary of the embedding, and a Doc2vec representation is constructed by treating a trace as a document. The control-flow level (Fig. 1 without event and case attributes) has been introduced as Trace2vec [4]. One drawback of this approach is the focus on the control-flow. Therefore, we introduce Case2vec, which incorporates the different kinds of attributes by concatenating them with the corresponding activity name. The key idea is to incorporate attributes in addition to activity names to enlarge the vocabulary and induce a better separation of cases. If attributes are taken into account, the concatenation of the activity and its respective attributes becomes an additional word and, therefore, includes the process context.

Figure 1 shows the architecture with the attribute extension, where the words of the vocabulary are constructed by concatenating `Activity`, `Resource`, and `Vendor`. We also evaluate the approach either using event or case attributes.

4   Experimental Evaluation

We implemented[1] the described representation learning techniques using *gensim*, *scikit-learn* and *fastcluster* in Python to evaluate their performance. We use two Business Process Intelligence Challenge (BPIC) event logs, an amended version of them, and a fully synthetic paper writing process to evaluate on the following objectives: Trace clustering, event abstraction, and interpretability through vector arithmetic operations in the vector space.

In the following, we describe the event logs, the experimental setup and report the results.

4.1   Datasets

We use real-life and artificial event logs to evaluate the different objectives.

Real-Life Event Logs. We use the BPIC15 [5] and BPIC19 [6] event logs to compare the applicability of the different approaches. We select a case attribute for both event logs that can be considered as the ground truth label for clustering. Although we do not know in advance if this process provides features that will lead to good clustering results with this label, we are not necessarily interested in the best clustering result, but rather how incorporating different attributes can influence the clustering performance.

For BPIC15, event logs are already split into five different municipalities. In BPIC19, the case attribute `Item Type` is used as the cluster label without the `Standard` cases to obtain evenly distributed clusters.

During the experiments for event abstraction we amended the real-life event logs with noise or additional attributes. For the event abstraction task, we amended activity names with random numbers in a certain amount of traces to show that the method is robust to small changes in activity names.

Artificial Event Log: Paper Writing Process. The artificial example event log is based on a synthetically generated process depicted in Fig. 2. It describes the main steps in a scientific paper writing process from identifying a problem to the submission of the paper. The activities are dependent on each other according to their sequential order. This event log is more comprehensible for interpreting the results of the vector arithmetic experiment. For the experiments we sampled 5,000 traces of this process according to [9].

4.2   Real-Life Event Logs: Trace Clustering

In the first experiment, we use the case representations for clustering cases into their classes to show applicability for process context separation.

---

[1] Source code publicly available at: https://github.com/alexsee/case2vec.

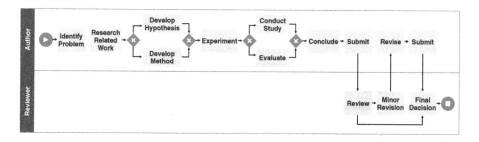

Fig. 2. Overview of the paper writing process [9].

Experimental Setup. Each event log is used individually to train the network according to the description in Sect. 3. For training, activities and attributes are used and none of the sequences are trimmed. Afterwards, we obtain the internal representation of each case and use the feature vectors as input for clustering.

Taking into account that process data in comparison to natural language has shorter sentence length and substantially smaller vocabulary, we employ a hyperparameter strategy to overfit the dataset for the clustering task. This is done using the ground truth label as an attribute with the goal to reach a Normalized Mutual Information (NMI) measure of 1.0 to ensure that the trained vector space has the capacity to model the underlying processes. This step is important before running the actual experiments to exclude weak results because of an impaired modeling capability of the underlying neural network. After a set of parameters is found that can overfit the dataset, the same optimization strategy can be employed during the actual experiments to maximize the NMI without the ground truth label.

In our parameter optimization strategy, we first optimize the vector size. We vary the vector size of the hidden and the embedding layer (2, 3, 4, 8, 16, 32, 64, 128, 256), and the number of epochs (10, 25, 50). Next, we optimize the window size of the embedding which determines how many activities before and after the current activity are considered. A value 5 or 7 seems optimal, and similar to the vector size, larger values do not improve the result and only run the risk of overfitting. Training epochs are varied between 10 and 50. The other parameters were standard parameters according to [4]. We trained the embedding with $sg = 0$ for the CBOW model, a learning rate set constant with $lr = 0.025$ for both Trace2vec and Case2vec and the decay factor alpha to 0.002. The number of inference epochs is set to 50. For clustering we opted to use hierarchical clustering. As a distance metric we use cosine distance to avoid a bias when dealing with traces of different length.

As an evaluation metric, we measure the NMI. We analyze the results using the non-parametric Friedman test. The Bonferroni corrected pairwise Wilcoxon signed-rank test is used for post-hoc analysis. We further report Kendall's W effect size.

Table 1. Best clustering performance grouped by approach, configuration, and event log.

Log	Approach	Vector size	Epochs	NMI
BPIC15	Trace2vec (original)	64	40	0.080
	Trace2vec (optimized)	4	50	0.132
	Case2vec (org:resource)	8	25	**0.980**
	Case2vec (Case Type)	3	25	0.010
	Case2vec (org:resource + Case Type)	8	50	**0.983**
	Case2vec (Responsible Actor)	128	25	0.398
	Case2vec (org:resource + Responsible Actor)	4	50	0.424
BPIC19	Trace2vec	32	10	0.560
	Case2vec (org:resource)	128	50	0.657
	Case2vec (org:resource + Document Type)	16	50	0.566
	Case2vec (Document Type)	128	25	0.591
	Case2vec (org:resource + Item Category)	16	25	0.626
	Case2vec (Item Category)	256	50	**0.805**
	Case2vec (org:resource + Vendor)	128	25	0.330
	Case2vec (Vendor)	2	50	0.296

Results. As a first step, we recreated the results by De Koninck et al. using the BPIC15 event log. As depicted in Table 1, Trace2vec reaches an NMI of 0.080 and increases to 0.132 after hyperparameter optimization. Using Case2vec with the case attribute `Responsible Actor` leads to a significant performance increase up to 0.398. The event attribute `org:resource`, which refers to the executing user, shows a performance of 0.980. Combining `org:resource` with one of the case attributes `Case Type` or `Responsible Actor` decreased the performance.

For the BPIC19 event log, Trace2vec reaches a performance up to 0.560. The case attribute `Item Category` reached the highest results with 0.805. However, using the `Vendor` results in a lower NMI than the control-flow only. Also, combining attributes also does not guarantee better results. Used separately, `org:resource` results in an NMI of 0.657 and `Document Type` in an NMI of 0.591. Combining the two leads to an NMI of 0.566, which results in a lower NMI than used separately.

Detailed results regarding the vector size are depicted in Fig. 3. The analysis of the results confirmed significant differences ($\chi^2(2) = 108$, $p < .001$, $W = 1$) between the approaches with a large effect. Post-hoc tests confirmed differences ($p < .001$) between all approaches with Case2vec performing better than Trace2vec. Incorporating `org:resource` lead to a significant better performance ($p < .001$) for BPIC15. For BPIC19, we discovered consistent results across the different parameter configurations, still there are significant differences ($\chi^2(2) = 24.111$, $p < .001$, $W = .223$) between the approaches. Similar to BPIC15, post-hoc tests confirm significant ($p < .001$) differences between all approaches.

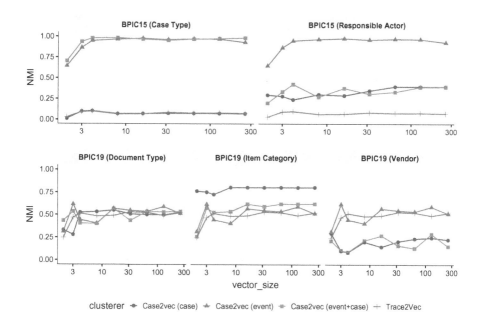

Fig. 3. Clustering results grouped by vector size, approach, configuration, and event log with 50 epochs.

4.3 Amended Real-Life Event Logs: Event Abstraction

The goal of event abstraction is to identify similar traces although activity names are slightly different. Eventually, these activity names can be adjusted to clean the event log. An example would be the activity name PR_created and Create_PR, which describes the same action with just a different name. The idea is to identify activities with similar function within the process so that a vector representation will allocate both activities close together in the vector space despite their different names.

The level of distorted activity names ranges from 0%, which is the normal case, up to 50%. The number of variations indicates the number of noise which is added to the activity name, e.g., letters or numerals. For example, if there are 2 variations and a 20% distortion level, the same random number is added to 20% of the traces, and the remaining 80% describe the unmodified variation. In case of 6 variations, besides the undistorted traces, there are traces distorted with 5 different random numbers to further increase uncertainty. Figure 4 shows results for different levels for different variations of the activity names. Case2vec with event attribute shows consistent results with deviations of ≤ 0.1 in NMI for distortion levels from 0% to 40%. A larger deviation of ~ 0.1 can only be seen with Trace2vec for 20% distortion.

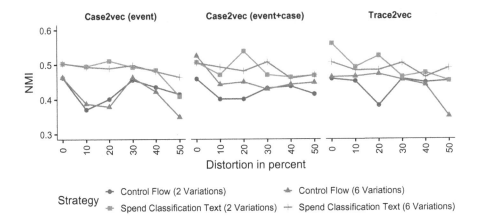

Fig. 4. Overview of the results of the event abstraction task.

Table 2. More paper process interpretability tasks

Add	Subtract	Top result
Experiment, Develop Method, Submit	Final Decision	Conduct Study
Conclude, Review, Submit	Final Decision	Develop Method
Develop Method, Submit	Final Decision	Conclude
Final Decision	Submit	Review
Experiment, Conclude	Submit	Develop Method

4.4   Synthetic Paper Process: Vector Arithmetic Interpretability

Since representation vectors are spanned within a vector space, arithmetic operations can be performed between vectors. The famous `king - man + woman = queen` example from Word2vec showed that representations can contain important semantic relationships. In this experiment, we investigate if vector arithmetic operations can also be used with process data. For testing the interpretability task, we have to come up with a certain scenario which allows a semantic interpretation. We use the paper writing process (see Fig. 2) because it is not pseudonymized and the activities can be read and understood by an analyst.

The first scenario is that the experiment was done and the paper was submitted, but the final decision has not taken place, because something is still missing that fulfills the criteria for an accepted paper. A possible composition would be to add the experiment and submission but subtract the final decision. When performing `Experiment + Submit - Final Decision` we would expect that something between `Experiment` and `Submit` is missing so that the `Final Decision` is still pending. The result of this computation returns `Evaluate` as the top result. The second top result is `Conduct Study` and the third top result is `Review`. Table 2 shows more example interpretability tasks. The result column shows the top result.

5 Discussion

In this section, we elaborate on the results from the experiment section and follow the order presented there.

5.1 Trace Clustering

BPIC15. Our experiments showed that hyperparameter optimization increases the performance of the original approach, but did not lead to useful results. However, it is not known if the separation by municipality solely based on the control-flow is possible. The process of a building permit application may be presumably highly standardized and, therefore, not a useful criterion for separating by municipality.

The results of our experiments show that the user of an activity is an attribute that is able to separate the cases into the five municipalities. This may be an obvious observation because persons may only work for a specific municipality. However, the event log also contains several persons that work across multiple municipalities. Case2vec, which includes the control-flow and the attributes, is able to find case representations for clustering that discriminate between the municipalities.

BPIC19. For the BPIC19 experiments, we found that Trace2vec performed significantly better compared to the BPIC15 event log. The best result was achieved by incorporating the `Item Category` case attribute, which seems to be strongly related to the `Item Type`. Interestingly, not all attributes improve the control-flow performance. For instance, the case attribute `Vendor` decreased the performance down to 0.290. This could be explained by the fact that a vendor is not a good separating attribute when categorizing according to an item type a company purchases. This would be the case if the company acquires most of its items from the same vendor regardless of the category of the item. Hence, even if the control-flow is able to separate by item type to some extent, an attribute, which is identical for most items, like a vendor, can obfuscate the results. This means that we cannot arbitrarily add more attributes for better results.

Case2vec seems to be sensitive to the selection of the attributes. Even though we showed that those methods provide good results after hyperparameter tuning, applying them to real-life event logs can be difficult because the quality of the result can usually not be determined since the ground truth is unknown. Attributes that contain random values or do not contribute to the desired clustering result lead to a significant drop in performance. However, when selecting appropriate attributes, Case2vec can outperform Trace2vec significantly. Still, finding good attributes can be difficult without prior knowledge.

5.2 Event Abstraction

For event abstraction, we ran several experiments with different amounts of traces including random numbers. We also changed the amount of different random numbers. Every attribute including a different random number will increase the vocabulary size. Still, Case2vec was able to identify and group traces according to their function despite them being amended. An even more interesting application than finding functionally similar traces with different names would be finding functionally similar traces with different performance metrics like cost or time. An analyst could study why these traces are similar in their role but differ in cost or time.

5.3 Interpretability Task

The example computation `Experiment + Submit − Final Decision` returns `Evaluate` as the top result and `Conduct Study` as the second top result. Both are sensible choices when we assume that `Evaluate` has already taken place and both are performed before `Submit`. The third top result is `Review`, which takes place directly after `Submit` and also shows a sensible reason assuming `Evaluate` and `Conduct Study` have been taken place and therefore cannot be the reason the submission is still blocked. Further results shown in Table 2 can be interpreted with similar reasoning.

However, in real-life event logs the interpretation of activities is not as clear because often they do not have interpretable names and even if, these names do not necessarily relate to a role in the process its name might suggest. Additionally, the developers of the Word2vec framework remark that vector arithmetic is not guaranteed to always produce sensible results. It is still interesting to see that on a small and well-defined event log the vector representation can deliver these results.

6 Conclusion

In this paper, we presented Case2vec, a representation learning technique based on a vector space model. It is trained using a neural network in an unsupervised fashion by using the sequence of activities including event attributes. It does not rely on any prior knowledge about the process and is able to learn robust and compact representations automatically.

The results of the evaluation in a trace clustering task showed that Case2vec is able to learn a good representation given useful control-flow or case attributes. When selecting appropriate attributes Case2vec can outperform Trace2vec significantly as shown in our real-life evaluation. The experiments on the additional tasks like event abstraction or arithmetic operations in the constructed vector space support that the learned representation is able to capture semantic characteristics of the process. However, Trace2vec and Case2vec seem to be sensitive to the selection of the attributes and finding good attributes can be difficult without prior knowledge. Feature selection methods from machine learning may help to identify attributes with a high information value, helping analysts to select useful attributes. Another limitation is that Case2vec only supports categorical attributes. Numerical values could be incorporated by grouping them into bins beforehand.

In conclusion, the internal representation of Case2vec is highly useful for trace clustering, finding functionally similar traces or executing vector space arithmetic operations for interpretability tasks.

Acknowledgment. This work is funded by the German Federal Ministry of Education and Research (BMBF) research project KI.RPA [01IS18022D].

References

1. Bose, R.P.J.C., van der Aalst, W.M.P.: Context aware trace clustering: towards improving process mining results. In: International Conference on Data Mining (SIAM) (2009)
2. Bui, H.-N., Vu, T.-S., Nguyen, H.-H., Nguyen, T.-T., Ha, Q.-T.: Exploiting CBOW and LSTM models to generate trace representation for process mining. In: Sitek, P., Pietranik, M., Krótkiewicz, M., Srinilta, C. (eds.) ACIIDS 2020. CCIS, vol. 1178, pp. 35–46. Springer, Singapore (2020). https://doi.org/10.1007/978-981-15-3380-8_4
3. Camargo, M., Dumas, M., González-Rojas, O.: Learning accurate LSTM models of business processes. In: Hildebrandt, T., van Dongen, B.F., Röglinger, M., Mendling, J. (eds.) BPM 2019. LNCS, vol. 11675, pp. 286–302. Springer, Cham (2019). https://doi.org/10.1007/978-3-030-26619-6_19
4. De Koninck, P., vanden Broucke, S., De Weerdt, J.: act2vec, trace2vec, log2vec, and model2vec: representation learning for business processes. In: Weske, M., Montali, M., Weber, I., vom Brocke, J. (eds.) BPM 2018. LNCS, vol. 11080, pp. 305–321. Springer, Cham (2018). https://doi.org/10.1007/978-3-319-98648-7_18
5. van Dongen, B.: BPI Challenge 2015. https://doi.org/10.4121/uuid:31a308ef-c844-48da-948c-305d167a0ec1
6. van Dongen, B.: BPI Challenge 2019. https://doi.org/10.4121/uuid:d06aff4b-79f0-45e6-8ec8-e19730c248f1
7. Le, Q., Mikolov, T.: Distributed representations of sentences and documents. In: International Conference on Machine Learning (2014)
8. Mikolov, T., Sutskever, I., Chen, K., Corrado, G., Dean, J.: Distributed representations of words and phrases and their compositionality. In: Advances in Neural Information Processing Systems (2013)
9. Nolle, T., Luettgen, S., Seeliger, A., Mühlhäuser, M.: BINet: multi-perspective business process anomaly classification. Inf. Syst. (2019)
10. Nolle, T., Seeliger, A., Mühlhäuser, M.: BINet: multivariate business process anomaly detection using deep learning. In: Weske, M., Montali, M., Weber, I., vom Brocke, J. (eds.) BPM 2018. LNCS, vol. 11080, pp. 271–287. Springer, Cham (2018). https://doi.org/10.1007/978-3-319-98648-7_16
11. Song, M., Yang, H., Siadat, S.H., Pechenizkiy, M.: A comparative study of dimensionality reduction techniques to enhance trace clustering performances. Expert Syst. Appl. **40**(9), 3722–3737 (2013)

12. Song, M., Günther, C.W., van der Aalst, W.M.P.: Trace clustering in process mining. In: Ardagna, D., Mecella, M., Yang, J. (eds.) BPM 2008. LNBIP, vol. 17, pp. 109–120. Springer, Heidelberg (2009). https://doi.org/10.1007/978-3-642-00328-8_11
13. Tavares, G.M., Barbon, S.: Analysis of language inspired trace representation for anomaly detection. In: Bellatreche, L., et al. (eds.) TPDL/ADBIS -2020. CCIS, vol. 1260, pp. 296–308. Springer, Cham (2020). https://doi.org/10.1007/978-3-030-55814-7_25
14. Tax, N., Sidorova, N., Haakma, R., van der Aalst, W.M.P.: Event abstraction for process mining using supervised learning techniques. In: Bi, Y., Kapoor, S., Bhatia, R. (eds.) IntelliSys 2016. LNNS, vol. 15, pp. 251–269. Springer, Cham (2018). https://doi.org/10.1007/978-3-319-56994-9_18

Supervised Conformance Checking Using Recurrent Neural Network Classifiers

Jari Peeperkorn[1]($\boxtimes$), Seppe vanden Broucke[1,2], and Jochen De Weerdt[1]

[1] Research Center for Information Systems Engineering (LIRIS), KU Leuven, Leuven, Belgium
{jari.peeperkorn,seppe.vandenbroucke,jochen.deweerdt}@kuleuven.be
[2] Department of Business Informatics and Operations Management, Ghent University, Ghent, Belgium

Abstract. Conformance checking is concerned with the task of assessing the quality of process models describing actual behavior captured in an event log across different dimensions. In this paper, a novel approach for obtaining the degree of recall and precision between a process model and event log is introduced. The approach relies on the generation of a so-called "antilog", randomly constructed from the activity vocabulary, on one hand, and a simulated "model log", which is played-out from the given model. In the case of recall the antilog and model log are used to train a recurrent neural network classifier. This network allows for calculating the probability of a trace being part of the model log or the antilog. If thereupon the event log is fed to the neural network, a value for recall can be obtained. In the case of precision the neural network is trained using a given event log and the antilog, and the model log is fed to it afterwards. We show that this new method can be used to measure global recall and precision correctly in some common examples.

Keywords: Process mining · Conformance checking · Machine learning · Neural networks · RNN

1 Introduction

Conformance checking covers different process mining techniques to compare event logs with process models. The latter can either be normative or an automatically discovered model. Conformance checking techniques can include both global conformance analysis, typically represented in the form of metrics, as well as local diagnostics, i.e. pinpointing conformance problems at a more fine-granular level, either in the log or in the model. Global conformance metrics typically measure conformance across one of the following quality dimensions: recall (or fitness), precision, generalisation and simplicity.

In this work, we propose a novel, fully data driven technique that can be used to measure recall and precision between process models and event logs and thus works at a global level, although it also has potential to provide insights into

© Springer Nature Switzerland AG 2021
S. Leemans and H. Leopold (Eds.): ICPM 2020 Workshops, LNBIP 406, pp. 175–187, 2021.
https://doi.org/10.1007/978-3-030-72693-5_14

local conformance diagnostics in future work. For now, the technique relies on the generation of a model log and comparing this with the given event log. This comparison is carried out by first generating a so-called "antilog" of either one of these two logs (the given event log or the model log). The antilog represents behavior not present in the log and could be generated in multiple ways. In this work, we investigate the potential of using the simplest of strategies to produce such an antilog, i.e. traces consisting of events representing activity executions randomly selected from the activity vocabulary. Once we have such an antilog, a recurrent neural network is trained to discriminate whether process instances belong to the (model or event) log or the antilog. Using this network we can obtain scores for the instances of the other (model or event) log. These scores represent how well an instance is described by the behavior in the log used for training.

Our technique has several compelling advantages. First of all, it proposes an new alternative to the common alignment or replay based algorithms, by utilizing a recurrent neural network classifier (RNN) model. Second, the RNN models are intrinsically probabilistic, thus giving a fine-grained analysis of model-log conformance. Furthermore, they are able to automatically detect temporal relations in sequences, making them a fitting tool to assess process instances. Moreover, by increasing the sample size of the antilog, we can investigate convergence and stability of metrics. Third, while being a black-box model, the RNN model could be complemented with visualizations that can pinpoint conformance problems at a local level, indicating at which timestep (activity) the prediction changes to antilog. Fourth, our technique allows for incremental updating of the model. Once the RNN is trained, conformance analysis can happen very fast, e.g. when checking the precision of different models. Fifth, despite the fact that we rely on model simulation to obtain a representation of the model behaviour, our technique is intrinsically more model-agnostic than other techniques which assume a certain model representation, and can be applied on any model that defines execution semantics. Finally, our technique links very well with predictive process monitoring techniques and has a lot of potential to extend conformance analysis towards other dimensions than control-flow, i.e. also including the resource and data dimensions. The use of RNNs in process mining is not new and has been applied in several areas such as predictive process monitoring [11,22]. However, the application of RNNs to conformance checking has not been investigated before.

The remainder of our paper is structured as follows: first the technique and its different components are introduced in Sect. 2. In Sect. 3 a set of initial experiments are detailed, which show our approach's potential for global conformance analysis. The paper is concluded by a section discussing previous related work (Sect. 4) and conclusions and directions for future work in Sect. 5.

2   RNN-Based Conformance Checking

2.1   Overview

Figure 1 presents the key idea behind our proposed technique. RNN based conformance checking relies on first "playing-out" the process model to obtain a model log. From this point onward the approaches to obtain either precision or recall differ. If we want to calculate recall, we take the model log and generate its model antilog. Again, we need to generate the antilog to capture non-conforming behaviour (not present in the model), in order to train the RNN. A recurrent neural network classifier is then trained to discriminate whether a certain process instance comes from the model log or its antilog. In practice this discriminator uses a sequence of activities as input, and outputs the probability of the instance belonging to either antilog or log (a value between 0 and 1). If we now use a new process instance as input in this network, we obtain a score on how well the model log (and therefore the model) describes it. Using the event log as input grants us an opportunity to obtain a global recall score. We can either use the average predicted value over all instances or we can use the fraction of instances with a score higher than a certain threshold (e.g. 0.5).

When precision of the process model is of interest, the technique starts with the event log and trains an RNN-classifier by combining it with its event antilog. By then training the RNN-classifier in a similar fashion and letting it classify the traces in the model log, we can obtain precision scores for every trace and thus for the entire model. Because the method as a whole provides labels for the RNN to train with automatically, it can be regarded as "self-supervising".

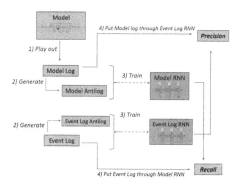

Fig. 1. Overview of the technique.

2.2   Model Log Generation

For the time being, the method relies on capturing the model behavior by means of a generated model log. In this work, we opted to "play-out" the model stochastically. This means that not necessarily all possible execution sequences allowed by the model are captured by the model log. However by making the model log sizeable enough, this problem can be partially solved. Only in extreme cases (like e.g. the flower model [9]), where the model entails a very large (infinite) set of potential process variants, this may lead to lower recall values than imposed by ground truth. However, we argue that this should not be seen as a major downside of our technique because even for models with a very large behavioural size (thus allowing a large number of variants), we can continually increase the size of the model log by simulating more cases. Moreover, because of the probabilistic nature of the approach, it will be possible to assess stability and convergence of recall and precision scores when increasing the size of the model log. In addition, the technique itself is fairly robust to infinite behaviour due to loops, given that the RNN discriminator in itself has sufficient generalization power to deal with such constructs. One advantage of using a model log is that the method is model agnostic and can be applied on any model that defines play-out execution semantics. For the time being, we have chosen to use the play-out functionality of the Python Process Mining library PM4PY [4].

2.3   Antilog Generation

Multiple methods to obtain the antilog were considered before settling on the intuitive approach proposed here. That is, experiments demonstrated that a strategy as straightforward as random generation produced satisfactory results, which indeed is an appealing proposition towards end-users. More specifically, we generate process instances with a sequence length uniformly selected between the minimal and maximal sequence length found in the log it is based on. Each activity is randomly selected from the activity vocabulary present in both event log and model log. This choice of vocabulary is not set in stone, as one could also opt to use only the vocabulary of the log in question. Therefore the model antilog and the event log antilog are likely to be very similar, only the array of different potential sizes of the instances might be different. Finally, a check is performed whether the newly generated antilog instance does not correspond to a certain instance variant from the log it is based on. In this way another possible difference between the model antilog and the event log antilog is created. However if the event log or simulated model log is large enough (i.e. has enough instances of each possible variant), this check is not strictly necessary and can be ignored. After all, the classifier will usually have significantly more examples of this particular instance with the correct (real) label than with the incorrect (antilog) label in its training set.

Observe that, despite the fact that in the current configuration the two antilogs will typically not differ much, we decided to still make a distinction between model antilog and event log antilog, given that we investigated some alternative antilog generation methods, which might lead to more significant differences between the two types of antilog. For instance, we considered a strategy that involved the addition of noise to the log. More sophisticated antilog generation strategies, e.g. relying on artificial negative event generation or generating the model antilog directly from the model also seem worthwhile. Nonetheless, given both the satisfactory results detailed below, as well as the fact that the random model log generation and random antilog generation provide the technique with a clear probabilistic nature, we argue that this is a proper choice. The antilog size was taken to be equal to the log it is based on. This was chosen in order to not introduce class imbalance in the training set of the classifier. If the training set would be wildly imbalanced, the likelihood of the classifier getting stuck in a local minimum whilst training would increase, with the end result of simply predicting the majority label with a near-certain probability, ultimately resulting in recall/precision scores of either 0 or 1.

2.4   Recurrent Neural Network Classifier

Recurrent neural networks (RNNs) are a type of artificial neural network specifically designed to handle sequential data. RNNs can be seen as a combination of multiple feedforward neural networks (one for each time step in the sequence). The hidden layers (recurrent layers) passing on messages, either one directional forward or bi-directional back and forth. Due to the vanishing/exploding gradient problem, simple recurrent neural networks do not handle long term dependencies well. Multiple solutions for this have been proposed, the most popular being Long Short-term Memory (LSTM) [15] and Gated Recurrent Units (GRU) [5]. Using such networks provides several advantages when there is a reason to incorporate these long term dependencies. However in this particular setting, we noticed that (non-fitting) switching of two subsequent activities was not being discriminated as nonconforming by the network when using an LSTM, which might be a problem in a conformance checking context. For example, when "sign contract" occurs before "check contract", this did not lead to a large change in predictions by the network, but could still indicate a problematic conformance error. Therefore we opted to use a bi-directional simple recurrent layer. The full architecture can be found in Fig. 2.

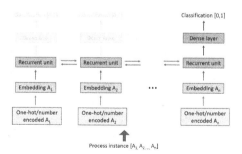

Fig. 2. Overview of the Recurrent Neural Network discriminator architecture.

The input is a process instance, i.e. a sequence of activities. These activities are presented one by one to the network, in an integer encoded fashion. In a first hidden layer this integer vector is converted into an embedding vector. An embedding is a vector representation, trained to be meaningful in the sense of constructing a lower-dimensional vector which retains as much of the original topology of the input space as possible. Embeddings can be trained in self-supervised fashion on a big corpus (e.g. word2vec [18]) or in a supervised fashion, added as an extra layer, as done here. The one-hot encoded vectors are multiplied by a weight matrix with dimensions *dimension embedding × activity vocabulary size*. This weight matrix thus contains the embeddings of the different activities in each column and is trained while training the entire network. You could also extract these embeddings for other purposes afterwards. Or use pre-trained embeddings, trained in a word2vec like fashion, as done by [7], albeit in different contexts. The activity embeddings are fed to the bi-directional simple RNN layer. The output of this layer is then send through a dense layer which predicts the label of the sequence. This output is a number between 0 and 1 and can be interpreted as a probability. This label is taken from the output of the last time step in the sequence (the final activity), but in theory the label could be predicted at each time step. This could provide extra insights into at which specific activity the nonconforming behavior takes place and could provide interesting visualizations. For now, however, the focus lies on predicting one overall label, i.e. whether a trace stems from a log or its antilog. In this research setting, the neural network implementation was executed by means of the Python library Keras[1] and sequences were padded to the maximal length, such that all eventual input sequences had the same size. The dimension of the embedding layer was set to 4 and the dimension of the recurrent layer was set to 8. The RNN uses a sigmoid activation function and was trained to optimize the binary cross entropy loss using RmsProp [24] for 40 epochs with a minibatch size of 64. In order to counteract overfitting a dropout of 0.5 was added between the recurrent layer and the dense output layer.

3   Experimental Evaluation

In this section the newly introduced technique will be tested[2]. In this experiment, we focus on global recall and precision calculation. Hereto, we use the event log in Table 1 together with the models 1–10 in Fig. 3, obtained from van Dongen et al. [9]. This set of models was supplemented with model 11, a model with low recall and high precision. We also use a model discovered from the event log by the Alpha miner [25] and 3 models discovered by the inductive miner [17], with the noise parameter set to 0, 0.5 and 1 respectively. These models can also be found in Fig. 3. The discovery algorithms were used as implemented in the ProM framework [10].

For each of the models we calculate the global fitness and precision, both by averaging the probability values as well as by counting the fraction of instances with a probability above the 0.5 threshold. This was done using the

---

[1] https://keras.io.

[2] The implementation of the technique, tests and the synthetic data used can be found on https://github.com/jaripeeperkorn/Supervised-Conformance-Checking-using-Recurrent-Neural-Network-Classifiers.

Table 1. The test log used for the experimental setup [9].

Instance	#
$\langle A, B, D, E, I \rangle$	1207
$\langle A, C, D, G, H, F, I \rangle$	145
$\langle A, C, G, D, H, F, I \rangle$	56
$\langle A, C, H, D, F, I \rangle$	23
$\langle A, C, D, H, F, I \rangle$	28

setting described above. We each time generated a model log with 1500 process instances, which is not sufficient to capture all possible behavior in some of the extreme models (e.g. the flower model). In most cases however, this was more than sufficient. The technique was performed 10 times for each model, and the median is taken to be the eventual value as well as the standard error $\sigma/\sqrt{10}$. These values are compered with different methods from literature: alignment based recall and precision [1], behavioral recall [13] and precision [27] and Markovian recall and precision with $k = 3$ [3]. We used the implementations in CoBeFra [26] for the alignment and behavioral based metrics. The results can be found in Table 2.

Table 2. Resulting recall and precision values.

Model	Recall					Precision				
	Prob.	Count	[1]	[13]	[3]	Prob.	Count	[1]	[27]	[3]
1	1.00 ± 0.00	1.00 ± 0.00	1.00	1.00	1.00	1.00 ± 0.04	1.00 ± 0.03	0.98	1.00	0.88
2	0.83 ± 0.00	0.83 ± 0.00	0.92	0.81	0.23	1.00 ± 0.00	1.00 ± 0.00	1.00	0.89	1.00
3	0.79 ± 0.01	1.00 ± 0.00	1.00	1.00	1.00	0.00 ± 0.00	0.00 ± 0.00	0.14	0.12	0.00
4	1.00 ± 0.00	1.00 ± 0.00	1.00	0.99	1.00	1.00 ± 0.01	1.00 ± 0.00	1.00	0.94	1.00
5	1.00 ± 0.00	1.00 ± 0.05	1.00	1.00	1.00	0.91 ± 0.02	0.89 ± 0.02	0.95	0.95	0.56
6	1.00 ± 0.00	1.00 ± 0.00	1.00	1.00	1.00	0.77 ± 0.01	0.77 ± 0.02	0.95	0.87	0.18
7	1.00 ± 0.00	1.00 ± 0.00	1.00	1.00	1.00	0.58 ± 0.04	0.60 ± 0.06	0.80	0.72	0.35
8	0.52 ± 0.02	0.52 ± 0.14	0.74	1.00	1.00	0.00 ± 0.00	0.00 ± 0.00	0.34	0.16	0.01
9	1.00 ± 0.00	1.00 ± 0.00	1.00	1.00	—	0.50 ± 0.01	0.50 ± 0.01	0.84	0.60	—
10	0.43 ± 0.11	0.45 ± 0.15	0.62	0.59	0.09	0.00 ± 0.00	0.00 ± 0.00	0.89	0.19	0.06
11	0.15 ± 0.01	0.17 ± 0.02	0.62	0.35	0.64	1.00 ± 0.01	1.00 ± 0.00	1.00	0.36	1.00
Alpha	0.96 ± 0.00	0.97 ± 0.00	1.00	0.99	0.77	0.73 ± 0.01	0.72 ± 0.01	0.96	0.92	0.38
Ind. 0	1.00 ± 0.00	1.00 ± 0.00	1.00	1.00	1.00	0.86 ± 0.04	0.89 ± 0.05	0.72	0.59	0.42
Ind. 0.5	1.00 ± 0.00	1.00 ± 0.00	1.00	0.99	0.77	0.84 ± 0.03	0.86 ± 0.04	0.79	0.69	0.44
Ind. 1	0.18 ± 0.11	0.17 ± 0.12	0.86	0.84	0.68	0.82 ± 0.06	0.81 ± 0.07	0.87	0.64	0.48

Our new technique approximately agrees with existing metrics from the literature for most of the models. The recall values obtained for model 2 can be explained by looking at the fraction of instances in the event log corresponding to the most frequent trace $1207/1459 \approx 0.83$. Similarly, for model 11, the

Model 1: good recall and good precision [9].

Model 2: single most frequent trace [9].

Model 3: flower model [9].

Model 4: all traces in parallel [9].

Model 5: G and H in parallel [9].

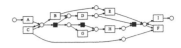

Model 6: G and H in self-loops [9].

Model 7: D in a self-loop [9].

Model 8: all transitions in parallel [9].

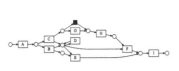

Model 9: C and F are in a loop and need to be executed the same amount to reach the final marking [9].

Model 10: Round-robin model [9].

Model 11: 3 least common traces in parallel.

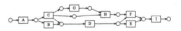

Model alpha: discovered by alpha miner [25] using the log in Table 1.

Model Ind. 0: discovered by inductive miner with noise set to 0 [17] using the log in Table 1.

Model Ind. 0.5: discovered by inductive miner with noise set to 0.5 [17] using the log in Table 1.

Model Ind. 1: discovered by inductive miner with noise set to 1 [17] using the log in Table 1.

Fig. 3. Models used for the experimental setup.

fraction of the least frequent traces in the event log $(56 + 23 + 28)/1459 \approx 0.07$ can be obtained. The recall score provided by the RNN is slightly higher, but not as high as the scores provided by alignment or Markovian based recall. For both model 2 and model 11 the precision is 1, as it should be. Noticeable differences in recall and/or precision appear in the special cases, i.e. model 3 (flower), model 8 (all parallel) and model 10 (round-robin). Whereas the extremely low precision values can be seen as correct, the incorrect low recall values can be attributed to an (incomplete) random model log generation. Due to the infinite (for the flower model) or high number of possible traces that can be generated by these models, the chance that not all or even few traces present in the event log are not simulated is high. This leads to incorrectly low recall values. This might be solvable by generating the model log in a different way (e.g. complete log generation with a maximum on the number of times a particular marking can be seen). However, for these overly general models, simply generating a bigger model log will not work, as the random antilog grows simultaneously (we chose to keep them the same size). The chance for a flower model to generate something close to the instances in the event log, is similar to the chance of the random antilog to generate it. In a way, an extra punishment on the recall value is given to overly general and imprecise models. It however also clear that each of the other method from literature have different examples they cannot handle properly. Another important difference in recall values can be seen in the model discovered by the inductive miner with noise parameter set to 1. Our newly proposed technique outputs a recall value significantly lower than the ones obtained by the methods from the literature. When looking at the model, you can actually see that it is not able to replay the most frequent instance from the event log $\langle A, B, D, E, I \rangle$, but is able to replay the other ones. The value obtained by our technique corresponds approximately with the fraction of the replayable traces in the event log $(145 + 56 + 23 + 28)/1459 \approx 0.17$, while the other techniques return values significantly higher. Apart from the extreme case in model 8, the method using the average of the probabilities and the method using the counts do not differ much.

We further investigated the convergence of the recall and precision values with increasing model log size (and therefore with increasing model antilog size as well). It was observed that convergence happens at a significantly higher number of instances than the amount of different variants the model can produce. This indicates that not necessarily the model log generation, but rather the random antilog generation requires enough examples in order to obtain a stable result. Because the RNN outputs the probability of an instance belonging to either the event log or the antilog, we can also show a probability distribution over all instances. This might provide end-users with a valuable visualisation. We also manually confirmed the method was able to pick up on small (unwanted) change. There is off course a trade-off. If it is necessary that small changes (e.g. switch "sign contract" before "check contract") not yet seen, result in a nonconforming label, having a RNN which is trained to generalize its predictions maximally, might provide an issue in combination with the current antilog generation. By reducing possible overfitting measures (e.g. dropout between layers) the shape of the distribution becomes more bimodal with two peaks (at 0 and 1) and no or little mass in between. On the other hand, if small changes are not a problem and we are more interested in the distribution of the fitness values, it is necessary to make sure the neural network is still generalizing enough. If you already know which specific behavior is certainly not desired, you can add this behavior manually to the antilog. This means adding multiple (different) traces containing the unwanted behavior to the antilog. The time it takes to run the method (precision or recall) with two event logs (and one antilog) of 1500 instances (size 20 activities, vocabulary size of 10) one time (including) training is around $9s$, as performed on an Intel(R) Core(TM) i7-9850h CPU @ 2.60 ghz. If the RNN is already trained and you would only need to put through one log this reduces to $0.2s$.

4   Related Work

One of the earliest research on Conformance Checking can be found in Cook et al. (1999) [6]. Multiple techniques for obtaining recall or precision have been proposed ever since. One example of an early precision metric was introduced by Greco et al. (2006) [14] which calculates a "set difference", between a set of traces representing the log's behavior and one representing the model's behavior. Other noteworthy earlier contributions are proper completion, token based sequence replay and the advanced behavioral appropriateness [20]. Behavioral recall (using a percentage of correctly classified positive events) and behavioral specificity (replaying the sequences and taking the percentage of correctly classified negative events) were introduced by Goedertier et al. (2009) [13] and was later supplemented with a similar method using the amount of "false positives" (behavior allowed by model, but labeled a negative event based on the log) [8] and by the behavioral precision [27]. Another method using behavioral profile based metrics, based on different constraints a process model can impose on a log, was introduced by Weidlich et al. (2011) [28]. Another metric is ETC precision which uses log prefix automatons and the number of "escaping" edges [19]. This was later altered in projected conformance checking (PCC), better scaled to real-life logs [16]. A lot of focus has been on the (average) alignment based trace recall and precision approach introduced in [1], supplemented by the one align precision and best align precision [2]. Van Dongen et al. (2016) proposed the use of "anti-alignments" to obtain a model's precision [9]. Recently promising work comparing Markovian abstractions of both event log and model has been shown as a potential efficient alternative to alignments based methods [3]. In recent years different axioms were proposed as well, describing (un)wanted behavior that conformance checking metrics should (not) adhere to [21].

The method introduced in this work draws some resemblance to Classifier-Adjusted Density Estimation for Anomaly Detection and One-Class Classification (or CADE) [12], which is an anomaly detection method that uses a classifier trained on discriminating between real data and synthetic data generated uniformly over all features. RNN's have earlier been used in predictive process monitoring [11,22]. Recent work in predictive monitoring has also introduced the use of Generative Adverserial Nets [23]: a self-supervising machine learning technique based on training a discriminator and a generator simultaneously, which draws some similarities to the technique introduced here.

5   Conclusion and Future Work

We have presented a new technique to obtain recall and precision of event logs and process models. The technique was tested on a small example log and models to show its potential for global conformance analysis. In future work, it should be interesting to hold our new technique against the recently proposed conformance checking axioms [21]. However, due to the intrinsic probabilistic nature of the technique, doing this theoretically might be hard and require empirical backing. Another interesting avenue would be to directly rely on the given process model, omitting the model log generation step, although this would require serious alterations to the technique. A less drastic improvement could be found in a more sophisticated antilog generation, though the simplicity combined with good results obtained by the current approach is nonetheless appealing. Another method could e.g. use negative events [8,13,27] or other smart usage of different data features. Next, it could be interesting to use the output at each time step (see Fig. 2) for explanatory purposes. Since we could then inspect at which activity exactly the model starts to classify the instance as being nonconforming. Preliminary results show that such an approach is viable. Finally, since the model could in theory be extended to include additional data attributes (e.g. the resource or other perspectives), including this in future implementations might provide us with additional advantages compared to competitive conformance checking techniques.

Acknowledgement. This research has been financed in part by the NeEDS research project, an EC H2020 MSCA RISE project with Grant agreement No. 822214.

References

1. van der Aalst, W., Adriansyah, A., van Dongen, B.: Replaying history on process models for conformance checking and performance analysis. WIREs Data Min. Knowl. Discov. **2**(2), 182–192 (2012). https://doi.org/10.1002/widm.1045
2. Adriansyah, A., Munoz-Gama, J., Carmona, J., van Dongen, B.F., van der Aalst, W.M.P.: Alignment based precision checking. In: La Rosa, M., Soffer, P. (eds.) BPM 2012. LNBIP, vol. 132, pp. 137–149. Springer, Heidelberg (2013). https://doi.org/10.1007/978-3-642-36285-9_15
3. Augusto, A., Conforti, R., Armas-Cervantes, A., Dumas, M., La Rosa, M.: Measuring fitness and precision of automatically discovered process models: a principled and scalable approach. IEEE Trans. Knowl. Data Eng. **PP**, 1 (2020)
4. Berti, A., van Zelst, S.J., van der Aalst, W.: Process mining for python (PM4Py): bridging the gap between process-and data science. In: Proceedings of the ICPM Demo Track 2019, co-located with 1st International Conference on Process Mining (ICPM 2019), Aachen, Germany, 24–26 June 2019, pp. 13–16 (2019)
5. Cho, K., van Merrienboer, B., Gülçehre, Ç., Bougares, F., Schwenk, H., Bengio, Y.: Learning phrase representations using RNN encoder-decoder for statistical machine translation. CoRR abs/1406.1078 (2014)
6. Cook, J.E., Wolf, A.L.: Software process validation: Quantitatively measuring the correspondence of a process to a model. ACM Trans. Softw. Eng. Methodol. **8**(2), 147–176 (1999)

7. De Koninck, P., vanden Broucke, S., De Weerdt, J.: act2vec, trace2vec, log2vec, and model2vec: representation learning for business processes. In: Weske, M., Montali, M., Weber, I., vom Brocke, J. (eds.) BPM 2018. LNCS, vol. 11080, pp. 305–321. Springer, Cham (2018). https://doi.org/10.1007/978-3-319-98648-7_18
8. De Weerdt, J., De Backer, M., Vanthienen, J., Baesens, B.: A robust F-measure for evaluating discovered process models. In: IEEE Symposium Series on Computational Intelligence, pp. 148–155 (2011)
9. van Dongen, B.F., Carmona, J., Chatain, T.: A unified approach for measuring precision and generalization based on anti-alignments. In: La Rosa, M., Loos, P., Pastor, O. (eds.) BPM 2016. LNCS, vol. 9850, pp. 39–56. Springer, Cham (2016). https://doi.org/10.1007/978-3-319-45348-4_3
10. van Dongen, B.F., de Medeiros, A.K.A., Verbeek, H.M.W., Weijters, A.J.M.M., van der Aalst, W.M.P.: The ProM framework: a new era in process mining tool support. In: Ciardo, G., Darondeau, P. (eds.) ICATPN 2005. LNCS, vol. 3536, pp. 444–454. Springer, Heidelberg (2005). https://doi.org/10.1007/11494744_25
11. Evermann, J., Rehse, J.R., Fettke, P.: Predicting process behaviour using deep learning. Decis. Support Syst. **100**, 129–140 (2017)
12. Friedland, L., Gentzel, A., Jensen, D.: Classifier-adjusted density estimation for anomaly detection and one-class classification. In: SDM, pp. 578–586, April 2014. https://doi.org/10.1137/1.9781611973440.67
13. Goedertier, S., Martens, D., Vanthienen, J., Baesens, B.: Robust process discovery with artificial negative events. J. Mach. Learn. Res. **10**, 1305–1340 (2009)
14. Greco, G., Guzzo, A., Pontieri, L., Sacca, D.: Discovering expressive process models by clustering log traces. IEEE Trans. Knowl. Data Eng. **18**(8), 1010–1027 (2006)
15. Hochreiter, S., Schmidhuber, J.: Long short-term memory. Neural Comput. **9**(8), 1735–1780 (1997). https://doi.org/10.1162/neco.1997.9.8.1735
16. Leemans, S.J.J., Fahland, D., van der Aalst, W.M.P.: Scalable process discovery and conformance checking. Softw. Syste. Model. **17**(2), 599–631 (2016). https://doi.org/10.1007/s10270-016-0545-x
17. Leemans, S.J.J., Fahland, D., van der Aalst, W.M.P.: Discovering block-structured process models from incomplete event logs. In: Ciardo, G., Kindler, E. (eds.) PETRI NETS 2014. LNCS, vol. 8489, pp. 91–110. Springer, Cham (2014). https://doi.org/10.1007/978-3-319-07734-5_6
18. Mikolov, T., Chen, K., Corrado, G., Dean, J.: Efficient estimation of word representations in vector space. In: 1st International Conference on Learning Representations, ICLR 2013 (2013)
19. Muñoz-Gama, J., Carmona, J.: A fresh look at precision in process conformance. In: Hull, R., Mendling, J., Tai, S. (eds.) BPM 2010. LNCS, vol. 6336, pp. 211–226. Springer, Heidelberg (2010). https://doi.org/10.1007/978-3-642-15618-2_16
20. Muñoz-Gama, J., Carmona, J.: A fresh look at precision in process conformance. In: Hull, R., Mendling, J., Tai, S. (eds.) BPM 2010. LNCS, vol. 6336, pp. 211–226. Springer, Heidelberg (2010). https://doi.org/10.1007/978-3-642-15618-2_16
21. Syring, A.F., Tax, N., van der Aalst, W.M.P.: Evaluating conformance measures in process mining using conformance propositions. In: Koutny, M., Pomello, L., Kristensen, L.M. (eds.) Transactions on Petri Nets and Other Models of Concurrency XIV. LNCS, vol. 11790, pp. 192–221. Springer, Heidelberg (2019). https://doi.org/10.1007/978-3-662-60651-3_8
22. Tax, N., Verenich, I., La Rosa, M., Dumas, M.: Predictive business process monitoring with LSTM neural networks. In: Dubois, E., Pohl, K. (eds.) CAiSE 2017. LNCS, vol. 10253, pp. 477–492. Springer, Cham (2017). https://doi.org/10.1007/978-3-319-59536-8_30

23. Taymouri, F., Rosa, M.L., Erfani, S., Bozorgi, Z.D., Verenich, I.: Predictive business process monitoring via generative adversarial nets: the case of next event prediction. In: Fahland, D., Ghidini, C., Becker, J., Dumas, M. (eds.) BPM 2020. LNCS, vol. 12168, pp. 237–256. Springer, Cham (2020). https://doi.org/10.1007/978-3-030-58666-9_14

24. Tieleman, T., Hinton, G.: Lecture 6.5–RmsProp: divide the gradient by a running average of its recent magnitude. COURSERA Neural Netw. Mach. Learn. **4**, 26–31 (2012)

25. van der Aalst, W., Weijters, T., Maruster, L.: Workflow mining: discovering process models from event logs. IEEE Trans. Knowl. Data Eng. **16**(9), 1128–1142 (2004)

26. vanden Broucke, S.K.L.M., De Weerdt, J., Vanthienen, J., Baesens, B.: A comprehensive benchmarking framework (CoBeFra) for conformance analysis between procedural process models and event logs in ProM. In: 2013 IEEE Symposium on Computational Intelligence and Data Mining (CIDM), pp. 254–261 (2013)

27. vanden Broucke, S.K.L.M., De Weerdt, J., Vanthienen, J., Baesens, B.: Determining process model precision and generalization with weighted artificial negative events. IEEE Trans. Knowl. Data Eng. **26**(8), 1877–1889 (2014)

28. Weidlich, M., Polyvyanyy, A., Desai, N., Mendling, J., Weske, M.: Process compliance analysis based on behavioural profiles. Inf. Syst. **36**, 1009–1025 (2011)

1st International Workshop on Streaming Analytics for Process Mining (SA4PM'20)

1st International Workshop on Streaming Analytics for Process Mining (SA4PM)

Streaming Process Mining is an emerging area in process mining that spans data mining (e.g. stream data mining; mining time series; evolving graph mining), process mining (e.g. process discovery; conformance checking; predictive analytics; efficient mining of big log data; online feature selection; online outlier detection; concept drift detection; online recommender systems for processes), scalable big data solutions for process mining and the general scope of online event mining, in addition to many other techniques that are all gaining interest and importance in industry and academia.

The SA4PM workshop aims at promoting the use and the development of new techniques to support the analysis of streaming-based processes. We aim at bringing together practitioners and researchers from different communities, e.g. Process Mining, Stream Data Mining, Case Management, Business Process Management, Database Systems and Information Systems who share an interest in online analysis and optimization of business processes and process-aware information systems with time, storage or complexity restrictions. The workshop aims at discussing the current state of ongoing research and sharing practical experiences, exchanging ideas and setting up future research directions.

The workshop started with an interesting invited talk by *Albert Bifet* on "Adaptive Machine Learning for Data Streams" with an extensive overview on recent contributions in the data stream mining field and on existing open-source tools. The invited talk also highlighted open issues in the field.

This 1st edition of the workshop attracted 8 international submissions. Each paper was reviewed by at least three members of the Program Committee. From these submissions, the top 4 were accepted as full papers for presentation at the workshop. The workshop was held in an online format due to the COVID-19 pandemic. The papers presented at the workshop provide a mix of novel research ideas and focus on online anomaly detection, concept drift detection, trace ordering and performance mining.

Jonghyeon Ko and Marco Comuzzi focus on online anomaly detection on the trace-level in event logs, in contrast to existing solutions that mainly addressed this problem in the offline setting. The online setting is crucial in this task for discovering anomalies in process execution as soon as they occur and, consequently, allowing early corrective actions to be promptly taken. This paper describes a novel approach to event log anomaly detection on event streams that uses statistical leverage. Leverage has been used extensively in statistics to develop measures to identify outliers and it has been adapted in this paper to the specific scenario of event stream data.

Next, *Ludwig Zellner et al.* address the problem of concept drift detection on streaming data by aggregating a considerable number of outliers in the stream. After considering various application of the proposed method, the authors concentrate on non-conforming traces of a stream on which they compute a local outlier factor and identify drifts as considerably changing outlier scores.

In a third contribution, *Florian Richter et al.* focus on online trace ordering by proposing an anytime structure visualizer called OTOSO, which is a monitoring tool based on OPTICS. The tool plots representations for density-based trace clusters in process event streams and identifies temporal deviation clusters as a time-dependent graph. The aim is to provide an on-demand overview of the temporal deviation structure during the process execution. Additionally, the work offers insights about temporally limited occurrences of trace clusters, which are usually difficult to detect when using a global clustering approach.

Finally, *Andrea Maldonado et al.* address the performance mining task by incorporating time interval information. The paper proposes the performance skyline approach to discover events that are crucial to the overall duration of real process executions. The authors then contribute three techniques for statistical analysis of performance skylines and process trace sets enabling more accurate process discovery, conformance checking and process enhancement.

We hope that the reader will find this selection of papers useful to keep track of the latest advances in the stream process mining area. We are looking forward to continuing to present new advances in future editions of the SA4PM workshop.

February 2021

Organization

Workshop Chairs

Andrea BurattinTechnical	University of Denmark, Denmark
Marwan Hassani	Eindhoven Univ. of Technology, The Netherlands
Sebastiaan van Zelst	Fraunhofer Inst. for Appl. Inf. Tech., Germany
Thomas Seidl	Ludwig-Maximilians-Univ. München, Germany

Program Committee

Agnes Koschmider	Kiel University, Germany
Ahmed Awad	University of Tartu, Estonia
Boudewijn van Dongen	Eindhoven Univ. of Technology, The Netherlands
Eric Verbeek	Eindhoven Univ. of Technology, The Netherlands
Felix Mannhardt	Eindhoven University of Technology, The Netherlands
Florian Richter	Ludwig-Maximilians-Univ., München, Germany
Francesco Folino	ICAR -CNR, Italy
Frederic Stahl	German Research Center for AI (DFKI), Germany
Jochen De Weerdt	KU Leuven, Belgium
Matthias Weidlich	Humboldt-Universität zu Berlin, Germany
Mohamed Medhat Gaber	Birmingham City University, UK
Sherif Sakr	University of Tartu, Estonia
Stefanie Rinderle-Ma	Technical University of Munich, Germany

Online Anomaly Detection Using Statistical Leverage for Streaming Business Process Events

Jonghyeon Ko and Marco Comuzzi[✉]

Department of Industrial Engineering, Ulsan National Institute of Science
and Technology (UNIST), Ulsan, Republic of Korea
{whd1gus2,mcomuzzi}@unist.ac.kr

Abstract. While several techniques for detecting trace-level anomalies in event logs in offline settings have appeared recently in the literature, such techniques are currently lacking for online settings. Event log anomaly detection in online settings can be crucial for discovering anomalies in process execution as soon as they occur and, consequently, allowing to promptly take early corrective actions. This paper describes a novel approach to event log anomaly detection on event streams that uses statistical leverage. Leverage has been used extensively in statistics to develop measures to identify outliers and it has been adapted in this paper to the specific scenario of event stream data. The proposed approach has been evaluated on both artificial and real event streams.

Keywords: Process mining · Online anomaly detection · Event streams · Information measure · Statistical leverage

1 Introduction

Information logged during the execution of business processes is available in so-called event logs, which contain events belonging to different process instances (or *cases*). Each event is described by multiple attributes, such as a timestamp and a label capturing the activity in the process that was executed.

Event logs are prone to errors, which can stem from a variety of root causes [1,2], such as system malfunctioning or sub-optimal resource behaviour. For instance, sloppy human resources may forget to log the execution of specific activities in a process, or a system reboot may assign a different case id to all the new events recorded after rebooting. Errors in event log hamper the possibility of extracting useful process insights from event log analysis, and should therefore be fixed as early as possible [20].

To this end, the research field of event log anomaly detection (or event log cleaning) has emerged recently, providing methods to detect anomalies at trace level [1,2,9,10,13], i.e., concerning the order and occurrence of activities in a process, and at event level [15,16,18], i.e., concerning the value of attributes

© Springer Nature Switzerland AG 2021
S. Leemans and H. Leopold (Eds.): ICPM 2020 Workshops, LNBIP 406, pp. 193–205, 2021.
https://doi.org/10.1007/978-3-030-72693-5_15

of events, using a variety of different approaches. Note that event log anomaly detection is normally (process) model-agnostic, that is, it does not assume the existence of a process model or clean traces from which a model can be extracted. This aspect separates this research field from traditional process mining research on compliance checking [14].

In the specific case of online settings, i.e., event streams, while research has recently emerged in the field of online compliance checking, only the work by Tavares et al. [19] addresses the issue of anomaly detection. In particular, the authors propose a method to detect point anomalies specified by Principal Component Analysis (PCA). These point anomalies, however, do not normally reflect real-life anomaly patterns, such as inserting, skipping or switching events, commonly considered by event log anomaly detection in offline settings. Therefore, we argue that there is a lot of potential for new research in this area.

More in general, event log anomaly detection in online settings can be crucial for discovering anomalies in process execution as soon as they occur and, consequently, allowing to promptly take early corrective actions. The online settings, however, obviously introduce additional challenges to the design of an event log anomaly detection method. In particular, owing to the finite memory assumption of online settings [4–6,19,20], only a limited number of (recent) events are available at any given time to take a decision. This prevents to apply effectively some of the approaches that have been proposed in the literature for event log anomaly detection in offline settings. Probabilistic methods that detect anomalies after having created an intermediate model of frequent process behaviour [1,10,13] are hampered by the fact that only a limited number of events may be available to create such models. Online settings also prevent the application of machine learning reconstructive techniques for anomaly detection, e.g. [16,17]. These, in fact, normally rely on deep learning models, which require a high number of data points (complete process traces in this scenario) to be trained effectively. Also, any update of these models may require a long training time.

In this paper we propose an information-theoretic approach to online event log anomaly detection at trace level. Specifically, we devise an anomaly score based on statistical leverage [11]. The leverage is a relative measure of the information content of observations in a dataset that has been used extensively in statistics to develop observation distance measures and outlier detection techniques. Since leverage captures the information content of one observation in respect of all others in a dataset, the anomaly score proposed in this paper can always be calculated reliably based on the information available at any point in time, resulting in an anomaly detection method that does not require extensive amount of data to be executed effectively.

After having presented the related work (in Sect. 2), Sect. 3 presents a trace anomaly score based on the notion of statistical leverage. Then, we discuss how this score can be applied to anomaly detection of streams of events, addressing issues such as the grace period, the finite memory assumption, and the identification of anomaly detection thresholds. The proposed method is evaluated (in Sect. 4) on both artificial and real event logs injected with trace-level anomalies. Conclusions finally are drawn in Sect. 5.

2   Related Work

While there is only limited literature regarding online event log anomaly detection, a number of recent contributions have focused on online conformance checking. To some extent, conformance checking can be seen as model-aware anomaly detection, since process models, given or extracted from clean traces, can be seen as signatures of positive behaviour to detect anomalies. As referred by [6], there are currently two research lines in online conformance checking: the prefix-alignment approach [20] and the model-based approach [4–6].

Conformance checking/alignment of streaming events tends to overestimate the computation of optimal alignments. In order to avoid this issue, [20] provides the first incremental/online conformance checking technique that uses prefix-alignment. Prefix-alignment [20] is characterised by high computational complexity and prevents to define a warming up period. Alternatively, Online Conformance Transition Systema (OCTS) [4,5] can partially check compliance on regions of a process. This technique also suffers from high computational complexity and prevents to consider the warm start scenario. In [6], the first solution to achieve a warm start with streaming events has been proposed by introducing weak order relations, that have reduced computational complexity.

Regarding event log anomaly detection, as mentioned in the Introduction, Tavares et al. [19] have first applied the online clustering algorithm DenStream [7] to detect anomalies on event streams. DenStream clusters cases into two groups, normal and anomalous, using histogram-based frequency of activities contained in each case. Since the histogram-based frequency ignores the sequence of events in traces, DenStream detects point anomalies in event logs defined by Principal Component Analysis (PCA) [21].

3   Research Framework

There are two different elements in the proposed framework: the anomaly score and the anomaly detection method. The former (presented in Sect. 3.1) concerns the definition of a trace anomaly score based on statistical leverage. The latter (Sect. 3.2) concerns setting a threshold value above which a trace is considered anomalous based on its anomaly score.

3.1   Anomaly Score

Statistical *leverage* [11] is a measure indicating how far away each observation is scattered from other observations in a dataset. It has been used as a key support measure for developing different observation distance metrics, such as Cook's distance, the Welsch-Kuh distance, and the Welsch's distance.

Given a matrix X, with $X \in \mathbb{R}^{J \times I}$, of a dataset with J observations and I numerical attributes (or variables), the leverage of the observations in X are the diagonal elements of the projection matrix $H = X(X^T X)^{-1} X^T$. Specifically, the leverage of the j-th observation in X is the diagonal element $h_{j,j} \in H$, which

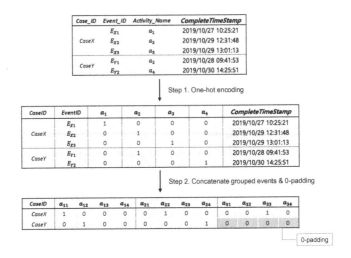

Fig. 1. One-hot encoding and 0-padding

is comprised by definition between 0 and 1. The higher its leverage, the more likely an observation to be an anomaly.

Our objective is to detect anomalies at the level of occurrence and order of events in traces. Therefore, we can abstract an event log E as a set of J traces $\{\sigma_j\}_{j=1,\ldots,J}$. Each trace is a sequence of events $e_{i,j}$ of variable length N_j, i.e., $\sigma_j = \{e_{1,j}, \ldots, e_{N_j,j}\}$. Events are ordered in a trace by timestamp in ascending order and are defined by the activity that they represent, which is one in a set $A = \{a_1 \ldots, a_K\}$ of K possible activity labels.

In order to define a leverage-based anomaly measure of traces in an event log, two pre-processing steps are necessary. The first one is an integer encoding step. This is necessary because the attributes of a dataset X must be numerical to calculate H, while the activity attribute in event logs is categorical. Second, events in an event log must be aggregated at trace level, such that the resulting matrix X has J rows, i.e., one for each trace. In conclusion, the projection matrix $H(E)$ can be calculated by considering an observation matrix $X(E)$ obtained from E applying the following pre-processing steps.

In the first pre-processing step (see Fig. 1), similarly to [16], we apply one-hot encoding, that is, each event $e_{i,j}$ is encoded into a set K dummy attributes $d_{i,j,k}$ such that:

$$d_{i,j,k} = \begin{cases} 1 & \text{if } e_{i,j} = a_k \\ 0 & \text{otherwise} \end{cases}$$

Then, for trace-level aggregation, the one-hot encoded events are horizontally concatenated for each trace. Since traces have different length, for the traces shorter than the longest one(s) in E, i.e., with less events than $N^{max} = \max_{\sigma_j \in E}\{N_j\}$, zero padding is applied. For example, given a case consisting of

4 events and $N^{max} = 5$, the fifth event of this case is zero padded, therefore, $d_{5,j,k} = 0$, $\forall k$. Based on this pre-processing, an event log E is encoded into an observation matrix $X(E)$ with J rows (traces) and $I = N^{max} \times K$ columns (attributes).

Using $X(E)$, we can now define a first leverage-based anomaly score $\hat{l}(\sigma_j)$ by extracting the diagonal elements of $H(E) = X(E) \cdot (X(E)^T \cdot X(E))^{-1} \cdot X(E)^T$:

$$\hat{l}(\sigma_j) = h_{j,j}$$

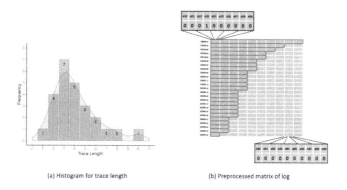

(a) Histogram for trace length   (b) Preprocessed matrix of log

Fig. 2. Seesaw effect of zero-padding

This first anomaly score is likely to be biased by the zero-padded attributes in the aggregation pre-processing step. Normally, these zero-padded attributes should be treated as `null` values by any statistical method and therefore not considered in the analysis. However, this is not the case when calculating $\hat{l}(\sigma_j)$. The presence of 0-padded values, as shown in Fig. 2, creates a a *seesaw* effect that increases the leverage of longer traces and decrease the one of shorter traces. Shorter traces, in fact, are more likely to be considered similar to each other, and therefore not anomalous, because they are encoded into a higher number of zero-padded values.

In order to counter this issue, we introduce a weighting factor w_j as a function of the trace length to increase/decrease the leverage $\hat{l}(\sigma_j)$ of shorter/longer traces σ_j. This weighting factor is calculated by first normalising the trace length N_j in the range $[0,1]$. This is done by applying the Z transformation to normalize N_j to the average $mean[N_j]$, followed by the application of a sigmoid function. The sigmoid-based normalisation is generally used to improve the fit accuracy and decrease the computational complexity of the fitting model [12]:

$$sig(Z_j) = \frac{1}{1 + e^{-Z_j}}, \text{ with } Z_j = \frac{\{N_j - mean[N]\}}{stdev[N]} \tag{1}$$

The weighting factor w_j is then defined as:

$$w_j = \left[1 - sig(Z_j)\right]^{c(N^{max})} \tag{2}$$

The power coefficient $c(N^{max})$ is required to adjust the strength of the weighting factor for different event logs. Intuitively, if all traces in a log have similar length, then this adjustment factor should be low, approaching 1; if trace length variance is very high, then the adjustment should be higher.

To define an appropriate value of the power coefficient $c(N^{max})$, a relation between N^{max} and the anomaly detection performance bias should be first found. However, this relation can only be estimated and not optimised because trace length has no upper bound, which would lead to a non-finite state optimisation problem. Therefore, we have estimated the value of $c(N^{max})$ by fitting a non-linear regression using 6 real-life event logs[1] . To model a non-linear regression function, we use the values of $c(N^{max})$ that achieve the highest F1-score in anomaly detection using the 6 different logs in offline settings, i.e., considering all the traces in an event log at the same time in the observation matrix X. Under significance level 0.01, the non-linear equation in Eq. 3 has been fitted with two coefficient parameters a and b as in Table 1.

$$c(N^{max}) = \begin{cases} -2.2822 + (N^{max})^{0.3422} & \text{if } N^{max} > \frac{2.2822}{0.3422} \\ 0 & \text{otherwise} \end{cases} \tag{3}$$

Table 1. Result of fitted non-linear regression model: $f(x) = a + x^b$

Parameter	Estimate	Standard Error	t value	p-value
a	−2.2822	0.3533	−6.46	0.0030
b	0.3422	0.0191	17.96	0.0001

In the end, using the estimated power coefficient $c(N^{max})$, we define the weighted *leverage*-based anomaly score as:

$$\hat{l}_w(\sigma_l) = w_j \cdot \hat{l}(\sigma_l). \tag{4}$$

3.2   Online Anomaly Detection

After having defined an anomaly score, the proposed anomaly detection method is complemented by the following four aspects: (i) grace period, (ii) finite memory usage, (iii) update of leverage scores, and (iv) anomaly threshold setting. These are described in detail next.

---

[1] These event logs belong to the ones made available by the Business Process Intelligence Challenge in 2012, 2013 and 2017.

Grace Period. Similarly to other online anomaly detection methods in the literature [8,19], for practical reasons it makes sense to begin taking decision on trace anomaly only after having received a sufficient number of events. For this purpose, we introduce the parameter Grace Period (GP), which specifies the minimum number of traces and events per trace that must be received before trace anomaly decisions can begin to be taken. In other words, the GP prevents to run the anomaly detection model at early stages, when a sufficient number of events has not been received yet. In this paper, the GP parameter is defined as the number of traces for which at least 2 events have been received. For example, if GP = 100, the anomaly detection starts from the first event after having received at least the first 2 events of 100 different traces.

Finite Memory Usage. Another condition to be satisfied by an anomaly detection method in online settings is the one of finite memory usage. In principle, events may be infinitely received as time goes by. However, handling an infinite number of events would require infinite memory, which is impossible in practical settings [19,20]. Therefore, to calculate leverage using always a finite number of events received from a stream, we introduce the parameter windows size (W), defined as the number of recent cases that are considered to determine anomalies when a new event is received. More formally, at a given time t, let us refer to E_t as the set of events received until t. Now, if the number of (possibly incomplete) traces represented by events in E_t is more than W, then the earliest traces are removed from the set of traces considered to calculate the projection matrix H. Specifically, events of the trace whose first event is the earliest in E_t is first removed and so on until the number of traces represented in E_t is W.

Update of Leverage Scores. Each time a new event is received, the leverage of the case to which this event belongs is updated (see Fig. 3). More in detail, let us assume that an event $e_{i,j}$ is received at time t. Then, after having possibly removed some cases represented by events in E_t to maintain the finite memory usage assumption, the leverage of the remaining traces $\hat{l}_w(\sigma_j)$ is calculated. The result obtained determines, based on the value of the considered anomaly detection threshold, whether the trace σ_j is considered anomalous after the arrival of event $e_{i,j}$, or not. This procedure is replicated each time a new event is received.

Anomaly Threshold Setting. The objective of anomaly detection is ultimately to determine whether traces are anomalous or not. Therefore, the problem of anomaly detection can be reduced to a binary classification problem. Based on the anomaly score $\hat{l}_w(\sigma_j)$, a decision should be made whether the trace σ_j is anomalous or not. This is normally done by setting the value of an anomaly detection threshold T for $\hat{l}_w(\sigma_j)$, such that a trace is anomalous if $\hat{l}_w(\sigma_j) > T$. In this work, we consider three constant thresholds and one variable threshold. We consider the constant values $T_{c1} = 0.1$, $T_{c2} = 0.15$, and $T_{c3} = 0.2$. These values are based on our experience in experiments in online and offline settings, where anomalous traces tend to have an anomaly score $\hat{l}_w(\sigma_j)$ greater than 0.1. As variable threshold, we consider the value

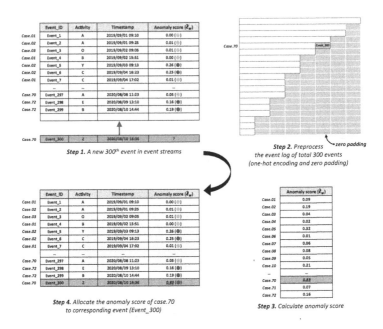

Fig. 3. Example of anomaly detection using a fixed anomaly threshold $T = 0.1$

$T_v = mean_{\sigma_j \subseteq E_t}[\hat{l}_w(\sigma_j)] + stdev_{\sigma_j \subseteq E_t}(\hat{l}_w(\sigma_j))$, which calculates the threshold based on the mean and standard deviation of the leverage scores of the traces considered in the observation matrix X. A similar principle to set the anomaly detection threshold is used by [16] for the timestamp anomaly detection threshold.

4   Evaluation

This section describes first the datasets that we used for evaluating the proposed online anomaly detection framework. Then, we present the evaluation metrics and experiment settings and, finally, we discuss the performance and computational cost of the proposed framework.

We consider two artificial logs used by [18] and one real-life log publicly available. The artificial logs are generated by simulating 2 process models (Small and Medium in [18]) using the PLG2 tool. Regarding the real log, we consider the *Helpdesk* event log, which contains events logged by a ticketing management system of the help desk of an Italian software company. These logs have been chosen because they have been considered by previous work in anomaly detection and they are also sufficiently small to control the running time of experiments. Descriptive statistics of these logs are reported in Table 2.

Evaluating an unsupervised approach of anomaly detection like the one that we propose requires event logs with labelled traces (normal v. anomalous), which are generally unavailable in practice. Therefore, a common practice in this

Table 2. Descriptive statistics of event logs (Log statistics are counted after injecting anomalies)

Type	Data	Statistics	Value
Artificial log	Small	- Number of cases	5,000
		- Number of events	44,811
		- Number of activities	20
		- Average # of cases per day	5,000
		- Average # of events per day	44,811
	Medium	- Number of cases	5,000
		- Number of events	29,683
		- Number of activities	32
		- Average # of cases per day	5,000
		- Average # of events per day	29,683
Real log	Helpdesk	- Number of cases	3,804
		- Number of events	13,901
		- Number of activities	9
		- Average # of cases per day	10.94
		- Average # of events per day	18.06

research field is to inject anomalies using different types of anomaly patterns in event log and creating labels during the anomaly injection process [1–3,16,18]. We consider the 5 anomaly patterns *Skip*, *Insert*, *Early*, *Late*, and *Rework* as defined in [18]: in *Skip*, a sequence of events is skipped in some cases; in *Insert*, one or more events are generated in random positions within existing traces; in *Early/Late*, timestamps of events are manipulated such that a sequence of events is moved earlier/later in a trace; in *Rework*, a sequence of events is repeated after its occurrence. Anomalies are randomly injected in an event log until 10% of the traces in the log have become anomalous. As far as performance measures are concerned, we consider the typical measures for classification problems obtained from the confusion matrix, i.e., precision, recall and F1-score. The datasets used in this paper and the code to reproduce the experiments discussed next are available at https://github.com/jonghyeonk/OnlineAnomalyDetection.

We set the GP to 1,000 cases and consider 3 values of window size W, i.e., $W \in [1000, 2000, 3000]$. A larger value of GP and W is likely to lead to better and more stable performance, while also implying a higher computational cost. The experiments are implemented in R on an Intel i7 Linux machine using a single CPU and 5 GB memory limit.

Table 3 shows the performance of anomaly detection in event streams for different anomaly detection threshold values and different values of W. Note that the 4 columns T_{c1} to T_v report average performance measures calculated from the start of the streaming (after the GP condition has been reached). It can be noticed that the three constant thresholds show on average better perfor-

Table 3. Performance of online anomaly detection (average from the start of the stream and calculated only on first/last 100 events)

Data	Window size	Time cost (average sec)	Metric	Threshold					
				T_{c1}	T_{c2}	T_{c3}	T_v	$T_v^{F:100}$	$T_v^{L:100}$
Small	1,000	1.09	Precision	0.22	0.22	0.22	0.21	0.06	1.00
			Recall	0.63	0.62	0.59	0.62	1.00	0.63
			F1-score	0.33	0.32	0.32	0.31	0.11	0.77
	2,000	1.07	Precision	0.26	0.25	0.25	0.25	0.06	1.00
			Recall	0.61	0.60	0.57	0.61	1.00	0.63
			F1-score	0.36	0.36	0.35	0.35	0.11	0.77
	3,000	1.23	Precision	0.75	0.79	0.82	0.75	0.20	1.00
			Recall	0.57	0.55	0.53	0.57	0.17	0.63
			F1-score	0.65	0.65	0.64	0.65	0.18	0.77
Medium	1,000	1.42	Precision	0.16	0.17	0.17	0.15	0.50	1.00
			Recall	0.72	0.71	0.71	0.73	0.29	0.50
			F1-score	0.26	0.27	0.27	0.25	0.36	0.67
	2,000	1.46	Precision	0.37	0.45	0.52	0.26	0.50	1.00
			Recall	0.65	0.64	0.63	0.68	0.29	0.50
			F1-score	0.47	0.53	0.57	0.38	0.36	0.67
	3,000	2.06	Precision	0.28	0.30	0.31	0.25	0.50	1.00
			Recall	0.67	0.67	0.66	0.67	0.29	0.50
			F1-score	0.39	0.41	0.42	0.37	0.36	0.67
Helpdesk	1,000	0.34	Precision	0.06	0.06	0.06	0.06	0.00	0.50
			Recall	0.99	0.98	0.96	1.00	0.00	0.96
			F1-score	0.12	0.12	0.12	0.12	0.00	0.66
	2,000	0.37	Precision	0.08	0.08	0.09	0.08	0.00	0.51
			Recall	0.80	0.74	0.68	0.77	0.00	0.96
			F1-score	0.15	0.15	0.15	0.14	0.00	0.67
	3,000	0.39	Precision	0.09	0.11	0.13	0.10	0.00	0.61
			Recall	0.59	0.49	0.39	0.60	0.00	0.93
			F1-score	0.16	0.19	0.20	0.17	0.00	0.74

mance than the variable threshold T_v. To better observe a trend of performance improvement as more events are received, the last two columns $T_v^{F:100}$ and $T_v^{L:100}$ show the average performance values calculated on the first 100 events received (after the GP condition has been met) and last 100 events received, respectively. The result shows a clear tendency of increasing performance. The performance is low at the initial stage, and it increases remarkably for the last 100 events received. Particularly in the case of the *Helpdesk* log, the proposed framework could not detect any anomalous cases in the first 100 events, while the performance clearly increases for the last 100 events. It should be noted that, as the number of events received increases, the performance of the proposed framework is likely to converge to the one showed by the average on the last 100 events.

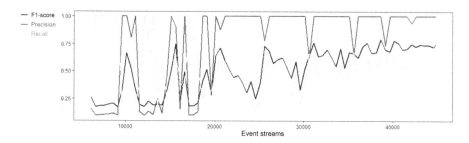

Fig. 4. Performance values as number of events received increases (Small log is applied, using T_v, W = 3000)

Regarding the window size W, the average time cost increases with the value of W. A larger window size also leads to better anomaly detection performance.

As an example, Fig. 4 breaks down the performance of the proposed framework along time, counted as the number of events received, in the case of the Small event log with W = 3000. It can be noted that the performance oscillates wildly until 20,000 events are received. After that, the performance tends to stabilise and, while the precision remains high, the recall also begins increasing more regularly. After 30,000 events are received, all the performance metrics appear to have become stable.

5   Conclusions

This paper has presented an approach to detect trace level anomalies in business process event streams using an anomaly score based on statistical leverage. A preliminary evaluation on artificial and real event logs also has been presented. The results obtained in this paper are important to determine the future work in this line of research.

First, the performance in the case of the *Helpdesk* log highlights an issue with anomaly threshold setting. The constant values chosen for the experiment (between 0.1 and 0.2) appear to be too low for this event log, which results in low precision and F1-score and only high recall. This points to the need for developing an advanced variable threshold that can adapt to the characteristics of different event logs.

Another limitation of the proposed approach, which also impacts the performance, is the fact that it does not distinguish between incomplete and completed traces when calculating the anomaly score. Therefore, many traces may be considered anomalous because they are incomplete, even though they will turn into normal at some point in the future as more events are received. A possible strategy to prevent this is to organise the events received into batches by different prefix length before calculating the anomaly score. This is inspired by [20] that, in the case of online compliance checking, addresses the issue of trace incompleteness using prefix-alignment.

Finally, considering word-embedding instead of one-hot encoding and zero-padding during pre-processing may be likely to reduce the size of the observation matrix X and, therefore, speed up the calculation of anomaly scores.

References

1. Bezerra, F., Wainer, J.: Algorithms for anomaly detection of traces in logs of process aware information systems. Inf. Syst. **38**(1), 33–44 (2013)
2. Böhmer, K., Rinderle-Ma, S.: Multi-perspective anomaly detection in business process execution events. In: Debruyne, C., et al. (eds.) OTM 2016. LNCS, vol. 10033, pp. 80–98. Springer, Cham (2016). https://doi.org/10.1007/978-3-319-48472-3_5
3. Böhmer, K., Rinderle-Ma, S.: Multi instance anomaly detection in business process executions. In: Carmona, J., Engels, G., Kumar, A. (eds.) BPM 2017. LNCS, vol. 10445, pp. 77–93. Springer, Cham (2017). https://doi.org/10.1007/978-3-319-65000-5_5
4. Burattin, A.: Online conformance checking for petri nets and event streams. In: 15th International Conference on Business Process Management (BPM 2017) (2017)
5. Burattin, A., Carmona, J.: A framework for online conformance checking. In: Teniente, E., Weidlich, M. (eds.) BPM 2017. LNBIP, vol. 308, pp. 165–177. Springer, Cham (2018). https://doi.org/10.1007/978-3-319-74030-0_12
6. Burattin, A., van Zelst, S.J., Armas-Cervantes, A., van Dongen, B.F., Carmona, J.: Online conformance checking using behavioural patterns. In: Weske, M., Montali, M., Weber, I., vom Brocke, J. (eds.) BPM 2018. LNCS, vol. 11080, pp. 250–267. Springer, Cham (2018). https://doi.org/10.1007/978-3-319-98648-7_15
7. Cao, F., Estert, M., Qian, W., Zhou, A.: Density-based clustering over an evolving data stream with noise. In: Proceedings of the 2006 SIAM International Conference on Data Mining, pp. 328–339. SIAM (2006)
8. Domingos, P., Hulten, G.: Mining high-speed data streams. In: Proceedings of the sixth ACM SIGKDD International Conference on Knowledge Discovery and Data Mining, pp. 71–80 (2000)
9. Genga, L., Alizadeh, M., Potena, D., Diamantini, C., Zannone, N.: Discovering anomalous frequent patterns from partially ordered event logs. J. Intell. Inf. Syst. **51**(2), 257–300 (2018)
10. Ghionna, L., Greco, G., Guzzo, A., Pontieri, L.: Outlier detection techniques for process mining applications. In: An, A., Matwin, S., Raś, Z.W., Ślęzak, D. (eds.) ISMIS 2008. LNCS (LNAI), vol. 4994, pp. 150–159. Springer, Heidelberg (2008). https://doi.org/10.1007/978-3-540-68123-6_17
11. Hoaglin, D.C., Welsch, R.E.: The hat matrix in regression and Anova. Am. Stat. **32**(1), 17–22 (1978)
12. Klimstra, M., Zehr, E.P.: A sigmoid function is the best fit for the ascending limb of the hoffmann reflex recruitment curve. Exp. Brain Res. **186**(1), 93–105 (2008)
13. Leemans, S.J.J., Fahland, D., van der Aalst, W.M.P.: Discovering block-structured process models from event logs containing infrequent behaviour. In: Lohmann, N., Song, M., Wohed, P. (eds.) BPM 2013. LNBIP, vol. 171, pp. 66–78. Springer, Cham (2014). https://doi.org/10.1007/978-3-319-06257-0_6
14. Leemans, S.J., Fahland, D., Van der Aalst, W.M.: Scalable process discovery and conformance checking. Softw. Syst. Model. **17**(2), 599–631 (2018)

15. Lu, X., Fahland, D., van den Biggelaar, F.J.H.M., van der Aalst, W.M.P.: Detecting deviating behaviors without models. In: Reichert, M., Reijers, H.A. (eds.) BPM 2015. LNBIP, vol. 256, pp. 126–139. Springer, Cham (2016). https://doi.org/10.1007/978-3-319-42887-1_11
16. Nguyen, H.T.C., Lee, S., Kim, J., Ko, J., Comuzzi, M.: Autoencoders for improving quality of process event logs. Expert Syst. Appl. **131**, 132–147 (2019)
17. Nolle, T., Luettgen, S., Seeliger, A., Mühlhäuser, M.: Analyzing business process anomalies using autoencoders. Mach. Learn. **107**(11), 1875–1893 (2018)
18. Nolle, T., Luettgen, S., Seeliger, A., Mühlhäuser, M.: BiNet: multi-perspective business process anomaly classification. Inf. Sci. 101458 (2019)
19. Tavares, G.M., da Costa, V.G.T., Martins, V.E., Ceravolo, P., Barbon Jr, S.: Leveraging anomaly detection in business process with data stream mining. iSys-Revista Brasileira de Sistemas de Informação **12**(1), 54–75 (2019)
20. van Zelst, S.J., Bolt, A., Hassani, M., van Dongen, B.F., van der Aalst, W.M.: Online conformance checking: relating event streams to process models using prefix-alignments. Int. J. Data. Sci. Analytics **8**(3), 269–284 (2019)
21. Wold, S., Esbensen, K., Geladi, P.: Principal component analysis. Chemom. Intell. Lab. Syst. **2**(1–3), 37–52 (1987)

Concept Drift Detection on Streaming Data with Dynamic Outlier Aggregation

Ludwig Zellner[(✉)], Florian Richter, Janina Sontheim, Andrea Maldonado, and Thomas Seidl

LMU Munich, Munich, Germany
{zellner,richter,sontheim,maldonado,seidl}@dbs.ifi.lmu.de

Abstract. Many processes no matter what kind are regularly changing over time, adapting themselves to external and internal circumstances. Analyzing them in a streaming context is a very demanding task. Particularly the detection and classification of significant deviations is important to be able to re-integrate these possible micro-processes. Assuming a deviation of a certain process, the significance is implicitly given when a high number of instances contain this deviation similarly. To enhance a process the integration of or preventive measures against those anomalies is of high interest for all stakeholders as the actual process core gets discovered more and more in detail. Considering various areas of application, we focus on previously neglected but potentially significant anomalies like small changes in the disease process of a virus infection that has to be discovered to develop an appropriate reaction mechanism. We concentrate on non-conforming traces of a stream on which we compute a local outlier factor. This allows us to detect relations between traces based on changing outlier scores. Hence, hereby connected traces are clusters with which we achieve the detection of concept drift. We evaluate our approach on a synthetic event log and a real-world dataset corresponding to a process representing building permit applications which emphasizes the extensive applicability.

Keywords: Concept drift detection · Local outlier factor · Micro-clusters

1 Introduction and Motivation

Nearly every established process consists of deviations, which potentially contain valuable information impelling its context. E.g. by considering the Internet of Things as highly attractive for botnet attacks, it is a very interesting task to distinguish, detect and classify attack traces from operations of the ordinary process. The classification task to identify device-threatening traces is pursued by attack triage systems, in which our approach can provide outlier scores. A second example is the analysis of disease spreading processes. Regarding the regulations of the German public health department in the 2020's pandemic, symptomatic people have to keep health journals during the course of an infection to record

S. Leemans and H. Leopold (Eds.): ICPM 2020 Workshops, LNBIP 406, pp. 206–217, 2021.
https://doi.org/10.1007/978-3-030-72693-5_16

the timeline of symptoms. These journals are eventually gathered to construct the typical disease process. Fever, dry cough and fatigue count as common symptoms for Covid-19 and represent the aforementioned process. Extraordinary and more rare symptoms are: Loss of the senses of taste and smell, conjunctivitis and the so called covid toes. Having these symptoms occur for some patients, their disease process is non-conforming and a concept drift emerges. To extend the process, the course of treatment and its effects can also be incorporated. In general, the detection and classification of new types of infectious diseases can be done independently of the deeper understanding of its nature and, therefore, this procedure can also be mapped to various other diseases. Regarding business processes, in case of structural changes in the business itself, the question is, which deviations occur in reaction to initial changes. In particular, using a building permit application process, we apply our approach to detect deviating subprocesses as candidates for reintegration after a split of departments.

At large, process mining has its focus on conformance checking for a long time. Particularly considering an anomalous group of traces which nevertheless belongs to the main process, can help to enhance the main process by reintegrating this process deviation. The deviation itself qualifies for further analysis by consisting of multiple similar instances. Hereafter, we refer to an anomaly as a micro-cluster of non-conforming traces i.e. anomalous traces. This microcluster is defined by multiple related process instances. The relation between traces can manifest itself in different ways. Therefore, a specific distance between these instances is crucial, which will be defined for our specific case in Sect. 3.

In this work, we introduce a novel approach to classify non-conforming traces into micro-clusters on trace streams. Based on these clusters we then achieve a more fine-grained concept drift detection. Specifically, the basis of our approach is a stream of traces. As a combination of state-of-the-art methods from process mining and an established algorithm from the field of machine learning and data mining, our approach is very flexible due to its modularity. As mentioned before the field of application is very wide concerning current emergency cases but can also be applied to every other process which consists of interesting and critical deviations. An evaluation on a synthetic event log and the building permit application dataset from the BPI Challenge of 2015 demonstrates the considerable effects of our method.

In Sect. 2 we analyze the existing work related to clustering traces to groups of anomalies. Section 3 describes the background of our work. Our framework and our algorithmic methods are explained in Sect. 4. Sections 5 and 6 address the performed experiments and corresponding evaluation. We conclude our work with Sect. 7.

2   Related Work

The aggregation of outliers by leveraging local outlier factor serves as means to cluster data. In addition, *LOF* also incorporates concepts from density-based algorithms like *reachability* or *core distance* from the field of data mining [2].

However, a major difference to its original field of application is the goal, which is not defined by finding local outliers, but is to find reductions of local outlierness, which eventually yield a new micro-cluster. First and foremost, the focus lies on finding groups of outliers, which define an anomaly of the underlying process, and detect concept drifts of different degree. Because this anomaly consists of a number of traces, it is highly probable, that there is a hidden micro-process driving those process instances. For that reason, it is an important task to analyze those micro-clusters and recognize the drift of the main process into one of those micro-processes. The underlying idea to our approach is trace clustering [5], which is based on an embedding of traces into vector space. The overall purpose is to split highly diverse processes into homogeneous subsets. Thus, the complexity is reduced, while the interpretability is increased.

In comparison to our approach Richter et al. [9] also use the idea of leveraging a reference model for distance computation between traces. They aim at clustering traces based on their distances to each other. However, the approach requires manual user knowledge to define the agglomerative clustering manually. It is not developed for stream application and does not provide further analysis like concept drift detection as we do.

With TESSERACT [8], concept drifts regarding the time dimension are focused. Interim-times between events of a stream are used to derive a drift indicator regarding sudden and incremental drifts. However, the change of completion times of an event often can have other reasons than indicating a drift. Therefore, we use the conformance of a trace by activity labels as a more direct indication.

Using statistical hypothesis testing Maaradji et al. [6] are detecting drifts between consecutive batches of traces by comparing the distribution of runs statistically. A subsequent paper by Ostovar et al. [7] focuses on the detection of drift within a trace rather than between traces. By using sliding windows and applying the G-test of independence this method even can be applied to event logs which are highly variable. In our approach high variability also is supported resulting in many micro-clusters which are given as means of drift categorization.

3 Preliminaries

In this section we describe the background of the methods, which will be harnessed in Sect. 4. These include a definition of trace streams, conformance checking, the method to compute distances between traces and the procedure behind the popular local outlier factor from the field of machine learning. *Case identifiers*, *activity labels* and *timestamps* are the minimal components of event logs. Their counterpart on streams can be defined similarly. In an event stream S an event e is emitted continuously in a specific order, duration and by reference to a certain *case identifier*. That means, that the context, in which the event is emitted, is known. However, the differences between an event log and an event stream is (1) the potentially *infinite* sequence of events and (2) the *incompleteness* of a case, which means that at a certain point in time a case does not have to be necessarily completed, since a new event could possibly be executed in context of this case [11]. Additionally the timestamp at which an event is emitted depends on the stream. Thus, it can be neglected and we define $\#_{case}(e) = c$ and $\#_{act}(e) = a$ for $e = (c, a)$. Because of the different structure this type of data needs a different handling compared to conventional process discovery. Hence, we refer hereby to state-of-the-art methods in the literature and assume incoming data to be given as traces $\#_{trace}(c) = \hat{c}$ consisting of occurring events. Furthermore, we check every trace for conformance against the reference model and additionally against every micro-cluster model in each iteration. Thus, we utilize a preferably fast but efficient conformance checking method. We decide to apply token-based replay [10] as it has an advantage in speed compared to alignments [1]. Furthermore, to concentrate on the applicability of our approach, we focus on incremental and recurring drifts in Sect. 5 and Sect. 6.

To be able to provide every trace with a comparable distance property we utilize the reference model by following the approach of Richter et al. [9]. By interpreting the model as a map, we define the *geodesic distance* between two traces as the average number of edges on the shortest paths between every vertex of two traces. A transition is only counted as a vertex, if it is one of the transitions causing problems in the conformance checking procedure. The distance computation follows the approach in [9] and is defined by the following formula where X and Y are traces and x and y are transitions within:

$$dist(X, Y) = \frac{1}{|X| \cdot |Y|} \cdot \sum_{x \in X} \sum_{y \in Y} shortest\_path(x, y)$$

Essentially, every trace is reduced to its events, which cause problems on the reference model. Thereafter, the reduced traces are compared pairwise by computing the geodesic distance between every contained event. After we average the result we get the required distance between two traces, which is also known as average-linkage. The ulterior motive is, that for deviating traces one or multiple transitions can be determined, which lead to the low fitness value. These transitions posses different distances to each other when the missing transitions are filled up by the reference model and the amount of edges between the transitions are counted. Thus, traces deviating with similar problematic transitions have a lower distance as deviating traces with very different problematic transitions. This part as another module can be substituted by any other computation method as well, which possibly leads to different results. The main part of our novel method is the usage of the popular *Local Outlier Factor* developed by Breunig et al. [3] from the field of machine learning. This algorithm is constructed to detect local outliers in a dataset. Therefore, every trace is assigned a degree (LOF), which describes its outlierness with respect to the surrounding neighborhood. The problem of different types of outliers, i.e. global versus local is described in [3]. Local outlierness, therefore, is defined relatively to the neighborhood of a data point. We leverage this perspective in our work by analyzing traces with respect to their surroundings. These neighboring traces can be arbitrarily distant to each other. If the average reachability distance, described below, between the neighbors is similar to the reachability distance of the trace itself, it gets reflected in a LOF of ~ 1.0. The LOF of a trace σ and the local reachability distance lrd is computed in the following way:

$$LOF_k(\sigma) = \frac{\sum\limits_{\gamma \in N_k(\sigma)} \frac{lrd_k(\gamma)}{lrd_k(\sigma)}}{|N_k(\sigma)|}, \quad lrd_k(\sigma) = \frac{|N_k(\sigma)|}{\sum\limits_{\gamma \in N_k(\sigma)} reach - dist_k(\sigma, \gamma)}$$

The variable k denotes the number of nearest neighbors surrounding σ, where the set of these neighbors is called k-neighborhood $(N_k(\sigma))$. and lrd refers to the local reachability distance: where *reachability–distance* is either the distance of γ to its k-th nearest neighbor or the real distance between γ and σ depending on which distance is greater. Analogously, if $LOF_k(\sigma) \gg 1.0$ the trace is an outlier.

4   Dynamic Outlier Aggregation

Our novel approach combines conformance checking from Process Mining with Local Outlier Factoring from the field of Data Mining to achieve the automated detection and classification of deviating traces in streams. The emerging micro-clusters are then used to detect concept drift on the aforementioned streams. This procedure differs from [9] to the extent that we aim for concept drift detection on streams. Richter et al. aim at clustering deviating traces by using a reference model given by the main process. The resulting micro-clusters represent deviations with increased potential due to its density. Furthermore, they do not provide means to the reader to automatically cluster or even classify new upcoming data. Trace clustering on deviating traces could also be achieved by our approach but it is not subject to this paper. After conformance checking of incoming traces to both the reference and micro-models, only the non-conforming traces are analyzed.

Our approach can be split into four modules (see Fig. 1), i.e. initial process discovery, conformance checking, LOF computation and micro-cluster aggregation. The former has to be done once for initiation to prepare a process model as a reference for further computations. The latter three steps are processed repeatedly after initiation. At first, in the initiation phase, a main model is discovered, which

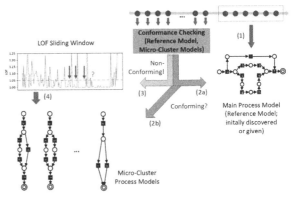

Fig. 1. Overview of our approach. Steps 1–4 denote the successive procedure to detect anomalous clusters.

serves as a reference model (Fig. 1 (1)). This reference model can be normative or declarative but the premise is that it does not yet hold traces, which actually should be considered non-conforming. Afterwards, in the iteration phase, the conformance of the incoming traces is checked and filtered by non-conformance, which means that we only analyze non-conforming traces in the following (Fig. 1 (2a and b)). Therefore, we use the popular method of token-replay as a conformance checking step. Here, the reference model is used to compute the fitness of a trace. For this purpose, the four counters *produced, consumed, missing* and *remaining* are used as described in Sect. 3 to determine, if a trace fits a model or not. Every conforming trace then is removed and the next steps are only based on the set of non-conforming traces. In the next step, we compute a local outlier factor (LOF) on every trace (Fig. 1 (3)). Gradually, the latest incoming traces are filtered in the aforementioned way and the local outlier factor of every stored trace is re-computed. Those computations serve as a snapshot, at which we decide, if there is a relation between two or more traces. This decision is based on the fact, that the scores of those traces drop below a given clustering threshold at a specific snapshot. For those corresponding traces a process model is discovered (Fig. 1 (4)). These micro-cluster models then are used as additional reference models, against which every incoming trace is checked before it is labelled non-conforming (Fig. 1 (2b)). Since the non-conforming traces are not removed from the set of local outlier factors, the computation of it has an increase in accuracy over time until it reaches the size of a sliding window. This sliding window is used to limit the complexity of re-computations. Algorithm 1 depicts the described procedure in pseudocode.

We assume that the streaming events are already gathered to traces. This means, that the procedure, in which we wait for a trace to start and end, is transferred to already existing efficient algorithms [8]. Furthermore, the outcome of this procedure does not have to be completely accurate since our approach can cope with inaccuracies arising in this step. In addition, it is assumed that the process model, which we use as a reference, is already discovered, solely relies on conforming traces and only has to be prepared for further usage. Another possibility is to provide a normative model by a domain expert. The iteration phase comprises repeating steps mainly consisting of the novel approach we call *Dynamic Outlier Aggregation*. This aggregation refers to the grouping of traces. Thus, if there is a number of traces, of which the LOF scores are below a constant threshold, which T exceeds a given constant number K, these traces are aggregated to micro-clusters. To compute the LOF for each trace we utilize a matrix holding all distances between every trace in the window. We start with the first trace having a distance of 0.0 to itself. For every trace being emitted afterwards by the stream we compute the distance between the current and every other trace in the sliding window based on their transitions causing problems on the reference model. The pairwise combination of the traces and the cartesian product between all these problematic transitions of a pair form the basis for further computation. The average of the overall hop count of the shortest paths between every aforementioned transition yields our resulting lower triangular matrix, which then is extended to a symmetric matrix for further computation. Besides, if in the next iteration step the number of distances exceeds W, the

---

Algorithm 1. Dynamic Outlier Aggregation

Input: The emission interface of a stream SI of single traces i, a lower bound L and a sliding window size W, MinPts K for LOF computation and a threshold T, below which a trace is assigned to a micro-cluster

Output: A collection of micro-clusters represented by event logs and a LOF for each trace

1: $R \leftarrow discover\_reference\_model()$ ▷ Assumption: Traces are conforming
2: $MC \leftarrow \emptyset$ ▷ MC denotes the set of micro-clusters
3: $IBT \leftarrow \emptyset$ ▷ IBT denotes the set of non-conforming traces with LOF below threshold
4: $M \leftarrow \emptyset$ ▷ M denotes the distance matrix
5: **while** $|SI| > 0$ **do**
6:     Get next i
7:     **if** $!fitsR(i) \wedge !fitsM(i)$ **then**   ▷ Check conformance with reference model and micro-cluster models
8:         $M \leftarrow compute\_distances(IBT, i)$
9:         $IBT \leftarrow token\_replay(R, i)$ ▷ Save activities with problems
10:         $M \leftarrow sym(M)$ ▷ Create symmetric matrix from M
11:         **if** $|rows(M)| > L$ **then**
12:             $IBT \leftarrow IBT[1:]$ ▷ Remove first element
13:             **if** $|rows(M)| > W$ **then**
14:                 $M \leftarrow delete\_first\_row\_col(M)$
15:             **end if**
16:             ▷ Separate traces forming a micro-cluster from outliers
17:             $micro\_cluster \leftarrow separate(IBT, lof, T)$
18:             **if** $|micro\_cluster| \geq K$ **then**
19:                 $MC \leftarrow MC \cup micro\_cluster$
20:             **end if**
21:         **end if**
22:     **end if**
23: **end while**

---

oldest trace is removed and so on. Thus, we can use the pre-computed distance matrix to return LOF for every trace. Because the LOF of every trace potentially changes in each iteration step, we have to separate the possible micro-clusters from the outliers. On that account, a threshold T is required, which indicates the affiliation of a trace to a micro-cluster. Hence, every trace, which possesses a LOF below T, is returned in a separate log. However, the connected trace and LOF is not yet removed from the matrix. The reason behind it is the increasing accuracy of LOF the more traces are located in a micro-cluster. Since the computational complexity of LOF is $\mathcal{O}(W^2)$, the overall complexity of *Dynamic Outlier Aggregation* is $\mathcal{O}(W^3)$ in a naive implementation. However, the theoretical complexity analysis based on stream data seems high, the actual complexity strongly depends on the window size, which covers only a small portion of the data. Thus, by choosing a fixed window size, which should be preferably small compared to the expected emitted data, the goal of our approach can be achieved in constant time ($\mathcal{O}(1)$).

5   Evaluation on a Synthetic Log

The proposed approach can be inspected and reproduced on GitHub[1]. The implementation is used to experiment with both synthetic and real-world logs. In

---

[1] https://github.com/zellnerlu/DOA.

addition to the applicability on real-world data, we focus on the detection of recurring drifts as an example. Furthermore, the results of the usage within a fixed parameter set are compared. Moreover, the limitations of our approach are discussed.

As a proof of concept we create an event log consisting of 1000 traces with the *Processes and Logs Generator* (PLG2) by Burattin [4] with which it is also possible to produce an evolution of a given process. Thus, we start with a rather small main process consisting of 6 activities and 2 XOR-gateways. Additionally, 3 deviation processes with the size of 1000 traces each are created by using the aforementioned option and the given parameters such as *depth, AND/XOR branches, AND/XOR weight etc.* are adjusted accordingly, such that the probability of occurrence is de- and increased. We inject the drifts, by splitting up and interleaving parts of the deviation processes with parts of the main process. Thus, an event log is produced, which consists of all 4 processes appearing successively. Here, we concentrate mainly on recurring drifts as an example to illustrate the process of detection. The scalability of our approach is evaluated by measuring the required time of the trace from comparison to the other traces in the sliding window until the clustering step. Furthermore, we utilize *F1-Score* as a quality measure, which is defined as the harmonic mean between precision and recall. Since our approach heavily relies on the creation of process models in different steps of the framework, we globally measure precision and recall instead of using a multi-class perspective.

5.1   Execution Times

We perform the experiments on an Intel Core i7 with 3.2 Ghz and 32 GB RAM. The operating system is Windows 10. The measured time required for the distance computation to the other traces in the sliding window as well as the clustering procedure varies between 1ms and 20 ms with an average duration of about 7 ms. In Fig. 2b the application of a sliding window of size 200 shows an outlier in terms of maximum duration, which indicates the increasing time requirement with larger sliding windows. Nevertheless, this shows the applicability at least on trace streams and the approach also qualifies for an extension to event streams due to the fast processing.

5.2   Impact of Inter-drift Distance and Sliding Window Size

At first we analyze the detection quality by varying the distance between drifts. This means, that we vary the size of the interleaved parts, namely inter-drift distance, by 5–75%, which means that 50, 100, 250, 500 and 750 traces of every event log are arranged in sequence. This is repeated until log completion. We choose *LOF* plots and Gantt Charts for visualization, where the x-axis shows the global trace identifiers, i.e. trace IDs within all emitted traces and the y-axis represents the *LOF* of a trace (*LOF* plot), i.e. the micro-cluster affiliation (Gantt Chart). It is also perceivable, that the *LOF* of some traces is highlighted with a certain color, i.e. the bars in the Gantt Charts are depicted with a certain color.

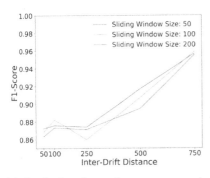

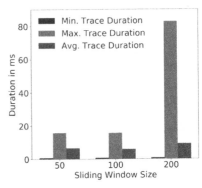

(a) Analyzing the performance on various inter-drift distances by number of interleaved traces and the application of different sliding window sizes.

(b) Execution times by using different sizes of sliding windows. Min. Trace Duration is between 30 and 60 microseconds.

Fig. 2. Quality measure and execution time depending on the sliding window size

This color represents the affiliation of the trace to a certain micro-cluster model. As it is shown in Fig. 2a, the resulting *F1-Score* increases in all three cases similarly, when increasing the inter-drift distance. This is due to the decreasing number of changes, which have to be detected. Thus, the probability of detecting *False Positives* decreases, as well. Figure 2a shows a continuously high detection quality, which also can be confirmed visually in Fig. 3.

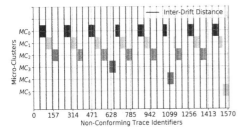

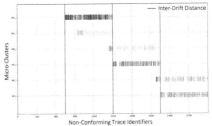

Fig. 3. Recurring drift detection on a a synthetic log with inter-drift distance of 50. The change points are emphasized with vertical black bars.

Fig. 4. Resulting Gantt Chart on an inter-drift distance of 750 and a sliding window size of 100.

One can see, that micro-cluster 0–2 are reliably representing the injected deviations from the main model. Interestingly, our approach regularly detects additional micro-clusters (3–5) beyond change points, which neither resemble recurring drifts nor were injected by design. This could be for reason of overly similar traces, which also neither do fit to the reference model nor to one of the micro-cluster models. Nonetheless, the visualization shows a clear detection of recurring drifts, where false micro-clusters can be simply excluded and the remainder can be analyzed by a domain expert further on. When we increase the size of the sliding window by keeping a relatively small lower bound L, we get a result as shown in Fig. 4. Here, the recurring drifts are still detected very well, but two micro-cluster models, e.g. green and yellow or purple and red, share one injected cluster each. In the latter pair another red cluster besides the first one is created, because the purple cluster was initially created before the change point at trace id 1500. This issue brings some limitations to light. On the one hand, to estimate a fitting sliding window size is difficult as a minimum of $K \leq W$ of LOF scores have to drop below T to be detected as a cluster. Therefore, a high difference between K and W leads to the detection of very small micro-clusters and a low difference leads to a very low sensitivity against concept drifts. On the other hand, some test execution have to be made to estimate a fitting threshold. In the experiments it appears, that T is preferably set right below the average LOF. Additionally, the current implementation poses the problem that concept drift detection is constrained to the control-flow dimension. To include other perspectives like time or resources the approach has to be altered.

6   Evaluation on the Event Log of BPIC 2015

As it is already introduced in Sect. 1, we use data from the Business Process Intelligence Challenge 2015[2]. It consists of building permit applications of five Dutch municipalities with a total number of 5649 traces. Due to the four year period, in which changes were regularly made to the rules and regulations, an incremental concept drift is expected in the data. For the first experiment we use the first log with 1199 traces. Figure 5 shows a snapshot of our analysis in LOF-plot format, which reveals the direction of the resulting Gantt-Chart. One has to be aware,

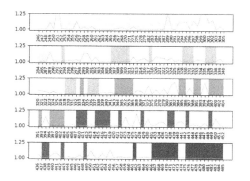

Fig. 5. Snapshot of five overlapping sliding window states. Every trace ID (x-axis) is assigned a LOF (y-axis). Background colors represent assignments to micro-clusters.

that appearing traces in this plot are already classified non-conforming to both the reference and all other micro-cluster models. The colors show, that clusters are aggregated, which are temporally overlapping. This implies the affinity for each outlier aggregation to a different, successive cluster. Thus, incremental drifts are detected as new micro-clusters are consecutively created.

Because BPIC 2015 is of high variability regarding the activity domain, traces inherently posses a rather high distance in between. Increasing the sliding window size does not lead to the expected micro-clusters in the data already holding

[2] https://doi.org/10.4121/uuid:31a308ef-c844-48da-948c-305d167a0ec1.

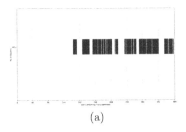 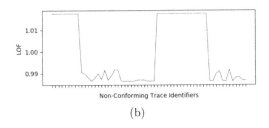

(a)   (b)

Fig. 6. Micro-cluster detection on hybrid log of BPIC 2015. The x-axis shows the non-conforming trace identifiers in both figures and the y-axis shows the micro-cluster affiliation (Fig. 6a) and the LOF (Fig. 6b).

an incremental drift but the loss in performance was relatively high. Thus, to show the applicability to on small recurring drifts, we extend the log to serve as a hybrid by repeatedly injecting data unrelated to the process. We randomly use 5–25 traces with an inter-drift distance between 25–100 to alter the log. Following this strategy, we simulate a changed process based on real-world data already comprising realistic noise and inherent drifts. The first 100 traces are used to create the initial reference model. We discover this model with a high dependency threshold to force many traces to be classified as non-conforming. In this setup, we set the sliding window size to 60, the lower bound to 35 and the offset to 0.005. The Gantt Chart in Fig. 6a marks all non-conforming traces, which are aggregated to a micro-cluster and the aggregation is derived by high difference between the LOF of each trace in Fig. 6b. By simulating the basic process as a recurring drift in otherwise noisy data, we cluster the BPIC 2015 as one micro-cluster. An important aspect, which has to be considered is, that real noisy traces in the basic process also get filtered. This leads to a more pure version of the BPIC 2015 as a micro-cluster.

7 Conclusion

With our novel *Dynamic Outlier Aggregation* we detect concept drifts of different extent. Trace clustering of non-conforming traces is used as a first step. Afterwards, we focus on the detection of recurring drifts as an example. Using a sliding window approach allows us to discover changes of different magnitude in the underlying process. Depending on the size of the traces these parameters provide control over the results as we exchange processing time for accuracy. In respect of a process like the spread of a disease, changes happen in a very different time span. For example, we are constantly in need of knowledge about mutating viruses like influenza or other more recent ones. Thus, the origin and classification of a newly mutated kind is of high interest for the development of a vaccine. Furthermore, if the sliding window is chosen with an appropriate size, and the underlying reference model consists of enough information of former viruses, even a rapid and significant change of a virus, is detected.

In future works, we will look at the application of this approach to single events instead of traces. The advantage of working on events instead of traces is the processing of incomplete traces with a higher rate. This involves the output of a micro-cluster probability for every emitted batch of connected events. In addition, we will be working on an appropriate solution to re-integrate clustered anomalies including a high amount of traces. This will also lead to great benefits for real-world applications, since unknown deviations are thoroughly analyzed, and the main process is purposefully extended.

References

1. Adriansyah, A., Sidorova, N., van Dongen, B.F.: Cost-based fitness in conformance checking. In: 2011 Eleventh International Conference on Application of Concurrency to System Design, pp. 57–66, June 2011
2. Ankerst, M., Breunig, M.M., Kriegel, H.P., Sander, J.: Optics: ordering points to identify the clustering structure. In: Proceedings of the 1999 ACM SIGMOD International Conference on Management of Data. SIGMOD 1999, pp. 49–60. Association for Computing Machinery, New York, NY, USA (1999)
3. Breunig, M.M., Kriegel, H.P., Ng, R.T., Sander, J.: LOF: identifying density-based local outliers. In: Proceedings of the 2000 ACM SIGMOD International Conference on Management of Data. SIGMOD 2000, pp. 93–104. Association for Computing Machinery, New York, NY, USA (2000)
4. Burattin, A.: PLG2: multiperspective process randomization with online and offline simulations. In: Azevedo, L., Cabanillas, C. (eds.) Proceedings of the BPM Demo Track Co-located with the 14th International Conference on Business Process Management, Rio de Janeiro, Brazil, vol. 1789, pp. 1–6 (2016)
5. Greco, G., Guzzo, A., Pontieri, L., Sacca, D.: Discovering expressive process models by clustering log traces. IEEE Trans. Knowl. Data Eng. **18**(8), 1010–1027 (2006)
6. Maaradji, A., Dumas, M., La Rosa, M., Ostovar, A.: Fast and accurate business process drift detection. In: Motahari-Nezhad, H.R., Recker, J., Weidlich, M. (eds.) BPM 2015. LNCS, vol. 9253, pp. 406–422. Springer, Cham (2015). https://doi.org/10.1007/978-3-319-23063-4_27
7. Ostovar, A., Maaradji, A., La Rosa, M., ter Hofstede, A.H.M., van Dongen, B.F.V.: Detecting drift from event streams of unpredictable business processes. In: Comyn-Wattiau, I., Tanaka, K., Song, I.-Y., Yamamoto, S., Saeki, M. (eds.) ER 2016. LNCS, vol. 9974, pp. 330–346. Springer, Cham (2016). https://doi.org/10.1007/978-3-319-46397-1_26
8. Richter, F., Seidl, T.: Tesseract: time-drifts in event streams using series of evolving rolling averages of completion times. Inf. Syst. **84**, November 2018
9. Richter, F., Zellner, L., Sontheim, J., Seidl, T.: Model-aware clustering of non-conforming traces. In: Panetto, H., Debruyne, C., Hepp, M., Lewis, D., Ardagna, C.A., Meersman, R. (eds.) On the Move to Meaningful Internet Systems: OTM 2019 Conferences, pp. 193–200. Springer International Publishing, Cham (2019)
10. Rozinat, A., Aalst, W.: Conformance checking of processes based on monitoring real behavior. Inf. Syst. **33**, 64–95 (2008)
11. van Zelst, S.J., van Dongen, B.F., van der Aalst, W.M.P.: Event stream-based process discovery using abstract representations. Knowl. Inf. Syst. **54**(2), 407–435 (2018)

OTOSO: Online Trace Ordering
for Structural Overviews

Florian Richter[(✉)], Andrea Maldonado, Ludwig Zellner, and Thomas Seidl

Ludwig-Maximilians-Universität München, Munich, Germany
{richter,maldonado,zellner,seidl}@dbs.ifi.lmu.de

Abstract. Identifying structures in data is an essential step to enhance insights and understand applications. Clusters and anomalies are the basic building blocks for those structures and occur in various types. Clusters vary in shape and density, while anomalies occur as single-point outliers, contextual or collective anomalies. In online applications, clusters even have a higher complexity. Besides static clusters, which represent a persistent structure throughout the whole data stream, many clusters are dynamic, tend to drift and are only observable in certain time frames. Here, we propose OTOSO, a monitoring tool based on OPTICS. OTOSO is an anytime structure visualizer, that plots representations for density-based trace clusters in process event streams. It identifies temporal deviation clusters and visualizes them as a time-dependent graph. Each node represents a cluster of traces by size and density. Edges yield information about merging and splitting trace clusters. The aim is to provide an on-demand overview over the temporal deviation structure during the process execution. Not only for online applications, but also for static datasets, our approach yields insights about temporally limited occurrences of trace clusters, which are difficult to detect using a global clustering approach.

Keywords: Trace clustering · Visualization · Operational support · Anytime clustering

1 Introduction

The ongoing digitalization of industries and social systems creates a strong demand for analysis tools to transform data into useful insights. Especially early warning systems for already known issues or still uncovered problems are highly requested. However, without a thorough exploration of the data, those systems cannot be developed, since we need to know what we are looking for beforehand.

In online applications, the time for analysis is always very precious and never sufficient. Therefore, an in-depth analysis has to be postponed, as interesting and promising aspects have been identified. A more shallow high-level analysis is more suitable as a time-efficient first exploration.

In the field of clustering, DBSCAN [4] is a prominent technique for density-based clustering. However, finding good parameters to generate results that

© Springer Nature Switzerland AG 2021
S. Leemans and H. Leopold (Eds.): ICPM 2020 Workshops, LNBIP 406, pp. 218–229, 2021.
https://doi.org/10.1007/978-3-030-72693-5_17

leverage the data into a given story is very tedious. Restarting clustering algorithms with arbitrary parameters is very different from output-driven experimentation. Therefore, OPTICS [1] was proposed as an extension, that offers a two-dimensional visualization for any multidimensional dataset. In OPTICS plots, the structure of the data is abstracted and parameters for density-based clusterings are visually determined.

In an online process mining application, we need to increase the abstraction level even further. Anytime variants for DBSCAN and OPTICS have been proposed in literature already. However, the structure of an online process is not covered by observing an event stream and building an up-to-date process model. The time perspective provides clusters with a further dimension of volatility.

In the context of processes, we differentiate between the major behavior, the baseline process, and process variants with deviation behavior. During the process execution, the baseline stays mostly static and rarely tends to shift its behavior. In contrast, variants often traverse different lifecycles dynamically. They emerge at certain points in time, merge with other variants, separate again and disappear eventually. In some time intervals, variants can remain inactive and reappear seasonally or randomly later.

In this work, we propose OTOSO, an on-demand temporal structure visualization of event streams. It is based on OPTICS and developed to cope with dynamic structure transformations. OTOSO collects trace data from an event stream as temporal deviation signatures, generates temporary OPTICS plots and aggregates their information into a graph plot. This plot shows relations between baseline and variant clusters. In a quick analysis, structure changes are identified visually. Each cluster is represented as a node of a specific size at a point in time. Relations between clusters are indicated by edges between nodes. The whole plot can then be interpreted as a map, that show the dynamic changes of the process during the event stream.

2   Related Work

To the best of our knowledge, there is no direct competitor that proposes an anytime structure overview for event streams. However, there are related methods that have to be mentioned here. There is a plethora of published techniques regarding process discovery, conformance checking and clustering. Due to space constraints, we only mention works that have a focus on temporal perspectives or which work on event streams.

Event stream monitoring emerges as a required preprocessing step for anytime analyses. Works in this field mainly prepare intermediate data for process discovery [3,7,9] and conformance checking [2,13]. These works propose methods to analyze event streams, which is the more complex task in comparison to trace stream analysis. The latter paradigm assumes that events are already grouped into traces, which is mostly a difficult requirement. In many practical scenarios, there is also a strong concurrency between cases. Cases can become inactive or are stopped without any further information. An approach based on event streams has to come up with a heuristic to deal with the lack of information.

In the area of temporal anomaly detection, Rogge et al. [12] analyzed interim times between events by applying kernel density estimation to identify outliers in the temporal perspective. In [11], the authors identify such outliers of event pairs online by using hashing for event collecting and applying z-scoring to define an in-control area for unsuspicious event relations. In [10], this idea is leveraged on the trace level to detect collective trace anomalies using density-based clustering on temporal deviation signatures. We adapt the presented clustering technique for OTOSO.

The area of event stream concept drift detection contains more established works. In [6], Hassani elaborated the idea of [7] to detect work-flow-based concept drifts using different structural metrics on process models. In [8], the authors present a technique to change forecasting models according to changed environments due to concept drifts. However, we are not aware of any concept drift detection approaches taking the temporal perspective into account.

3   Preliminaries

An event stream $S : \mathbb{N} \to \mathbb{N} \times A \times \mathbb{N}$ is a mapping from natural numbers to the event domain. Each event $e = (c, a, t)$ consists of an case identifier $c \in \mathbb{N}$, an activity label $a \in A$ and a timestamp $t \in \mathbb{N}$. For case identifiers from another domain, there is typically a canonical translation into the natural numbers. The same holds for the timestamps. In the following, we will not distinguish between cases and case identifiers, as the context provides enough clarification.

Since OTOSO can also be applied to event logs, we define an event log as a finite multiset of events. Although an event log is mostly grouped by case identifiers, for OTOSO the log should be sorted by timestamp. Additional event attributes like resources are ignored in this work, although they might enhance the results in future works.

Next, we call tuples of two activities $(a_1, a_2) \in A^2$ relations. A relation (a_1, a_2) exists in a case c, if there are two events $e_1 = (c, a_1, t_1)$ and $e_2 = (c, a_2, t_2)$ with $t_1 < t_2$. We canonically define the mean μ and variance σ of all time intervals in a finite set of cases for a certain relation. Using z-scoring as follows we account for the imbalance between all different relations and define the temporal deviation signature as:

$$TDS^c(a_1, a_2) = \begin{cases} \frac{|t_2 - t_1| - \mu_{(a_1, a_2)}}{\sigma_{(a_1, a_2)}} & , e_1 = (c, a_1, t_1), e_2 = (c, a_2, t_2) \in c \\ 0 & , \text{otherwise} \end{cases}$$

In case of multiple occurrences of a relation, the average z-score is used. A distance is a positive-definite function, that is symmetrical and fulfills the triangle inequality. In the following, we use the Euclidian distance due to its popularity and will not go into detail about other functions in this work. For the clustering step, we require a measure of density. Density is defined by a number of objects n in a certain area of radius ε. If an object, here a case represented by its temporal deviation signature, contains at least $MinPts$ many objects within a neighborhood $N_\varepsilon(c)$ of radius ε, this case is a core object. All cases within the neighborhood are at least border objects, if their neighborhood is not dense enough to be core objects themselves. All remaining cases are noise.

One of the most popular density-based methods is DBSCAN [4]. It selects objects and classifies them depending of their neighborhood as core, border or noise points. For a more in-depth description, we point to the corresponding work of Ester et al. A major drawback of DBSCAN is the difficulty to choose an appropriate value for the neighborhood distance ε. To overcome this issue, Ankerst et al. developed OPTICS [1]. Given $MinPts$, this method determines for each object its core distance, the minimal distance needed such that the ε-neighborhood contains $MinPts$ many objects. Derived from the core distance, the reachability distance between two objects is computed then. According to this distance, the processing order is depending on the nearest neighbor that has not been processed yet. This 2D reachability plot uses the ordering on the x-axis and the reachability distance on the y-axis. Since dense object clusters in the data space have low pairwise reachability distances, they are accumulated in the plot and clusters are identified as troughs in the reachability plot. Using a horizontal line as a density threshold, all troughs below this level represent clusters using the height as the according ε-value.

4 OTOSO

OPTICS visualizes the cluster structure of a static dataset. However, especially in process mining, process behaviors are dynamic and cluster structures are likely to change. To visualize not only a snapshot in a particular time frame, but the evolution of process variants and trace anomalies, we propose OTOSO, which is briefly summarized a visual time series of trace cluster structures. OTOSO consists of two phases. First, the event stream is observed and the necessary statistics are collected. By using a hashing data structure, the data is provided for the second module on-demand. At any point in time, the stored data can be queried as input for OPTICS to produce the current temporal cluster structure in the recent event stream. All those individual clustering snapshots are used to iteratively plot the clustering overview for the whole event stream.

4.1 Monitoring Temporal Deviations

OTOSO uses an event stream as input. In contrast to trace streams, the events have to be collected individually before case statistics can be extracted. A major problem of stream input is that we can never be sure that a case is still active. Therefore, we need an aging mechanism to discard old cases without the certainty that they are canceled or just paused and will be continued later.

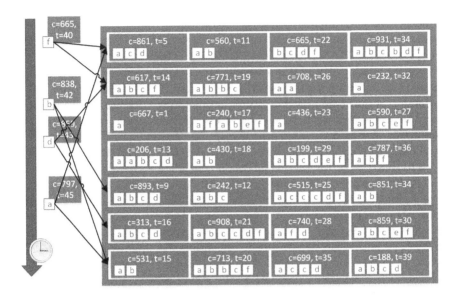

Fig. 1. Example hash table with $h = 7$ and $w = 4$. Each observed stream event has two potential rows to store it. Since the table is already full, either an event can be appended to its corresponding trace or an old trace has to be discarded. Do not be confused with the activity labels, since complete event information is stored.

We utilize Cuckoo-Hashing as it already provided a useful discarding technique for StrProM [7]. A hash table of height h is filled with case data, that is the last timestamp, the case identifier and all observed events. Two hash functions are applied on the case identifier to determine two potential hash table cells for each case. Instead of storing the case data directly in the hash table, we store a small and finite collection of cases in a cell. Technically, this width w of the table is implemented using arrays. Thus, the decaying factor can be adjusted without corrupting the operation complexity.

For each observed event, both hash functions are applied to identify all potential storage cells. If the case is already stored, it is updated by adding the event and setting the last-modified timestamp. In Fig. 1, the stream event in the top left corner belongs to case $c = 665$. A potential storage option is in the first hash table row. The case is already present in this row at the third position. We can update this cell by appending the event and updating the timestamp to $t = 40$. If the case has not been stored yet, we replace the case with the stored case, that has the oldest last-modified timestamp. The replaced case is the least recent one in this hash table cell. We try to insert it in the secondary position. Either, the secondary position has empty space, or we replace it again with the oldest case in this position. The procedure is recursively repeated until the secondary position has only more current entries and we discard the current item. Considering Fig. 1 again, the second stream event with $c = 838$ has storage options for row 5 and 7. Neither holds data for this case already. Using the first option, we attempt to store this new case in the fifth row, depending on the first of both hash functions. The oldest case here is case $c = 893$. We replace it with the new case and try to re-insert $c = 893$. The timestamp $t = 9$ tends to be already deprecated, however, since there are older entries in the table, there might be a chance to discard it after a series of replacements. This would be the case, if the alternative storage position is in row 1 or 3. Otherwise, the already existing timestamps in the remaining rows are newer and case $c = 893$ is discarded.

With this strategy, the hash table is always a finite representation of the recent cases, however some older behaviors potentially survive in the data structure since the swap operations regard the table only partially. Another drawback is that events in the beginning of cases are represented excessively, as the chance to be discarded is increased for longer cases. Alternatively, the length of the case can be included in the discarding mechanism. Nevertheless, this gives older cases an advantage to be kept stored, since smaller and recent cases are discarded. To the best of our knowledge, a perfectly fair sampling for event streams is still an open research topic, so we accept the drawbacks and discard by recency only.

Regarding hash functions, there are various ways to implement a set of two functions. Most programming languages provide at least one built-in hash function. To derive a second one, it is mostly sufficient to reverse the case identifier and use the same function again. Another strategy splits the identifier in two chunks and uses the hash value for the first and for the second chunk to determine both positions. We did not perform an in-depth evaluation on this topic here.

4.2   Structure Analysis

The hash table provides at most $h \cdot w$ many cases at any point in time t. The cases do not have to be completed already. The complete hash table is processed to extract the case data and to generate the z-scored temporal deviation signatures for all cases, which is used as input for OPTICS to cluster the traces. The output gives an impression on the recent temporal trace clustering structure. For the stream structure overview, we extract all clusters depending on the chosen density parameters $(\varepsilon, MinPts)$. For each cluster C, we create a node at position $x = t$ and $y = \sum_{c \in C} \text{core}_{\varepsilon, MinPts}(c)$ which is the occurrence time and aggregated cluster density. The size of each node is depending on the number of contained cases in the cluster respectively the number of cluster elements that are currently stored in the hash table.

In the basic variant, OTOSO connects cluster nodes if the distance between cluster centers is below the distance threshold Δ_{TDS} and the nodes occur in consecutive time slots. The extension connects cluster nodes of distant time slots. This allows to identify temporally limited clusters that reoccur after a period of inactivity. In Fig. 2, OTOSO is applied to an event stream producing OPTICS plots for various timestamps. At a tickrate of $10k$ events, further intermediate results are requested. For four of these intermediate queries, we show the OPTICS plots in the top row of the figure. For each OPTICS plot, a vertical slice in the OTOSO plot below is generated. Typically, a process produces one

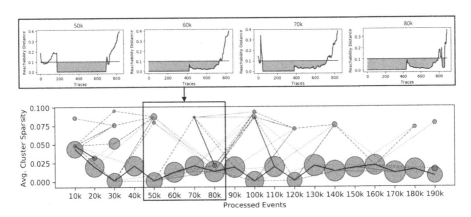

Fig. 2. OTOSO applied to an event stream. Each slice corresponds to a point in time and a hierarchy of clusterings at this timestamp.

major cluster containing cases that behave ordinary. These are the large spheres in each slice. For the $50k$ mark, besides the major cluster, two variants of low density are active. Both are related to previous queries, but disappear for the next two queries. Solid lines indicate a strong similarity between clusters of consecutive clusters. Dashed lines indicate similarity between slices over a larger timeframe. Here, we only include lines connecting slices within a timeframe of $30k$ events. In slice $60k$, all variants disappear. In $70k$, a small variant emerges. It has some similarity with the major cluster in $50k$, but no connection to the major cluster in $60k$. Hence, the temporal deviation profile first covered this deviation, but the variant did not occur in the succeeding process window. Interestingly, the small cluster in slice $80k$ grows slightly in size, but drastically in density. Regarding the solid line, we recognize a close similarity between both clusters, so their behavior represented by the temporal deviation signature is also similar.

This visualization allows to detect different structural changes in an event stream. Lifecycles of emerging and vanishing variants can be followed as illustrated before. The connections of a cluster node indicates, whether this variant has disappeared or has been inactive for some time. If a node emerges without initial connection, the corresponding variant starts suddenly. Otherwise, a connected new node hints towards a gradually emerging variant. These mechanisms are related to types of concept drift, however it is difficult to clearly label the effects according to sudden, gradual and incremental drifts due to the complexity of an event stream. Many activities and therefore activity relations are included in the temporal deviation profile. Nevertheless, the OTOSO plot gives an overview over the whole structure. A sudden drift, for instance, will likely affect a small number of traces and will maybe only affect some activities. The abstraction level of the visualization is to high to register concept drifts with a high confidence, except they appear as large-scale effects.

5   Evaluation

In the following, we evaluate the correlation between the size of the hash table and the currency of the collected event data. Afterwards, we show the benefits of applying OTOSO in comparison to using density-based clustering on the data as a static data chunk. Finally, we build a stream of a sequence of event logs to show the capability to detect the transitions between dissimilar event stream sections. We uploaded OTOSO into a GitHub project[1], thereby the experiments can be reproduced.

5.1   Datasets

Working with pure synthetic datasets causes some issues concerning the detection simplicity of anomalies or clusters in the data. We need datasets that are realistic, because synthetic datasets allow too much freedom and often are unfairly beneficial to the method's evaluation. Therefore, we utilize the BPI challenge datasets from 2015[2] and 2017[3], in the following abbreviated as BPIC15 and BPIC17. BPIC15 contains data of building permit applications over four years in five Dutch municipalities. Five partitions show the process of each municipality individually. Each sublog contains about a thousand cases. The challenge of this dataset lays in its about 400 activities and its resulting complexity from the large number of potential relations. The publications regarding this challenge show, that there is a high similarity between sublog 1, 2 and 5 while sublog 3 and 4 represent a slightly different behavior. In BPIC17, a loan application process of a Dutch financial institute over one year is logged. The offer log contains only a subset of 24 offer related activities. 128985 events are recorded in 42995 cases. Due to its larger size, we are able to simulate an online observation of the whole fiscal year.

5.2   Hash Table Size

We use BPIC17 to investigate the influence of the hash table size on the currency of the data. Each event log is transformed into an event stream. Observing the stream event by event, each recent event is inserted into the hash table. Every 1000 events, we determine the average time difference to the current event timestamp. In Fig. 3, we show the results. Starting with a small hash table, which only contains 1000 cases, we compare three different dimensions for the table. In the first case, a table of height 10 with 100 buckets in each position is used. The second hash table has height 100 and width 10, while the third is a one-dimensional table of height 1000. The average recency is below 10 days. Towards the end, no new cases are starting, so no old cases are discarded and the table gets slightly outdated.

---

[1] https://github.com/Skarvir/OTOSO.

[2] https://doi.org/10.4121/uuid:31a308ef-c844-48da-948c-305d167a0ec1.

[3] https://doi.org/10.4121/uuid:5f3067df-f10b-45da-b98b-86ae4c7a310b.

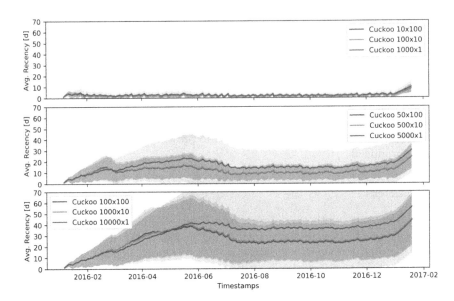

Fig. 3. Avg. recency and standard deviation is given for nine Cuckoo hash tables with different dimensions as *height* × *width*.

The second plot shows three hash tables of size 5000 having analogous changes regarding their dimensions. Due to the higher capacity, more cases can be stored and the table contains more obsolete items. Storing more items leads to a more stable clustering and following techniques are affected by noise or short-term outliers. There is no clear method to determine the best recency and the corresponding table size, since this is completely depending on the user-defined time window and the arrival frequency of events and cases. Finally, the application is also an important factor, since the detection of point-wise anomalies benefits from higher currency while the detection of long-term structures requires data with high stability. However, the important point we want to highlight is the advantage of using a two-dimensional hash table. The width allows shorter rehash cycles, which is already shown in [5,7]. The new insight here is the greater recency for small numbers of buckets in each position. Already in the second plot, but much clearer in the third one with a hash table of size 10000, the one-dimensional hash table has a delay of about 40 days, while both variants with few buckets have smaller temporal shifts. The difference between using 10 or 100 buckets is rather marginal. Therefore, we recommend using small numbers of buckets, since the iteration over a large list of buckets is more time-consuming than rehashing at another position.

5.3   Static Clustering vs. Dynamic Clustering

The BPIC17 dataset contains a significant cluster with deviating temporal behavior, that contains accepted offers with a delay in its execution. In Fig. 4a, we show the result of OPTICS applied to the whole event log using the temporal deviation signatures as a representation. Using a neighborhood size of 0.5, two major clusters are yielded. The largest one contains the majority of cases and represents the baseline of this process. The second largest one is shown in OPTICS as a thinner and deep trough on the right side. Since this method yields a static overview over the temporal clustering structure, we would assume that the cluster is omnipresent during the complete event stream.

In Fig. 4c, the final OTOSO plot is given. After all events in the stream have been processed, the clusters are nodes with radii according to the number of contained cases. The height is determined by their density. Lines indicate a strong similarity between consecutive clusters. Thus, by following a line we observe the lifecycle of a specific cluster.

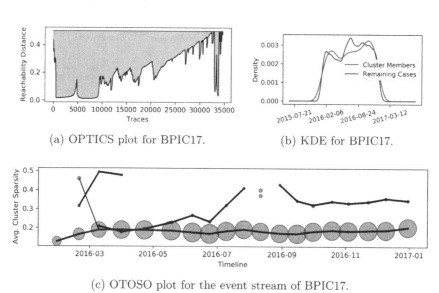

(a) OPTICS plot for BPIC17.

(b) KDE for BPIC17.

(c) OTOSO plot for the event stream of BPIC17.

Fig. 4. OPTICS and OTOSO applied on the BPIC17 datalog. $MinPts = 100$ and results are yielded each $10k$ events.

In the beginning, the results are not reliable. Many cases have been collected only partially yet. As a rule of thumb, we recommend to neglect insights from the first k cases if the hash table has size $k = h \cdot w$. Hence, starting with April, a baseline of large clusters has been emerged and retains an almost constant size for the remaining stream. More interesting is the other line above. It indicates a much smaller cluster, that still has a high density. During August the cluster vanishes but returns again in September. Instead, two new and dissimilar clusters emerge for this short period and vanish afterwards again. To show what OTOSO has highlighted there, we extract all cases contained in the previously mentioned deviating cluster. This set of cases corresponds to the thin and deep trough in Fig. 4a. For this cluster and also for the remaining cases, we plotted the starting times as a kernel density estimation in Fig. 4b. Here, we observe a peak in starting cases in August. The rising number of arriving cases, which do not belong to the variant cluster, shifts more weight towards the baseline cluster and the two new variants. The resulting loss in density for our previous variant cluster leads to its disappearance for one observation tick. While it is possible to detect such effects with static methods, this analysis is quite tedious. Besides, we already knew what we were looking for. OTOSO highlights this anomaly during the online observation of the event stream. In applications, that require short reaction times, observing the OTOSO visualization provides a very quick indication for an abnormal behavior.

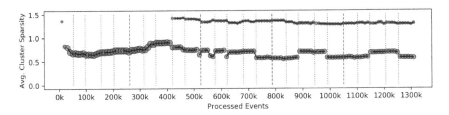

Fig. 5. OTOSO applied to a five-fold concatenation of all five BPIC15 sublogs. $MinPts = 100$ and an intermediate result is demanded every $10k$ events.

5.4   OTOSO on Event Stream with Concept Drifts

Finally, we use the BPIC15 dataset to how concept drifts affect the structural overview. The dataset is quite small, so we concatenate all five sublogs into one larger event log. Further, we concatenated this event log 5 times with itself to create an even larger log with five segments or 25 sublogs. This event log is then transformed into an event stream.

In Fig. 5, the OTOSO plot is given after processing the event stream. As discussed before, we neglect the results from the first two segments of the stream. After $500k$ events have been processed, the hash table is filled sufficiently and the structure of the data starts to appear. The red lines indicate the border points when a sublog ends and a new one starts. Especially in the last two segments, there is a significant similarity in BPIC15 between sublog 1, 2 and 5 and also between 3 and 4. The black similarity line indicates this relation. There is a much sparser and small cluster above. We do not have expert knowledge to verify or explain its meaning. On the one hand, it is possible to neglect it due to its sparsity. On the other hand, this cluster exists in all sublogs and it shows a strong similarity. In reality, we would recommend a thorough examination, but due to the lack of expert knowledge, we have to dispense with further speculations.

6   Conclusion

In a world of continuously emerging digitalization, it is very important to get preliminary insights early and with a high level of abstraction. OTOSO provides an online overview over structures in an event stream. Emerging or vanishing clusters are visually identified and lifecycles of those structures are tracked.

Although some structural dimensions are monitored like density, size and similarity of clusters, process data contains more information, which can be used to augment the structural overview plot. Also, the plot depends on suitable user-defined parameters. Estimating good parameters is a very difficult task. Thus, and because a data stream cannot be replayed, it is beneficial to enable on-demand parameter adaptations while results are visualized.

References

1. Ankerst, M., Breunig, M.M., Kriegel, H.P., Sander, J.: Optics: ordering points to identify the clustering structure. ACM SIGMOD Rec. **28**(2), 49–60 (1999)
2. Burattin, A., Carmona, J.: A framework for online conformance checking. In: Teniente, E., Weidlich, M. (eds.) BPM 2017. LNBIP, vol. 308, pp. 165–177. Springer, Cham (2018). https://doi.org/10.1007/978-3-319-74030-0_12
3. Burattin, A., Sperduti, A., van der Aalst, W.M.: Control-flow discovery from event streams. In: 2014 IEEE Congress on Evolutionary Computation (CEC), pp. 2420–2427. IEEE (2014)
4. Ester, M., Kriegel, H.P., Sander, J., Xu, X., et al.: A density-based algorithm for discovering clusters in large spatial databases with noise. In: KDD, vol. 96, pp. 226–231 (1996)
5. Fan, B., Andersen, D.G., Kaminsky, M., Mitzenmacher, M.D.: Cuckoo filter: practically better than bloom. In: Proceedings of the 10th ACM International on Conference on emerging Networking Experiments and Technologies, pp. 75–88 (2014)
6. Hassani, M.: Concept drift detection of event streams using an adaptive window. In: ECMS, pp. 230–239 (2019)
7. Hassani, M., Siccha, S., Richter, F., Seidl, T.: Efficient process discovery from event streams using sequential pattern mining. In: 2015 IEEE Symposium Series on Computational Intelligence, pp. 1366–1373. IEEE (2015)
8. Maisenbacher, M., Weidlich, M.: Handling concept drift in predictive process monitoring. In: 2017 IEEE International Conference on Services Computing (SCC), pp. 1–8. IEEE (2017)
9. Navarin, N., Cambiaso, M., Burattin, A., Maggi, F.M., Oneto, L., Sperduti, A.: Towards online discovery of data-aware declarative process models from event streams. In: 2020 International Joint Conference on Neural Networks. IEEE (2020)
10. Richter, F., Lu, Y., Sontheim, J., Zellner, L., Seidl, T.: TOAD: trace ordering for anomaly detection. In: 2020 International Conference on Process Mining (ICPM), pp. 1–8. IEEE (2020)
11. Richter, F., Seidl, T.: Looking into the tesseract: time-drifts in event streams using series of evolving rolling averages of completion times. Inf. Syst. **84**, 265–282 (2019)
12. Rogge-Solti, A., Kasneci, G.: Temporal anomaly detection in business processes. In: Sadiq, S., Soffer, P., Völzer, H. (eds.) BPM 2014. LNCS, vol. 8659, pp. 234–249. Springer, Cham (2014). https://doi.org/10.1007/978-3-319-10172-9_15
13. van Zelst, S.J., Bolt, A., Hassani, M., van Dongen, B.F., van der Aalst, W.M.: Online conformance checking: relating event streams to process models using prefix-alignments. Int. J. Data Sci. Anal. **8**(3), 269–284 (2019)

Performance Skyline: Inferring Process Performance Models from Interval Events

Andrea Maldonado[⊠], Janina Sontheim, Florian Richter, and Thomas Seidl

Ludwig-Maximilians-Universität München, Munich, Germany
{maldonado,sontheim,richter,seidl}@dbs.ifi.lmu.de

Abstract. Performance mining from event logs is a central task in managing and optimizing business processes. Established analysis techniques work with a single timestamp per event only. However, when available, time interval information enables proper analysis of the duration of individual activities as well as the overall execution runtime. Our novel approach, performance skyline, considers extended events, including start and end timestamps in log files, aiming at the discovery of events that are crucial to the overall duration of real process executions. As first contribution, our method gains a geometrical process representation for traces with interval events by using interval-based methods from sequence pattern mining and performance analysis. Secondly, we introduce the performance skyline, which discovers dominating events considering a given heuristic in this case, event duration. As a third contribution, we propose three techniques for statistical analysis of performance skylines and process trace sets, enabling more accurate process discovery, conformance checking, and process enhancement. Experiments on real event logs demonstrate that our contributions are highly suitable for detecting and analyzing the dominant events of a process.

Keywords: Interval events · Performance analysis · Process mining · Dominant duration path · Skyline operator

1 Introduction

To plan a process optimally, prevent mistakes as well as to answer questions about its performance, we need to know the process. The more substantial the knowledge, the better informed decisions users can take about their plan of action. Over a century ago Gantt charts [1] were introduced to schedule work according to resources in the manufacture industry. Since then data-centric process mining models aid to understand constantly changing processes by considering both, their prescription and the posterior description of run instances, in multiple fields. Often certain tasks in a process last long, without the user knowing whether that duration is expected or not. By taking performance indicators of the time dimension, such as the lead-, service- and waiting time into account, performance analysis examines event data over time [2].

© Springer Nature Switzerland AG 2021
S. Leemans and H. Leopold (Eds.): ICPM 2020 Workshops, LNBIP 406, pp. 230–242, 2021.
https://doi.org/10.1007/978-3-030-72693-5_18

This analysis enables businesses to discover performance patterns, optimize processes as well as to identify and prevent mistakes in them. In many cases, analyzing processes that contain big amounts of events often includes computational and visual overhead for the user, specially when only a subset of them might be interesting to asses the user's question. Furthermore including additional performance information for more substantial knowledge may worsen the visual charge. To alleviate this, we select a subset of events that may be specially interesting using the skyline operator [3], considering that events of dominant duration are crucial to process performance, and moreover optimization, resources usage and task prioritization.

Consider the following example: After a vacation visit to your favorite city, you write a review about the hotel you stayed in. This review is added to a platform's collection, from which hotel reputation companies forward customer feedback to hotels. Broadly speaking, this process ingests customer reviews from multiple sources, does multiple transformations, aggregates them and stores the result at a given location to provide hotels with an accurate rating. Since ratings may strongly influence guests booking decisions, an up-to-date result is essential to a host's reputation in the hospitality market. If all events of this process are executed sequentially, they all directly contribute to the overall duration of the process. For this reason it is often shorten by executing events parallely. Knowing events inter dependencies, service- and waiting times is advantageous to choose which activities should run in parallel. Independent events or events series of similar length can be run in parallel to make the process more efficient. Improving a process that already utilizes parallel event executions requires performance analysis of previously ran instances on a activity level. Focusing and speeding up activities that last considerably longer than others might have a higher impact on the overall duration than doing so on the ones that already perform relatively well on the same trace. Thus identifying these activities is key, specially for traces with a high amount of events and high service time deviation. Our performance skyline approach highlights events that dominate others in some given metric on one trace, in this occasion the metric is the service duration.

The next section, presents state-of-the art methods to analyze process performance as the performance spectrum miner and the critical path method. Subsequently, Sect. 3 presents interval events, the skyline operator and the geometric interval representation. These concepts form the basis of our methods. Section 4 and Sect. 5 introduces our contributions: The geometrical process representation, the performance skyline and three statistical analysis techniques. Following then, Sect. 6 demonstrates how our approaches work beyond theory by experimenting with them on real process logs from TrustYou GmbH, a German Guest Feedback and Hotel Reputation Software company. Lastly Sect. 7 closes listing achievements and further expansion possibilities as future work on this topic. Code and real log data for replicating our experiments are available on the open source project [https://github.com/andreamalhera/performanceskyline].

2   Related Work

Performance models regard temporal aspects of processes. The performance spectrum miner [2] maps all observed flows between activities together regarding their performance over time . Bringing a temporal perspective into process analysis, performance spectrum enables reliable pattern recognition for batching behavior. Nevertheless, using only one timestamp the performance spectrum forces models to overlook some aspects like waiting time, actual event duration and actual trace duration i.a. service time. Not extracting this knowledge restricts models and pattern detection methods derived from it. Other models, as for example PROM's dotted chart [4], use two-dimensional space projections with start time in the horizontal axis and case ids in the vertical axis to describe processes. Nevertheless it is also limited by using a single timestamp. Event intervals [5] have proven useful to extract insights about the idle periods of processes even from events of a single timestamp, but are limited by the assumption that all tasks occur sequentially.

When logs provide additional time interval information, performance insight may be mined using e.g. interval events. Heuristics miner for time intervals [6,7] uses interval events to mine the dependency relations among activities in a process more precisely. Similarly to transactional events [8] with *transaction types* like *start* and *complete*, interval events have already been defined by others [2,6,7,9,10], but slightly differently than in this paper. In the first one instance of an activity, which starts and ends, is described as two transactional events of the same activity. Other definitions of interval events implicitly assume that an event ends exactly when the next one begins. Another kind of multi-timestamped events are queue events in a single station queue log [11], which are used to predict delays in service processes online and thus improve customer experience. Being highly adapted to queues, queue events are not suitable to answer other performance questions nor to handle more complex processes, containing non-sequential activities as well. Disregarding complexity and in other cases provided multiple timestamps, hinders to identify gaps between two events.

Flow analysis [12] is a family of control-flow model based techniques to estimate the overall performance of a process given some knowledge about the performance of its tasks. These promising techniques can be extended [13] to also mine performance relations between a set of events and the overall process. Moreover another possible extension could focus on finer granularity for the dominating tasks regarding a given heuristics. Flow analysis is also restricted by process complexity, working properly for control-flow based models and furthermore block-structured models [8] only. Even when one disposes of logs containing interval events, it is a challenge to find suitable ways to integrate additional timestamps for events in a model including as much information as necessary but without generating an visual overload. Comparably the non-control-flow based model, multi-channel performance spectrum for predictive monitoring, [10] classifies cases using multiple performance-related dimensions, yet intra-case features of individual cases remain undiscovered due to its relatively coarse granularity and inter-case design.

The critical path method (CPM) [12,14–16] also filters interesting points from a process. The critical path comprehends the longest series of dependent events required from start to end, which add up to the overall trace duration [16]. Applying the CPM, identifies the critical path in petri-nets and precedence network process models [17] by noting *estimated/early* and *late*, start time and end time for each activity. Broadly this methods demonstrates how combining state-of-the-art process mining with other approaches results in suitable opportunities. Even though, real processes often present deviation in their duration. Since this approach only considers constant duration of activities across traces and focuses on precedence rather than process performance, analysis could be enriched by performance mining and statistical inspection of time deviating events.

3   Preliminaries

3.1   Interval Events

A process instance can be split into events. An **event** $e = (c, a, t) \in \mathbb{N} \times A \times \mathbb{N}$ is as an tuple consisting of case id c, an activity id a, and a timestamp t. If any two events contain the same case id, they belong to the same trace. A trace is an instance of a process, containing multiple events. An **interval event** $e = (c, a, t^+, t^-) \in \mathbb{N} \times A \times \mathbb{N} \times \mathbb{N}$ is as an tuple consisting of case id c, an activity id a, a start timestamp t^+ and an end timestamp t^-. The duration of an event e_i its duration can be computed as $(\pi_{t^-}(e_i) - \pi_{t^+}(e_i))$, where $\pi_k(e_i)$ is the value for key k in event e_i.

3.2   Skyline Operator

In the field of database queries, the Skyline Operator [3] filters out a set of interesting points from a potentially large set of data points. Whether or not a point is interesting depends on metrics given by the user and if a point is not dominated by any other. It can be used for example to find interesting hotel matches, meaning all hotels that are not worse than any other hotel in nearness to the beach and price. We call the line connecting the set of dominating points of interest: the dominant path. To the best of our knowledge skyline operators have not been used in the field of process mining before. Mostly because often control-flow models are used to analyze a process. For the purpose of performance analysis in processes we consider the dominant duration path of a trace, which comprehends the series of events, which last the longest in a trace or process from start to end and add up to the overall trace duration, similar to dominant points of interest presented in [3]. Consider the following example: A trace is composed of two events, A and B, both starting at t_1. Additionally, A ends at t_3, and B ends at t_2 with $t_3 > t_2$. Thus, the overall duration of the trace is $(t_3 - t_1)$ and only A is part of the dominant path. Consequently, to decrease the process duration based on this trace, event A needs to be sped up. Decreasing the duration of only event B would not improve the overall performance, it is not part of the dominant path and its duration $(t_2 - t_1)$ is lower than $(t_3 - t_1)$. For our approach the skyline operator was implemented using Allen's interval terms [18]: The performance skyline includes all events without *during* relationship to any other event in the same trace.

3.3 Geometric Interval Representation

In a geometric interval representation [20] temporal intervals are projected to points in a two-dimensional space, as in Fig. 1. At the top of both subfigures of Fig. 1, a schematic representation of a series containing 4 events A, B, C and D is depicted. Below it events are projected as points in a two-dimensional space. Using start and end time as axes. E.g. event A starts at t_1 and finishes at t_3; event B starts at t_2 and finishes at t_7; event C starts at t_4 and ends at t_5; and event D starts at t_6 and finishes at t_7, lasting for $(t_7 - t_6)$ time units.

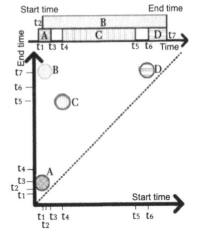

Fig. 1. Schematic representation of four example intervals depicted in the geometric interval representation with start time and end time as axes. Ad.

4 Performance Models for Interval Events

4.1 Geometrical Process Representation

Process interval events as defined previously in Sect. 3.1 contain temporal intervals, thus can be visualized in the geometric interval representation. Figure 2 describes an example trace, which entails four events similar to Fig. 1 with their corresponding start; end timestamps and a different activity each. Events A, B, C, and D are connected through a line marking they correspond to the same trace. Event B lasts the longest in this trace, since it is furthest away from the zero-duration diagonal. In contrast, the event with activity A appears to have the lowest duration. Events on the same vertical, start at the same time; those sharing one horizontal position end at the same time and if a line passing through two events is parallel to the zero-duration diagonal, they last the same.

4.2 Performance Skyline

The **performance skyline** ρ_c of a trace σ_c is the largest sub sequence of events $\rho_c = (e_1, \cdots, e_i \ldots, e_j, \ldots, e_n)$, where $\rho_c \subseteq \sigma_c$, and $\pi_{t-}(e_i) \leq \pi_{t-}(e_j)$ for all $1 \leq i \leq j \leq n$. Additionally, because a trace of interval events is ordered based on the start timestamps $\pi_{t+}(e_i) \leq \pi_{t+}(e_j) \Leftrightarrow i \leq j$. For the reason that events in the performance skyline in the case of start time and end time as axis are those which directly contribute to the overall duration of a process at any given point, they are equivalent to the set of events on the dominant path, previously presented in Sect. 3. If $e_i \in \rho_c$, there is no other event, which starts before $\pi_{t+}(e_i)$ and ends after $\pi_{t-}(e_i)$. The performance skyline of the trace in Fig. 2 is depicted in Fig. 3. In this example $\rho_{T_1} = \{A, B, D\}$ compose the performance skyline. C does not belong to the performance skyline, because even though $\{C, B\} \in \sigma_{T_1}$ and $\pi_{t-}(C) = t_5 \leq t_7 = \pi_{t-}(B)$, also $\pi_{t+}(C) = t_4 \nleq t_2 = \pi_{t+}(B)$.

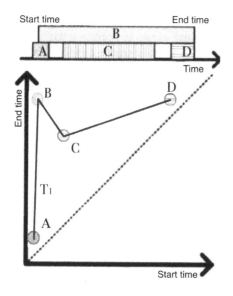

Fig. 2. Schematic representation of one example trace with four activities as intervals in [19]

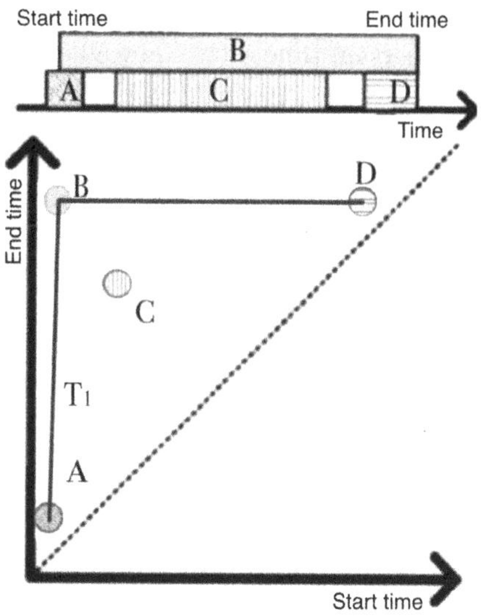

Fig. 3. Performance skyline for example trace T_1 in Fig. 2. Ad. [19].

To understand a process we analyze a representative set of traces. To include and compare events from several traces of the same process with each other, these are aligned to the left by subtracting the start time value $\pi_t^+(e_1)$ to all timestamps $\pi_t^+(e_i), \pi_t^-(e_i)$ for all depicted events in the same trace. This way for every trace ρ_c the first aligned event e_1' starts at $\pi_t^+(e_1') = \pi_t^+(e_1) - \pi_t^+(e_1) = 0$ and ends at $\pi_t^-(e_1') = \pi_t^-(e_1) - \pi_t^+(e_1)$. Any other aligned event e_i' starts and ends relatively to it, computed $\pi_t^+(e_i') = \pi_t^+(e_i) - \pi_t^+(e_1)$

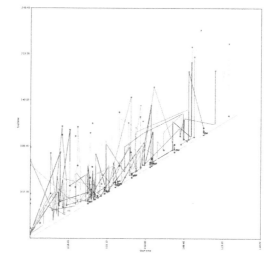

Fig. 4. Real log snippet from industry process with 390 event points of 30 activities, with corresponding point colors, from 13 traces represented by different line colors. Horizontal axis shows starting times within the first 01:06 h and vertical axis shows ending times between 00:05 h–02:00 h. (Color figure online)

and $\pi_t^-(e_i') = \pi_t^-(e_i) - \pi_t^+(e_1)$ correspondingly. Thus events that usually start at a certain time after the whole process starts are easier to compare with each other. From here on in this paper, events in all presented traces are aligned.

Figure 4 shows a sample of a real log containing 390 events describing 30 activities on 13 traces of a process. Points in similar positions and same color represent similar activities. Detected patterns between traces identifies behavior that could be expected from future traces of the same process. In this case similarities between traces can be observed in the peak often on the last event on most traces depicted as a blue point. Even so, contemplating all events of several traces simultaneously challenges recognizably in the visualization and burdens performance with computational overhead. Selecting only a subset of interesting events, e.g. those on the dominant path, to form a baseline of expected behavior for a trace set eases its comparison between traces as well as with future ones. For this purpose statistical analysis techniques will be introduced next.

5   Statistical Analysis Techniques

Methods in this section generalize the process analysis by considering multiple traces in the same performance skyline model and furthermore depicting stochastic summaries of these traces in the plot. Results visualized in this section originate from real logs.

5.1   Average Trace Skyline

The **average trace** $\bar{\sigma} = \{\bar{e}_{a_1}, \ldots, \bar{e}_{a_i}, \ldots, \bar{e}_{a_n}\}$ of a process trace set is the resulting trace of averaging all events start and end times for each activity on the trace set, i.e. for every activity the start time results in $\pi_t^+(\bar{e}_{a_i}) = \frac{1}{m}\sum_{j=1}^m \pi_t^+(e_j)$ and the end time in $\pi_t^-(\bar{e}_{a_i}) = \frac{1}{m}\sum_{j=1}^m \pi_t^-(e_j)$, where $\pi_a(e_j) = a_i$. An average trace is suitable as a comparable expectation for inquires that involve all activities of a trace set. It eases the view to gain representative knowledge about all activities start and end times as well as the relationships between consecutive activities. Figure 5 shows the average trace of the depicted trace set in Fig. 4. Comparing the form of the average trace to the trace set's, e.g. peaks in the 4th and last activities of the average trace with the ones of similar color in the trace set, a common behavior of the underlying process is revealed.

Moreover combining the average trace and the performance skyline in the *average trace skyline* results in a description of a representative dominant path for multiple traces of a process. For further details on the dominant path, see Sect. 3. The **average trace skyline** of a process is the performance skyline of the average trace. An average trace skyline is suitable to evaluate performance expectations for a trace set because it facilitates gaining knowledge about activities that are often part of the dominant path and their relations to each other. Figure 6 shows the average trace skyline of the trace set in Fig. 4, which is the performance skyline of the trace in Fig. 5. This average trace skyline highlights five out of thirty activities, which are part of the dominant path. With this information, the user can focus on those sparing them of unnecessary visual and computational overhead.

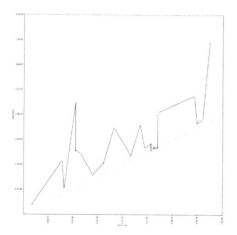

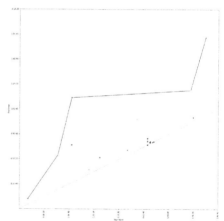

Fig. 5. Average trace from Fig. 4 with 30 activities. Horizontal axis shows starting times within the first 01:06 h and vertical axis shows ending times between 00:05 h–02:00 h.

Fig. 6. Performance skyline with five activities on the dominant path marked by line. Horizontal axis shows starting times within the first 01:06 h and vertical axis shows ending times between 00:05 h –02:00 h.

5.2 Average Skyline Trace

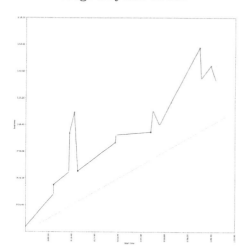

The **average skyline trace** $\bar{\rho} = \{\bar{e}_1, \ldots, \bar{e}_i, \ldots, \bar{e}_n\}$ of a process is the resulting skyline from averaging activities in performance skylines of all traces. $\pi_t^+(\bar{e}_{a_i}) = \frac{1}{m}\sum_{j=1}^m \pi_t^+(e_j)$ and the end time in $\pi_t^-(\bar{e}_{a_i}) = \frac{1}{m}\sum_{j=1}^m \pi_t^-(e_j)$, where $\pi_a(e_j) = a_i$ and $e_j \in \rho_k$, with ρ_k being a performance skyline from a trace in the trace set. An average skyline trace is suitable to consider performance aspects for a given trace set, because it offers knowledge about activities that might be part of any trace's dominant path, here dominant duration path, and their relations to each other. Consequently it is more inclusive regarding what activities to include than the average trace skyline. Figure 7

Fig. 7. Average skyline trace with 30 activities. Horizontal axis shows starting times within the first 01:06 h and vertical axis shows ending times between 00:05 h–02:00 h.

shows the average of the performance skylines of traces in Fig. 4. This average trace skyline highlights seventeen out of thirty activities, more than the average trace skyline in Fig. 6. With this information, the user can focus on duration dominant activities for any of the traces. This is useful in case traces contain a diverse set of duration dominant activities between them.

5.3   Expected Skyline Activity Set

Having computed performance skylines of a process trace set $R_\Sigma = \{\rho_1, \ldots, \rho_j, \ldots, \rho_n\}$, the **expected skyline activity set** R_A of that process is the set of activities that have a probability of appearance on performance skylines that is equal or higher than a given threshold t_A. The appearance probability for an activity a_k is computed as follows:

$$P(\pi_a(e_i) = a_k) = \frac{|\pi_a(e_i) = a_k|}{|e_i \in \rho_j|}, where\ e_i \in \rho_i$$

Furthermore $t_{a_k} \leq P(\pi_a(e_i) = a_k) \Rightarrow a_k \in R_\Sigma$. After computing appearance probabilities, an expected threshold value shall be chosen to define the expected skyline activity set of the presented process. The higher the chosen threshold value, the fewer activities will be part of the expected skyline activity set. Analogously the lower the chosen threshold value, the more activities will be part of the expected skyline activity set. An expected skyline activity set is suitable to analyze performance of a process across events and traces; as well as provide estimation and knowledge about deviation of events in a process dominant path. Figure 8 shows a bar chart, where each bar represents one of seventeen activities in the data set and their length corresponding appearance probabilities in R_Σ. Choosing e.g. $t_A = 60\%$ results in $|R_\Sigma| = 4$, containing following activities: `UpdateCrawlStartTask`, `SplitCrawlInputTask`, `DumpTask(target_filename=review_2yold)(chunk=prep)(sql_filename=review)` and `ConvertDumpTask(filename=review_2017_3)(chunk=prep)(sql_filename=review)`.

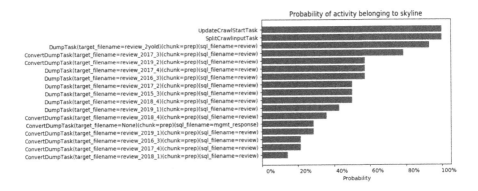

Fig. 8. Performance skyline activity set with corresponding probabilities of activity belonging to the performance skyline.

6   Discussion

Considering that performance skyline explicitly includes multiple traces, it offers a suitable method for performance analysis and thus provides the user to take more informed scheduling and planing decisions than the one offered by combining critical path method and process discovery models [17]. Furthermore introduced interval events bearing multiple timestamps extend performance spectrum [2] analysis techniques by including bi-dimensional information about relations between events within one trace or process while still offering a slim visualization. This way both time stamps can be used to take waiting times and events' duration into consideration.

The experimental dataset comprehends three months of logs for a process called `daily` at TrustYou GmbH, a German Guest Feedback and Hotel Reputation Software company. Broadly explained this process ingests customer reviews from multiple sources, does multiple transformations, aggregates them and stores the result at a given location. Being a data process, only computing resources are involved and thus control-flow deviations such as order of activities execution, do not vary without showing performance deviations on interdependent activities as well. The data collection compounds 62,074 interval events spread among 50 traces. It contains 261 different values for activity id. A trace has on average 1238 events. Most activities appear mostly once in a trace, except for six of them which correspond to a few hundred events.

Additionally taking only a subset of dominant activities of interest into account to describe a whole process reduces computational time an eases the search for potential improvement and answers to performance questions that might only concern a certain metric. For example finding events of dominant duration is useful when searching long lasting single activities that can be optimized while also regarding their order of execution and parallelization, which might be inflexible due to their inter-dependencies. For the experiment trace set an average of $5,22\%$ of its events are part of the performance skyline. In order to include the performance knowledge of multiple traces at once, statistical analysis techniques select duration dominant activities that concern any, most or some of the traces. Different techniques serve multiple purposes and data and show advantages to solve various matters: First, the average trace skyline includes only activities that belong to the dominant path. These are highly suited for comparing the average behavior of activities with each other. With this information independent non-dominant events can be paralleled to duration dominant events and thus performance of the whole process can be optimized. Nevertheless being very exclusive with average duration activities, which means that if there is an activity A, which due to high performance variance often appears on the performance skyline for some traces, but which is on average dominated by another activity B. Activity A will not be part of the average trace skyline. Furthermore, groups of traces that have different sets of activities on their skyline might be representative, and conforming, without appearing most often. For this reason the order of steps, averaging and computing a skyline, for a trace set leads to different expectation skylines. Second, the average skyline trace includes every activity on any skyline in the set to the average skyline trace. Even if this technique is advantageous to compare all duration dominant activities, it can produce an expectation skyline trace that is significantly sensitive to outliers because it includes activities that might only be part of a skyline computed from some traces or even the average trace. Lastly, as a trade off between only including most frequent dominant path's activities and all activities from any dominant path disregarding their relevance, an expected skyline activity set provides estimation and knowledge about performance deviation of events in a process dominant path, which can be used e.g. to identify trace anomalies. All of our presented statistical analysis techniques enable more informed decisions taking, e.g. how to best schedule events, without overflowing the visualization. Furthermore as a rather data-driven opposed to control-flow based approach, our performance skyline yields flow analysis [12] like results for mining non-block-structured models.

7   Conclusion and Future Work

With our new approach of *performance skyline* we introduce a novel approach for the performance mining of events containing multiple timestamps. Combining interval based sequence pattern mining and process mining techniques facilitates more accurate process discovery by integrating additional performance knowledge across traces and events. Applying statistical analysis techniques on the performance skyline enrich dominant path analysis with probabilistic performance knowledge enabling more complete conformance checking, detecting and discerning patterns, and thus adapting processes to be faster and more resourceful. Results from applying these methods to the real data set for a company exemplifies how combining performance mining and sequence pattern mining techniques is most suitable to identify and analyze the dominant path in a process model using a trace set containing interval events.

Future work involves further experiments with variations on implementation of the skyline operator, e.g. investigating minimal time duration as a dominant feature, which may be useful to optimize resources allocation. Furthermore it includes testing different alignment references, and inquiring skylines on other dominant features besides duration, as waiting time, or even non-time related aspects as memory usage. Moreover investigation of inferring models from streaming interval events as well as general research for suitability solving tasks in process discovery, conformance checking, and process enhancement could be expanded:

- Broader **process discovery**: identify loops and choices in the performance skyline aggregations, extending the model by adding more information or enriching further the visualization of already present items, researching more variations of this model, as adding a third dimension plotting information about resources, dependencies, further timestamps and `case id`.
- **Conformance checking** for recognition and prediction tasks: Anomaly detection on trace - activity and event level, drift recognition, and more

specifically predictive process monitoring for interval events, using e.g. skyline expectation maximization [21] for performance prediction on event level, as well as event anomaly detection through such predictions.
- **Process enhancement**: recognizing bottlenecks, or using detected probabilistic pattern knowledge for optimal networks queuing and resource allocation efficiently.

Acknowledgements. This work has been funded by the German Federal Ministry of Education and Research (BMBF) under Grant No. 01IS18036A. The authors of this work take full responsibility for its content.

References

1. Gantt, H.: A graphical daily balance in manufacture, vol. 24, pp. 1322–1336, June 1920
2. Denisov, V., Belkina, E., Fahland, D., van der Aalst, W.: The performance spectrum miner: visual analytics for fine-grained performance analysis of processes. In: Casati, F. (ed.) BPMTracks 2018, ser. CEUR Workshop Proceedings, pp. 96–100, September 2018. CEUR-WS.org
3. Borzsonyi, S., Kossmann, D., Stocker, K.: The skyline operator, December 2000
4. Robledo, P.: Process mining plays an essential role in digital transformation. https://medium.com/@pedrorobledobpm/process-mining-plays-an-essential-role-in-digital-transformation-384839236bbe
5. Suriadi, S., Ouyang, C., Van Der Aalst, W.M., Ter Hofstede, A.H.: Event interval analysis: why do processes take time? Decis. Support Syst. **79**, 77–98 (2015)
6. Burattin, A., Sperduti, A.: Heuristics miner for time intervals. In: ESANN (2010)
7. Burattin, A.: Heuristics miner for time interval. Process Mining Techniques in Business Environments. LNBIP, vol. 207, pp. 85–95. Springer, Cham (2015). https://doi.org/10.1007/978-3-319-17482-2_11
8. van der Aalst, W.M.P.: Process Mining: Data Science in Action, 2nd edn. Springer, Heidelberg (2016). https://doi.org/10.1007/978-3-662-49851-4
9. Richter, F., Seidl, T.: TESSERACT: time-drifts in event streams using series of evolving rolling averages of completion times. In: Carmona, J., Engels, G., Kumar, A. (eds.) BPM 2017. LNCS, vol. 10445, pp. 289–305. Springer, Cham (2017). https://doi.org/10.1007/978-3-319-65000-5_17
10. Denisov, V., Fahland, D., van der Aalst, W.M.: Predictive performance monitoring of material handling systems using the performance spectrum. In: 2019 International Conference on Process Mining (ICPM), pp. 137–144. IEEE (2019)
11. Senderovich, A., Weidlich, M., Gal, A., Mandelbaum, A.: Queue mining – predicting delays in service processes. In: Jarke, M., Mylopoulos, J., Quix, C., Rolland, C., Manolopoulos, Y., Mouratidis, H., Horkoff, J. (eds.) CAiSE 2014. LNCS, vol. 8484, pp. 42–57. Springer, Cham (2014). https://doi.org/10.1007/978-3-319-07881-6_4
12. Dumas, M., La Rosa, M., Mendling, J., Reijers, H.: Fundamentals of business process management (2013)
13. Senderovich, A., Weidlich, M., Gal, A.: Context-aware temporal network representation of event logs: model and methods for process performance analysis. Inf. Syst. **84**, 240–254 (2019)
14. Shaffer, L.R., Ritter, J., Meyer, W.L.: The Critical-Path Method. McGraw-Hill, New York (1965)

15. Fondahl, J.: A non-computer approach to the critical path method for the construction industry, June 1961
16. Santiago, J., Magallon, D.: Critical path method, CEE 320 - VDC SEMINAR, February 2009
17. Thomas, L., Kumar, M.M., Annappa, B.: Efficient process mining through critical path network analysis, February 2014
18. Allen, J.: Maintaining knowledge about temporal intervals, vol. 26, no. 11, pp. 832–843. ACM, New York (1983)
19. Marwan Hassani, J.W., Lu, Y., Seidl, T.: A geometric approach for mining sequential patterns in interval-based data streams. In: 2016 IEEE International Conference on Fuzzy Systems (FUZZ-IEEE), pp. 2128–2135 (2016)
20. Samet, H.: Hierarchical representations of collections of small rectangles. ACM Comput. Surv. **20**(4), 271–309 (1988). https://doi.org/10.1145/50020.50021
21. Moon, T.K.: The expectation-maximization algorithm. IEEE Signal Process. Mag. **13**(6), 47–60 (1996)

5th International Workshop on Process Querying, Manipulation, and Intelligence (PQMI 2020)

5th International Workshop on Process Querying, Manipulation, and Intelligence (PQMI 2020)

The aim of the fifth International Workshop on Process Querying, Manipulation, and Intelligence (PQMI 2020) was to provide a high-quality forum for researchers and practitioners to exchange research findings and ideas on methods and practices in the corresponding areas. *Process Querying* combines concepts from Big Data and Process Modeling and Analysis with Business Process Intelligence and Process Analytics to study techniques for retrieving and manipulating models of processes, both observed and recorded in the real world and envisioned and designed in conceptual models, to systematically organize and extract process-related information for subsequent use. *Process Manipulation* studies inferences from real-world observations for augmenting, enhancing, and redesigning models of processes with the ultimate goal of improving real-world business processes. *Process Intelligence* looks into application of the representation models and approaches in Artificial Intelligence (AI), such as knowledge representation, search, automated planning, reasoning, natural language processing, autonomous agents, and multi-agent systems, among others, for solving problems in process mining, that is automated process discovery, conformance checking, and process enhancement, and vice versa using process mining techniques to tackle problems in AI. Techniques, methods, and tools for process querying, manipulation, and intelligence have applications in Business Process Management and Process Mining. Examples of practical problems tackled by the themes of the workshop include business process compliance management, business process weakness detection, process variance management, process performance analysis, predictive process monitoring, process model translation, syntactical correctness checking, process model comparison, infrequent behavior detection, process instance migration, process reuse, and process standardization.

PQMI 2020 attracted seven high-quality submissions. Each paper was reviewed by at least three members of the Program Committee. The review process led to three accepted papers.

The invited talk of Ernesto Damiani, with which the workshop began, was about the opportunities and pitfalls of applying AI techniques to Business Process Management. The accepted paper by Daniel Schuster, Sebastiaan J. van Zelst, and Wil M. P. van der Aalst presents an approach that uses structural properties of a process model to speed up computations of alignments between the model and execution traces. The second accepted paper by Adam Burke, Sander Leemans, and Moe Wynn introduces several techniques to estimate the likelihoods of decisions encoded in process models based on the given collections of execution traces. Finally, the paper by Amin Jalali talks about the use of graph databases to solve various problems in process mining.

We hope that the reader will benefit from these proceedings to know more about the latest advances in research in the corresponding topics.

October 2020   PQMI Workshop Organizers

Organization

Workshop Organizers

Artem Polyvyanyy   The University of Melbourne
Claudio Di Ciccio   Sapienza University of Rome
Sebastian Sardina   RMIT University
Renuka Sindhgatta   Queensland University of Technology
Arthur ter Hofstede   Queensland University of Technology

Program Committee

Agnes Koschmider   Kiel University
Anna Kalenkova   The University of Melbourne
Catarina Moreira   Queensland University of Technology
Chiara Di Francescomarino   Fondazione Bruno Kessler-IRST
David Knuplesch   alphaQuest
Fabrizio Maggi   Free University of Bozen-Bolzano
Hagen Völzer   IBM Research – Zurich
Han van der Aa   University of Mannheim
Hyerim Bae   Pusan National University
Hye-Young Paik   The University of New South Wales
Jochen De Weerdt   Katholieke Universiteit Leuven
Jorge Munoz-Gama   Pontificia Universidad Católica de Chile
Kanika Goel   Queensland University of Technology
Marí Teresa Gómez-López   Universidad de Sevilla
Maurizio Proietti   CNR-IASI
Mieke Jans   Hasselt University
Minseok Song   Pohang University of Science and Technology
Pnina Soffer   University of Haifa
Rong Liu   Stevens Institute of Technology
Seppe vanden Broucke   Katholieke Universiteit Leuven
Shazia Sadiq   The University of Queensland

Alignment Approximation
for Process Trees

Daniel Schuster[1]([⊠]) [iD], Sebastiaan van Zelst[1,2] [iD],
and Wil M. P. van der Aalst[1,2] [iD]

[1] Fraunhofer Institute for Applied Information Technology FIT,
Sankt Augustin, Germany
{daniel.schuster,sebastiaan.van.zelst}@fit.fraunhofer.de
[2] RWTH Aachen University, Aachen, Germany
wvdaalst@pads.rwth-aachen.de

Abstract. Comparing observed behavior (event data generated during
process executions) with modeled behavior (process models), is an essential step in process mining analyses. Alignments are the de-facto standard
technique for calculating conformance checking statistics. However, the
calculation of alignments is computationally complex since a shortest
path problem must be solved on a state space which grows non-linearly
with the size of the model and the observed behavior, leading to the
well-known *state space explosion problem*. In this paper, we present a
novel framework to approximate alignments on process trees by exploiting their hierarchical structure. Process trees are an important process
model formalism used by state-of-the-art process mining techniques such
as the inductive mining approaches. Our approach exploits structural
properties of a given process tree and splits the alignment computation
problem into smaller sub-problems. Finally, sub-results are composed to
obtain an alignment. Our experiments show that our approach provides
a good balance between accuracy and computation time.

Keywords: Process mining · Conformance checking · Approximation

1 Introduction

Conformance checking is a key research area within process mining [1]. The
comparison of observed process behavior with reference process models is of crucial importance in process mining use cases. Nowadays, *alignments* [2] are the
de-facto standard technique to compute conformance checking statistics. However, the computation of alignments is complex since a shortest path problem
must be solved on a non-linear state space composed of the reference model and
the observed process behavior. This is known as the *state space explosion problem* [3]. Hence, various approximation techniques have been introduced. Most
techniques focus on decomposing Petri nets or reducing the number of alignments to be calculated when several need to be calculated for the same process
model [4–8].

In this paper, we focus on a specific class of process models, namely process trees (also called *block-structured* process models), which are an important

© Springer Nature Switzerland AG 2021
S. Leemans and H. Leopold (Eds.): ICPM 2020 Workshops, LNBIP 406, pp. 247–259, 2021.
https://doi.org/10.1007/978-3-030-72693-5_19

process model formalism that represent a subclass of sound *Workflow nets* [9]. For instance, various state-of-the-art process discovery algorithms return process trees [9–11]. In this paper, we introduce an alignment approximation approach for process trees that consists of two main phases. First, our approach splits the problem of alignments into smaller sub-problems along the tree hierarchy. Thereby, we exploit the hierarchical structure of process trees and their semantics. Moreover, the definition of sub-problems is based on a *gray-box view* on the corresponding subtrees since we use a simplified/abstract view on the subtrees to recursively define the sub-problems along the tree hierarchy. Such sub-problems can then be solved individually and in parallel. Secondly, we recursively compose an alignment from the sub-results for the given process tree and observed process behavior. Our experiments show that our approach provides a good balance between accuracy and computation effort.

The remainder is structured as follows. In Sect. 2, we present related work. In Sect. 3, we present preliminaries. In Sect. 4, we present the formal framework of our approach. In Sect. 5, we introduce our alignment approximation approach. In Sect. 6, we present an evaluation. Section 7 concludes the paper.

2   Related Work

In this section, we present related work regarding alignment computation and approximation. For a general overview of conformance checking, we refer to [3].

Alignments have been introduced in [2]. In [12] it was shown that the computation is reducible to a shortest path problem and the solution of the problem using the A* algorithm is presented. In [13], the authors present an improved heuristic that is used in the shortest path search. In [14], an alignment approximation approach based on approximating the shortest path is presented.

A generic approach to decompose Petri nets into multiple sub-nets is introduced in [15]. Further, the application of such decomposition to alignment computation is presented. In contrast to our approach, the technique does not return an alignment. Instead, only partial alignments are calculated, which are used, for example, to approximate an overall fitness value. In [4], an approach to calculate alignments based on Petri net decomposition [15] is presented that additionally guarantees optimal fitness values and optionally returns an alignment. Comparing both decomposition techniques with our approach, we do not calculate sub-nets because we simply use the given hierarchical structure of a process tree. Moreover, our approach always returns a valid alignment.

In [5], an approach is presented that approximates alignments for an event log by reducing the number of alignments being calculated based on event log sampling. Another technique based on event log sampling is presented in [8] where the authors explicitly approximate conformance results, e.g., fitness, rather than alignments. In contrast to our proposed approach, alignments are not returned. In [6] the authors present an approximation approach that explicitly focuses on approximating multiple optimal alignments. Finally, in [7], the authors present a technique to reduce a given process model and an event log s.t. the original

Table 1. Example of an event log from an order process

Event-id	Case-id	Activity name	Timestamp	$\cdots$
$\cdots$	$\cdots$	$\cdots$	$\cdots$	$\cdots$
200	13	create order (c)	2020-01-02 15:29	$\cdots$
201	27	receive payment (r)	2020-01-02 15:44	$\cdots$
202	43	dispatch order (d)	2020-01-02 16:29	$\cdots$
203	13	pack order (p)	2020-01-02 19:12	$\cdots$
$\cdots$	$\cdots$	$\cdots$	$\cdots$	$\cdots$

behavior of both is preserved as much as possible. In contrast, the proposed approach in this paper does not modify the given process model and event log.

3   Preliminaries

We denote the power set of a given set X by $\mathcal{P}(X)$. A multi-set over a set X allows multiple appearances of the same element. We denote the universe of multi-sets for a set X by $\mathcal{B}(X)$ and the set of all sequences over X as X^*, e.g., $\langle a, b, b \rangle \in \{a, b, c\}^*$. For a given sequence σ, we denote its length by $|\sigma|$. We denote the empty sequence by $\langle \rangle$. We denote the set of all possible permutations for given $\sigma \in X^*$ by $\mathbb{P}(\sigma) \subseteq X^*$. Given two sequences σ and σ', we denote the concatenation of these two sequences by $\sigma \cdot \sigma'$. We extend the $\cdot$ operator to sets of sequences, i.e., let $S_1, S_2 \subseteq X^*$ then $S_1 \cdot S_2 = \{\sigma_1 \cdot \sigma_2 \mid \sigma_1 \in S_1 \wedge \sigma_2 \in S_2\}$. For traces σ, σ', the set of all interleaved sequences is denoted by $\sigma \diamond \sigma'$, e.g., $\langle a, b \rangle \diamond \langle c \rangle = \{\langle a, b, c \rangle, \langle a, c, b \rangle, \langle c, a, b \rangle\}$. We extend the $\diamond$ operator to sets of sequences. Let $S_1, S_2 \subseteq X^*$, $S_1 \diamond S_2$ denotes the set of interleaved sequences, i.e., $S_1 \diamond S_2 = \bigcup_{\sigma_1 \in S_1, \sigma_2 \in S_2} \sigma_1 \diamond \sigma_2$.

For $\sigma \in X^*$ and $X' \subseteq X$, we recursively define the projection function $\sigma_{\downarrow_{X'}} : X^* \to (X')^*$ with: $\langle \rangle_{\downarrow_{X'}} = \langle \rangle$, $\left(\langle x \rangle \cdot \sigma \right)_{\downarrow_{X'}} = \langle x \rangle \cdot \sigma_{\downarrow_{X'}}$ if $x \in X'$ and $\left(\langle x \rangle \cdot \sigma \right)_{\downarrow_{X'}} = \sigma_{\downarrow_{X'}}$ else.

Let $t = (x_1, \ldots, x_n) \in X_1 \times \ldots \times X_n$ be an n-tuple over n sets. We define projection functions that extract a specific element of t, i.e., $\pi_1(t) = x_1, \ldots, \pi_n(t) = x_n$, e.g., $\pi_2((a, b, c)) = b$. Analogously, given a sequence of length m with n-tuples $\sigma = \langle (x_1^1, \ldots, x_n^1), \ldots, (x_1^m, \ldots, x_n^m) \rangle$, we define $\pi_1^*(\sigma) = \langle x_1^1, \ldots, x_1^m \rangle, \ldots, \pi_n^*(\sigma) = \langle x_n^1, \ldots, x_n^m \rangle$. For instance, $\pi_2^*(\langle (a, b), (a, c), (b, a) \rangle) = \langle b, c, a \rangle$.

3.1   Event Logs

Process executions leave *event data* in information systems. An *event* describes the execution of an activity for a particular *case*/process instance. Consider Table 1 for an example of an *event log* where each event contains the executed activity, a timestamp, a case-id and potentially further attributes. Since, in this paper, we are only interested in the sequence of activities executed, we define an event log as a multi-set of sequences. Such sequence is also referred to as a *trace*.

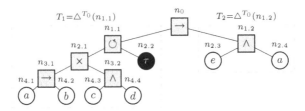

Fig. 1. Process tree $T_0 = (\{n_o, \ldots, n_{4.4}\}, \{(n_o, n_{1.1}), \ldots, (n_{3.2}, n_{4.4})\}, \lambda, n_0)$ with $\lambda(n_0) = \rightarrow, \ldots, \lambda(n_{4.4}) = d$

Definition 1 (Event log). *Let $\mathcal{A}$ be the universe of activities. $L \in \mathcal{B}(\mathcal{A}^*)$ is an event log.*

3.2 Process Trees

Next, we define the syntax and semantics of process trees.

Definition 2 (Process Tree Syntax). *Let $\mathcal{A}$ be the universe of activities and $\tau \notin \mathcal{A}$. Let $\bigoplus = \{\rightarrow, \times, \wedge, \circlearrowright\}$ be the set of process tree operators. We define a process tree $T = (V, E, \lambda, r)$ consisting of a totally ordered set of nodes V, a set of edges E, a labeling function $\lambda{:}V \rightarrow \mathcal{A} \cup \{\tau\} \cup \bigoplus$ and a root node $r \in V$.*

- *$(\{n\}, \{\}, \lambda, n)$ with $\lambda(n) \in \mathcal{A} \cup \{\tau\}$ is a process tree*
- *given $k > 1$ process trees $T_1 = (V_1, E_1, \lambda_1, r_1), \ldots, T_k = (V_k, E_k, \lambda_k, r_k)$, $T = (V, E, \lambda, r)$ is a process tree s.t.:*
  - *$V = V_1 \cup \ldots \cup V_k \cup \{r\}$ (assume $r \notin V_1 \cup \ldots \cup V_k$)*
  - *$E = E_1 \cup \ldots \cup E_k \cup \{(r, r_1), \ldots, (r, r_k)\}$*
  - *$\lambda(x) = \lambda_j(x)\ \forall j \in \{1, \ldots, k\} \forall x \in V_j, \lambda(r) \in \{\rightarrow, \wedge, \times\}$*
- *given two process trees $T_1 = (V_1, E_1, \lambda_1, r_1)$ and $T_2 = (V_2, E_2, \lambda_2, r_2)$, $T = (V, E, \lambda, r)$ is a process tree s.t.:*
  - *$V = V_1 \cup V_2 \cup \{r\}$ (assume $r \notin V_1 \cup V_2$)*
  - *$E = E_1 \cup E_2 \cup \{(r, r_1), (r, r_2)\}$*
  - *$\lambda(x) = \lambda_1(x)$ if $x \in V_1, \lambda(x) = \lambda_2(x)$ if $x \in V_2, \lambda(r) = \circlearrowright$*

In Fig. 1, we depict an example process tree T_0 that can alternatively be represented textually due to the totally ordered node set, i.e., $T_0 \hat{=} \rightarrow(\circlearrowright(\times(\rightarrow(a, b), \wedge(c, d)), \tau), \wedge(e, a))$. We denote the universe of process trees by $\mathcal{T}$. The degree d indicates the number of edges connected to a node. We distinguish between incoming d^+ and outgoing edges d^-, e.g., $d^+(n_{2.1}) = 1$ and $d^-(n_{2.1}) = 2$. For a tree $T = (V, E, \lambda, r)$, we denote its *leaf nodes* by $T^L = \{v \in V | d^-(v) = 0\}$. The child function $c^T{:}V \rightarrow V^*$ returns a sequence of child nodes according to the order of V, i.e., $c^T(v) = \langle v_1, \ldots, v_j \rangle$ s.t. $(v, v_1), \ldots, (v, v_j) \in E$. For instance, $c^T(n_{1.1}) = \langle n_{2.1}, n_{2.2} \rangle$. For $T = (V, E, \lambda, r)$ and a node $v \in V$, $\triangle^T(v)$ returns the corresponding tree T' s.t. v is the root node, i.e., $T' = (V', E', \lambda', v)$. Consider T_0, $\triangle^{T_0}(n_{1.1}) = T_1$ as highlighted in Fig. 1. For process tree $T \in \mathcal{T}$, we denote its height by $h(T) \in \mathbb{N}$.

trace part	a	b	$\gg$	$\gg$	c	f	$\gg$	$\gg$
model part	$n_{4.1}$	$n_{4.2}$	$n_{2.2}$	$n_{4.4}$	$n_{4.3}$		$n_{2.4}$	$n_{2.3}$
	$\lambda(n_{4.1})=a$	$\lambda(n_{4.2})=b$	$\lambda(n_{2.2})=\tau$	$\lambda(n_{4.4})=d$	$\lambda(n_{4.3})=c$		$\lambda(n_{2.4})=a$	$\lambda(n_{2.3})=e$

Fig. 2. Optimal alignment $\gamma = \langle (a, n_{4.1}), \ldots, (\gg, n_{2.3}) \rangle$ for $\langle a, b, c, f \rangle$ and T_0

Definition 3 (Process Tree Semantics). *For given* $T = (V, E, \lambda, r) \in \mathcal{T}$, *we define its language* $\mathcal{L}(T) \subseteq \mathcal{A}^*$.

- *if* $\lambda(r) = a \in \mathcal{A}$, $\mathcal{L}(T) = \{\langle a \rangle\}$
- *if* $\lambda(r) = \tau$, $\mathcal{L}(T) = \{\langle \rangle\}$
- *if* $\lambda(r) \in \{\rightarrow, \times, \wedge\}$ *with* $c^T(r) = \langle v_1, \ldots, v_k \rangle$
  - *with* $\lambda(r) = \rightarrow$, $\mathcal{L}(T) = \mathcal{L}(\triangle^T(v_1)) \cdot \ldots \cdot \mathcal{L}(\triangle^T(v_k))$
  - *with* $\lambda(r) = \wedge$, $\mathcal{L}(T) = \mathcal{L}(\triangle^T(v_1)) \diamond \ldots \diamond \mathcal{L}(\triangle^T(v_k))$
  - *with* $\lambda(r) = \times$, $\mathcal{L}(T) = \mathcal{L}(\triangle^T(v_1)) \cup \ldots \cup \mathcal{L}(\triangle^T(v_k))$
- *if* $\lambda(r) = \circlearrowright$ *with* $c^T(r) = \langle v_1, v_2 \rangle$, $\mathcal{L}(T) = \{\sigma_1 \cdot \sigma_1' \cdot \sigma_2 \cdot \sigma_2' \cdot \ldots \cdot \sigma_m \mid m \geq 1 \wedge \forall 1 \leq i \leq m (\sigma_i \in \mathcal{L}(\triangle^T(v_1))) \wedge \forall 1 \leq i \leq m-1 (\sigma_i' \in \mathcal{L}(\triangle^T(v_2)))\}$

In this paper, we assume binary process trees as input for our approach, i.e., every node has two or none child nodes, e.g., T_0. Note that every process tree can be easily converted into a language equivalent binary process tree [9].

3.3 Alignments

Alignments [12] map observed behavior onto modeled behavior specified by process models. Figure 2 visualizes an alignment for the trace $\langle a, b, c, f \rangle$ and T_0 (Fig. 1). The first row corresponds to the given trace ignoring the skip symbol $\gg$. The second row (ignoring $\gg$) corresponds to a sequence of leaf nodes s.t. the corresponding sequence of labels (ignoring τ) is in the language of the process tree, i.e., $\langle a, b, d, c, a, e \rangle \in \mathcal{L}(T_0)$. Each column represents an alignment move. The first two are *synchronous moves* since the activity and the leaf node label are equal. The third and fourth are *model moves* because $\gg$ is in the log part. Moreover, the third is an *invisible* model move since the leaf node label is τ and the fourth is a *visible* model move since the label represents an activity. Visible model moves indicate that an activity should have taken place w.r.t. the model. The sixth is a log move since the trace part contains $\gg$. Log moves indicate observed behavior that should not occur w.r.t. the model. Note that we alternatively write $\gamma \hat{=} \langle (a, a), \ldots, (\gg, e) \rangle$ using their labels instead of leaf nodes.

Definition 4 (Alignment). *Let* $\mathcal{A}$ *be the universe of activities,* $\sigma \in \mathcal{A}^*$ *be a trace and* $T = (V, E, \lambda, r) \in \mathcal{T}$ *be a process tree with leaf nodes* T^L. *Note that* $\gg, \tau \notin \mathcal{A}$. *A sequence* $\gamma \in ((\mathcal{A} \cup \{\gg\}) \times (T^L \cup \{\gg\}))^*$ *with length* $n = |\gamma|$ *is an alignment iff:*

1. $\sigma = \pi_1^*(\gamma)_{\downarrow_{\mathcal{A}}}$
2. $\left\langle \lambda \left(\pi_2 (\gamma(1)) \right), \ldots, \lambda \left(\pi_2 (\gamma(n)) \right) \right\rangle_{\downarrow_{\mathcal{A}}} \in \mathcal{L}(T)$
3. $(\gg, \gg) \notin \gamma$ *and* $(a, v) \notin \gamma \; \forall a \in \mathcal{A} \; \forall v \in T^L (a \neq \lambda(v))$

For a given process tree and a trace, many alignments exist. Thus, costs are assigned to alignment moves. In this paper, we assume the *standard cost function*. Synchronous and invisible model moves are assigned cost 0, other moves are assigned cost 1. An alignment with minimal costs is called *optimal*. For a process tree T and a trace σ, we denote the set of all possible alignments by $\Gamma(\sigma, T)$. In this paper, we assume a function α that returns for given $T \in \mathcal{T}$ and $\sigma \in \mathcal{A}^*$ an optimal alignment, i.e., $\alpha(\sigma, T) \in \Gamma(\sigma, T)$. Since process trees can be easily converted into Petri nets [1] and the computation of alignments for a Petri net was shown to be reducible to a shortest path problem [12], such function exists.

4   Formal Framework

In this section, we present a general framework that serves as the basis for the proposed approach. The core idea is to recursively divide the problem of alignment calculation into multiple sub-problems along the tree hierarchy. Subsequently, we recursively compose partial sub-results to an alignment.

Given a trace and tree, we recursively split the trace into sub-traces and assign these to subtrees along the tree hierarchy. During splitting/assigning, we regard the semantics of the current root node's operator. We recursively split until we can no longer split, e.g., we hit a leaf node. Once we stop splitting, we calculate optimal alignments for the defined sub-traces on the assigned subtrees, i.e., we obtain sub-alignments. Next, we recursively compose the sub-alignments to a single alignment for the parent subtree. Thereby, we consider the semantics of the current root process tree operator. Finally, we obtain a *valid*, but not necessarily optimal, alignment for the initial given tree and trace since we regard the semantics of the process tree during splitting/assigning and composing.

Formally, we can express the splitting/assigning as a function. Given a trace $\sigma \in \mathcal{A}^*$ and $T = (V, E, \lambda, r) \in \mathcal{T}$ with subtrees T_1 and T_2, ψ splits the trace σ into k sub-traces $\sigma_1, \ldots, \sigma_k$ and assigns each sub-trace to either T_1 or T_2.

$$\psi(\sigma, T) \in \left\{ \langle (\sigma_1, T_{i_1}), \ldots, (\sigma_k, T_{i_k}) \rangle \mid i_1, \ldots, i_k \in \{1, 2\} \wedge \sigma_1 \cdot \ldots \cdot \sigma_k \in \mathbb{P}(\sigma) \right\} \quad (1)$$

We call a splitting/assignment *valid* if the following additional conditions are satisfied depending on the process tree operator:

- if $\lambda(r) = \times$: $k = 1$
- if $\lambda(r) = \rightarrow$: $k = 2 \wedge \sigma_1 \cdot \sigma_2 = \sigma$
- if $\lambda(r) = \wedge$: $k = 2$
- if $\lambda(r) = \circlearrowleft$: $k \in \{1, 3, 5, \ldots\} \wedge \sigma_1 \cdot \ldots \cdot \sigma_k = \sigma \wedge i_1 = 1 \wedge \forall j \in \{1, \ldots, k - 1\}\left((i_j = 1 \Rightarrow i_{j+1} = 2) \wedge (i_j = 2 \Rightarrow i_{j+1} = 1)\right)$

Secondly, the calculated sub-alignments are recursively composed to an alignment for the respective parent tree. Assume a tree $T \in \mathcal{T}$ with subtrees T_1 and T_2, a trace $\sigma \in \mathcal{A}^*$, a valid splitting/assignment $\psi(\sigma, T)$,

---

Algorithm 1: Approximate alignment

input: $T = (V, E, \lambda, r) \in \mathcal{T}, \sigma \in \mathcal{A}^*, TL \geq TH \geq 1$
begin

1   if $|\sigma| \leq TL \vee h(T) \leq TH$ then
2   return $\alpha(\sigma, T)$;   // optimal alignment

3   else
4        $\psi(\sigma, T) = \langle (\sigma_1, T_{i_1}), \ldots, (\sigma_k, T_{i_k}) \rangle$;   // valid splitting
5   for $(\sigma_j, T_{i_j}) \in \langle (\sigma_1, T_{i_1}), \ldots, (\sigma_k, T_{i_k}) \rangle$ do
6            $\gamma_j \leftarrow$ approx. alignment for σ_j and T_{i_j};   // recursion
7        $\gamma \leftarrow \omega(\sigma, T, \langle \gamma_1, \ldots, \gamma_k \rangle)$;   // composing
8   return γ;

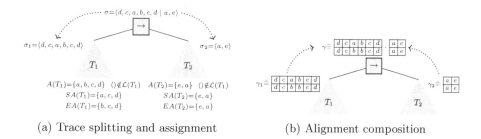

(a) Trace splitting and assignment   (b) Alignment composition

Fig. 3. Overview of the two main actions of the approximation approach

and a sequence of k sub-alignments $\langle \gamma_1, \ldots, \gamma_k \rangle$ s.t. $\gamma_j \in \Gamma(\sigma_j, T_{i_j})$ with $(\sigma_j, T_{i_j}) = \psi(\sigma, T)(j)$ *for all* $j \in \{1, \ldots, k\}$. The function ω composes an alignment for T and σ from the given sub-alignments.

$$\omega(\sigma, T, \langle \gamma_1, \ldots, \gamma_k \rangle) \in \{\gamma \mid \gamma \in \Gamma(\sigma, T) \wedge \gamma_1 \cdot \ldots \cdot \gamma_k \in \mathbb{P}(\gamma)\} \qquad (2)$$

By utilizing the definition of process tree semantics, it is easy to show that, given a valid splitting/assignment, such alignment γ returned by ω always exists.

The overall, recursive approach is sketched in Algorithm 1. For a given tree T and trace σ, we create a valid splitting/assignment (line 4). Next, we recursively call the algorithm on the determined sub-traces and subtrees (line 6). If given thresholds for trace length (TL) or tree height (TH) are reached, we stop splitting and return an optimal alignment (line 2). Hence, for the sub-traces created, we eventually obtain optimal sub-alignments, which we recursively compose to an alignment for the parent tree (line 7). Finally, we obtain a valid, but not necessarily optimal, alignment for T and σ.

5   Alignment Approximation Approach

Here, we describe our proposed approach, which is based on the formal framework introduced. First, we present an overview. Subsequently, we present specific strategies for splitting/assigning and composing for each process tree operator.

5.1   Overview

For splitting a trace and assigning sub-traces to subtrees many options exist. Moreover, it is inefficient to try out all possible options. Hence, we use a *heuristic* that guides the splitting/assigning. For each subtree, we calculate four characteristics: the activity labels A, if the empty trace is in the subtree's language, possible start-activities SA and end-activities EA of traces in the subtree's language. Thus, each subtree is a *gray-box* since only limited information is available.

Consider the trace to be aligned $\sigma = \langle d, c, a, b, c, d, a, e \rangle$ and the two subtrees of T_0 with corresponding characteristics depicted in Fig. 3a. Since T_0's root node is a sequence operator, we need to split σ once to obtain two subtraces according to the semantics. Thus, we have 9 potential splittings positions: $\langle |_1 \ d \ |_2 \ c \ |_3 \ a \ |_4 \ b \ |_5 \ c \ |_6 \ d \ |_7 \ a \ |_8 \ e \ |_9 \rangle$. If we split at position 1, we assign $\sigma_1 = \langle \rangle$ to the first subtree T_1 and the remaining trace $\sigma_2 = \sigma$ to T_2. Certainly, this is not a good decision since we know that $\langle \rangle \notin \mathcal{L}(T_1)$, the first activity of σ_2 is not a start activity of T_2 and the activities b, c, d occurring in σ_2 are not in T_2.

Assume we split at position 7 (Fig. 3a). Then we assign $\sigma_1 = \langle d, c, a, b, c, d \rangle$ to T_1. All activities in σ_1 are contained in T_1, σ_1 starts with $d \in SA(T_1)$ and ends with $d \in EA(T_1)$. Further, we obtain $\sigma_2 = \langle a, e \rangle$ whose activities can be replayed in T_2, and start- and end-activities match, too. Hence, according to the gray-box-view, splitting at position 7 is a good choice. Next, assume we receive two alignments γ_1 for T_1, σ_1 and γ_2 for T_2, σ_2 (Fig. 3b). Since T_1 is executed before T_2, we concatenate the sub-alignments $\gamma = \gamma_1 \cdot \gamma_2$ and obtain an alignment for T_0.

5.2   Calculation of Process Tree Characteristics

In this section, we formally define the computation of the four tree characteristics for a given process tree $T = (V, E, \lambda, r)$. We define the activity set A as a function, i.e., $A : T \to \mathcal{P}(\mathcal{A})$, with $A(T) = \{\lambda(n) \mid n \in T^L, \lambda(n) \neq \tau\}$. We recursively define the possible start- and end-activities as a function, i.e., $SA : T \to \mathcal{P}(\mathcal{A})$ and $EA : T \to \mathcal{P}(\mathcal{A})$. If T is not a leaf node, we refer to its two subtrees as T_1 and T_2.

$$
SA(T) = \begin{cases}
\{\lambda(r)\} & \text{if } \lambda(r) \in \mathcal{A} \\
\emptyset & \text{if } \lambda(r) = \tau \\
SA(T_1) & \text{if } \lambda(r) = \to \wedge \langle \rangle \notin \mathcal{L}(T_1) \\
SA(T_1) \cup SA(T_2) & \text{if } \lambda(r) = \to \wedge \langle \rangle \in \mathcal{L}(T_1) \\
SA(T_1) \cup SA(T_2) & \text{if } \lambda(r) \in \{\wedge, \times\} \\
SA(T_1) & \text{if } \lambda(r) = \circlearrowleft \wedge \langle \rangle \notin \mathcal{L}(T_1) \\
SA(T_1) \cup SA(T_2) & \text{if } \lambda(r) = \circlearrowleft \wedge \langle \rangle \in \mathcal{L}(T_1)
\end{cases}
$$

$$
EA(T) = \begin{cases}
\{\lambda(n)\} & \text{if } \lambda(r) \in \mathcal{A} \\
\emptyset & \text{if } \lambda(r) = \tau \\
EA(T_2) & \text{if } \lambda(r) = \to \wedge \langle \rangle \notin \mathcal{L}(T_2) \\
EA(T_1) \cup EA(T_2) & \text{if } \lambda(r) = \to \wedge \langle \rangle \in \mathcal{L}(T_2) \\
EA(T_1) \cup EA(T_2) & \text{if } \lambda(r) \in \{\wedge, \times\} \\
EA(T_1) & \text{if } \lambda(r) = \circlearrowleft \wedge \langle \rangle \notin \mathcal{L}(T_1) \\
EA(T_1) \cup EA(T_2) & \text{if } \lambda(r) = \circlearrowleft \wedge \langle \rangle \in \mathcal{L}(T_1)
\end{cases}
$$

The calculation whether the empty trace is accepted can also be done recursively.

- $\lambda(r) = \tau \Rightarrow \langle \rangle \in \mathcal{L}(T)$ and $\lambda(r) \in \mathcal{A} \Rightarrow \langle \rangle \notin \mathcal{L}(T)$
- $\lambda(r) \in \{\to, \wedge\} \Rightarrow \langle \rangle \in \mathcal{L}(T_1) \wedge \langle \rangle \in \mathcal{L}(T_2) \Leftrightarrow \langle \rangle \in \mathcal{L}(T)$
- $\lambda(r) \in \times \Rightarrow \langle \rangle \in \mathcal{L}(T_1) \vee \langle \rangle \in \mathcal{L}(T_2) \Leftrightarrow \langle \rangle \in \mathcal{L}(T)$
- $\lambda(r) = \circlearrowleft \Rightarrow \langle \rangle \in \mathcal{L}(T_1) \Leftrightarrow \langle \rangle \in \mathcal{L}(T)$

5.3   Interpretation of Process Tree Characteristics

The decision where to split a trace and the assignment of sub-traces to subtrees is based on the four characteristics per subtree and the process tree operator. Thus, each subtree is a gray-box for the approximation approach since only limited information is available. Subsequently, we explain how we interpret the subtree's characteristics and how we utilize them in the splitting/assigning decision.

Consider Fig. 4 showing how the approximation approach assumes a given subtree T behaves based on its four characteristics, i.e., $A(T), SA(T), EA(T),$ $\langle\rangle \in \mathcal{L}(T)$. The most liberal *interpretation* $\mathcal{I}(T)$ of a subtree T can be considered as a heuristic that guides the splitting/assigning. The interpretation $\mathcal{I}(T)$ depends on two conditions, i.e., if $\langle\rangle \in \mathcal{L}(T)$ and whether there is an activity that is both, a start- and end-activity, i.e., $SA(T) \cap EA(T) \neq \emptyset$. Note that $\mathcal{L}(T) \subseteq \mathcal{L}(\mathcal{I}(T))$ holds. Thus, the interpretation is an approximated view on the actual subtree.

In the next sections, we present for each tree operator a splitting/assigning and composing strategy based on the presented subtree interpretation. All strategies return a splitting per recursive call that minimizes the overall edit distance between the sub-traces and the closest trace in the language of the interpretation of the assigned subtrees. For $\sigma_1, \sigma_2 \in \mathcal{A}^*$, let $\updownarrow(\sigma_1, \sigma_2) \in \mathbb{N} \cup \{0\}$ be the Levenshtein distance [16]. For given $\sigma \in \mathcal{A}^*$ and $T \in \mathcal{T}$, we calculate a valid splitting $\psi(\sigma, T) = \langle(\sigma_1, T_{i_1}), \ldots, (\sigma_j, T_{i_k})\rangle$ w.r.t. Eq. (1) s.t. the sum depicted below is minimal.

$$\sum_{j \in \{1,\ldots,k\}} \left(\min_{\sigma' \in \mathcal{I}(T_{i_j})} \updownarrow(\sigma_j, \sigma') \right) \tag{3}$$

In the upcoming sections, we assume a given trace $\sigma = \langle a_1, \ldots, a_n\rangle$ and a process tree $T = (V, E, \lambda, r)$ with subtrees referred to as T_1 and T_2.

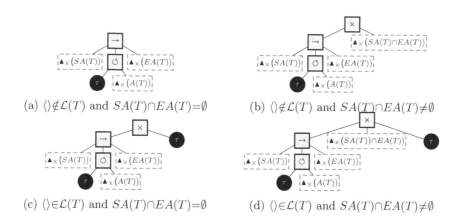

(a) $\langle\rangle \notin \mathcal{L}(T)$ and $SA(T) \cap EA(T) = \emptyset$

(b) $\langle\rangle \notin \mathcal{L}(T)$ and $SA(T) \cap EA(T) \neq \emptyset$

(c) $\langle\rangle \in \mathcal{L}(T)$ and $SA(T) \cap EA(T) = \emptyset$

(d) $\langle\rangle \in \mathcal{L}(T)$ and $SA(T) \cap EA(T) \neq \emptyset$

Fig. 4. Most liberal interpretation $\mathcal{I}(T)$ of the four characteristics of a process tree $T \in \mathcal{T}$. For a set $X = \{x_1, \ldots, x_n\}$, $\blacktriangle_\times(X)$ represents the tree $\times(x_1, \ldots, x_n)$

5.4   Approximating on Choice Operator

The choice operator is the most simple one since we just need to assign σ to one of the subtrees according to the semantics, i.e., assigning σ either to T_1 or T_2. We compute the edit distance of σ to the closest trace in $\mathcal{I}(T_1)$ and in $\mathcal{I}(T_2)$ and assign σ to the subtree with smallest edit distance according to Eq. (3).

Composing an alignment for the choice operator is trivial. Assume we eventually get an alignment γ for the chosen subtree, we just return γ for T.

5.5   Approximating on Sequence Operator

When splitting on a sequence operator, we must assign a sub-trace to each subtree according to the semantics. Hence, we calculate two sub-traces: $\langle(\sigma_1, T_1), (\sigma_2, T_2)\rangle$ s.t. $\sigma_1 \cdot \sigma_2 = \sigma$ according to Eq. (3). The optimal splitting/assigning can be defined as an optimization problem, i.e., Integer Linear Programming (ILP).

In general, for a trace with length n, $n+1$ possible splitting-positions exist: $\langle|_1\ a_1\ |_2\ a_2\ |_3\ \ldots\ |_n\ a_n\ |_{n+1}\rangle$. Assume we split at position 1, this results in $\langle(\langle\rangle, T_1), (\sigma, T_2)\rangle$, i.e., we assign $\langle\rangle$ to T_1 and the original trace σ to T_2.

Composing the alignment from sub-alignments is straightforward. In general, we eventually obtain two alignments, i.e., $\langle\gamma_1, \gamma_2\rangle$, for T_1 and T_2. We compose the alignment γ for T by concatenating the sub-alignments, i.e., $\gamma = \gamma_1 \cdot \gamma_2$.

5.6   Approximating on Parallel Operator

According to the semantics, we must define a sub-trace for each subtree, i.e., $\langle(T_1, \sigma_1), (T_2, \sigma_2)\rangle$. In contrast to the sequence operator, $\sigma_1 \cdot \sigma_2 = \sigma$ does *not* generally hold. The splitting/assignment w.r.t. Eq. (3) can be defined as an ILP. In general, each activity can be assigned to one of the subtrees independently.

For example, assume $\sigma = \langle c, a, d, c, b\rangle$ and $T \widehat{=} \wedge(\rightarrow(a, b), \circlearrowright(c, d))$ with subtree $T_1 \widehat{=} \rightarrow(a, b)$ and $T_2 \widehat{=} \circlearrowright(c, d)$. Below we assign the activities to subtrees.

$$\langle\ c,\ a,\ d,\ c,\ b\ \rangle$$
$$T_2\ T_1\ T_2\ T_2\ T_1$$

Based on the assignment, we create two sub-traces: $\sigma_1 = \langle a, b\rangle$ and $\sigma_2 = \langle c, d, c\rangle$. Assume that $\gamma_1 \widehat{=} \langle(a, a), (b, b)\rangle$ and $\gamma_2 \widehat{=} \langle(c, c), (d, d), (c, c)\rangle$ are the two alignments eventually obtained. To compose an alignment for T, we have to consider the assignment. Since the first activity c is assigned to T_2, we extract the corresponding alignment steps from γ_1 until we have explained c. The next activity in σ is an a assigned to T_1. We extract the alignment moves from γ_1 until we explained the a. We iteratively continue until all activities in σ are covered. Finally, we obtain an alignment for T and σ, i.e., $\gamma \widehat{=} \langle(c, c), (a, a), (d, d), (c, c), (b, b)\rangle$.

5.7 Approximating on Loop Operator

We calculate $m \in \{1, 3, 5, \dots\}$ sub-traces that are assigned alternately to the two subtrees: $\langle (\sigma_1, T_1), (\sigma_2, T_2), (\sigma_3, T_1), \dots, (\sigma_{m-1}, T_2), (\sigma_m, T_1) \rangle$ s.t. $\sigma = \sigma_1 \cdot \dots \cdot \sigma_m$. Thereby, σ_1 and σ_m are always assigned to T_1. Next, we visualize all possible splitting positions for the given trace: $\langle |_1\, a_1\, |_2\, |_3\, a_2\, |_4 \dots |_{2n-1}\, a_n\, |_{2n} \rangle$. If we split at each position, we obtain $\langle (\langle\rangle, T_1), (\langle a_1 \rangle, T_2), (\langle\rangle, T_1), \dots, (\langle a_n \rangle, T_2), (\langle\rangle, T_1) \rangle$. The optimal splitting/assignment w.r.t Eq. (3) can be defined as an ILP.

Composing an alignment is similar to the sequence operator. In general, we obtain m sub-alignments $\langle \gamma_1, \dots, \gamma_m \rangle$, which we concatenate, i.e., $\gamma = \gamma_1 \cdot \dots \cdot \gamma_m$.

6 Evaluation

This section presents an experimental evaluation of the proposed approach.

We implemented the proposed approach in PM4Py[1], an open-source process mining library. We conducted experiments on real event logs [17,18]. For each log, we discovered a process tree with the Inductive Miner infrequent algorithm [10].

In Figs. 5 and 6, we present the results. We observe that our approach is on average always faster than the optimal alignment algorithm for all tested parameter settings. Moreover, we observe that our approach never underestimates the optimal alignment costs, as our approach returns a valid alignment. W.r.t. optimization problems for optimal splittings/assignments, consider parameter setting TH:5 and TL:5 in Fig. 5. This parameter setting results in the highest splitting along the tree hierarchy and the computation time is the lowest compared to the other settings. Thus, we conclude that solving optimization problems for finding splittings/assignments is appropriate. In general, we observe a good balance between accuracy and computation time. We additionally conducted experiments with a decomposition approach [15] (available in ProM[2]) and compared the calculation time with the standard alignment implementation (LP-based) [12] in ProM. Consider Table 2. We observe that the decomposition approach does not yield a speed-up for [17] but for [18] we observe that the decomposition approach is about 5 times faster. In comparison to Fig. 6a, however, our approach yields a much higher speed-up.

Table 2. Results for decomposition based alignments

Approach	[17] (sample: 100 variants)	[18] (sample: 100 variants)
Decomposition [4]	25.22 s	20.96 s
Standard [12]	1.51 s	103.22 s

---

[1] https://pm4py.fit.fraunhofer.de/.
[2] http://www.promtools.org/.

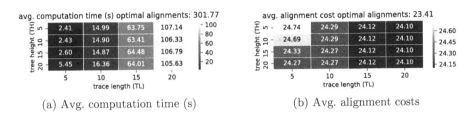

(a) Avg. computation time (s)   (b) Avg. alignment costs

Fig. 5. Results for [17], sample: 100 variants, tree height 24, avg. trace length 28

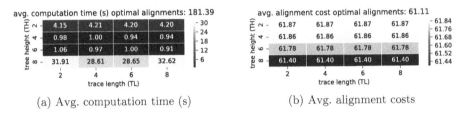

(a) Avg. computation time (s)   (b) Avg. alignment costs

Fig. 6. Results for [18], sample: 100 variants, tree height 10, avg. trace length 65

7   Conclusion

We introduced a novel approach to approximate alignments for process trees. First, we recursively split a trace into sub-traces along the tree hierarchy based on a gray-box view on the respective subtrees. After splitting, we compute optimal sub-alignments. Finally, we recursively compose a valid alignment from sub-alignments. Our experiments show that the approach provides a good balance between accuracy and calculation time. Apart from the specific approach proposed, the contribution of this paper is the formal framework describing how alignments can be approximated for process trees. Thus, many other strategies besides the one presented are conceivable.

References

1. van der Aalst, W.M.P.: Process Mining - Data Science in Action. Springer, Heidelberg (2016). https://doi.org/10.1007/978-3-662-49851-4
2. van der Aalst, W.M.P., Adriansyah, A., van Dongen, B.F.: Replaying history on process models for conformance checking and performance analysis. Wiley Interdiscip. Rev. Data Min. Knowl. Discov. **2**(2), 182–192 (2012)
3. Carmona, J., van Dongen, B.F., Solti, A., Weidlich, M.: Conformance Checking - Relating Processes and Models. Springer, Cham (2018). https://doi.org/10.1007/978-3-319-99414-7
4. Lee, W.L.J., Verbeek, H.M.W., Munoz-Gama, J., van der Aalst, W.M.P., Sepúlveda, M.: Recomposing conformance: closing the circle on decomposed alignment-based conformance checking in process mining. Inf. Sci. **466**, 55–91 (2018)

5. Fani Sani, M., van Zelst, S.J., van der Aalst, W.M.P.: Conformance checking approximation using subset selection and edit distance. In: Dustdar, S., Yu, E., Salinesi, C., Rieu, D., Pant, V. (eds.) CAiSE 2020. LNCS, vol. 12127, pp. 234–251. Springer, Cham (2020). https://doi.org/10.1007/978-3-030-49435-3_15

6. Taymouri, F., Carmona, J.: An evolutionary technique to approximate multiple optimal alignments. In: Weske, M., Montali, M., Weber, I., vom Brocke, J. (eds.) BPM 2018. LNCS, vol. 11080, pp. 215–232. Springer, Cham (2018). https://doi.org/10.1007/978-3-319-98648-7_13

7. Taymouri, F., Carmona, J.: Model and event log reductions to boost the computation of alignments. In: SIMPDA 2016, vol. 1757 (2016). CEUR-WS.org

8. Bauer, M., van der Aa, H., Weidlich, M.: Estimating process conformance by trace sampling and result approximation. In: Hildebrandt, T., van Dongen, B.F., Röglinger, M., Mendling, J. (eds.) BPM 2019. LNCS, vol. 11675, pp. 179–197. Springer, Cham (2019). https://doi.org/10.1007/978-3-030-26619-6_13

9. Leemans, S.J.J.: Robust process mining with guarantees. Ph.D. dissertation, Department of Mathematics and Computer Science (2017)

10. Leemans, S.J.J., Fahland, D., van der Aalst, W.M.P.: Discovering block-structured process models from event logs containing infrequent behaviour. In: Lohmann, N., Song, M., Wohed, P. (eds.) BPM 2013. LNBIP, vol. 171, pp. 66–78. Springer, Cham (2014). https://doi.org/10.1007/978-3-319-06257-0_6

11. Schuster, D., van Zelst, S.J., van der Aalst, W.M.P.: Incremental discovery of hierarchical process models. In: Dalpiaz, F., Zdravkovic, J., Loucopoulos, P. (eds.) RCIS 2020. LNBIP, vol. 385, pp. 417–433. Springer, Cham (2020). https://doi.org/10.1007/978-3-030-50316-1_25

12. Adriansyah, A.: Aligning observed and modeled behavior. Ph.D. dissertation, Eindhoven University of Technology (2014)

13. Dongen, B.F.: Efficiently computing alignments. In: Weske, M., Montali, M., Weber, I., vom Brocke, J. (eds.) BPM 2018. LNCS, vol. 11080, pp. 197–214. Springer, Cham (2018). https://doi.org/10.1007/978-3-319-98648-7_12

14. van Dongen, B., Carmona, J., Chatain, T., Taymouri, F.: Aligning modeled and observed behavior: a compromise between computation complexity and quality. In: Dubois, E., Pohl, K. (eds.) CAiSE 2017. LNCS, vol. 10253, pp. 94–109. Springer, Cham (2017). https://doi.org/10.1007/978-3-319-59536-8_7

15. van der Aalst, W.M.P.: Decomposing petri nets for process mining: a generic approach. Distrib. Parallel Databases, **31**(4), 471–507 (2013)

16. Levenshtein, V.I.: Binary codes capable of correcting deletions, insertions, and reversals. In: Soviet physics doklady, vol. 10, no. 8 (1966)

17. van Dongen, B.F.: BPI Challenge 2019. Dataset (2019)

18. van Dongen, B.F., Borchert, F.: BPI Challenge 2018. Dataset (2018)

Stochastic Process Discovery by Weight Estimation

Adam Burke[✉][iD], Sander J. J. Leemans[iD], and Moe Thandar Wynn[iD]

Queensland University of Technology, Brisbane, Australia
{at.burke,s.leemans,m.wynn}@qut.edu.au

Abstract. Many algorithms now exist for discovering process models from event logs. These models usually describe a control flow and are intended for use by people in analysing and improving real-world organizational processes. The relative likelihood of choices made while following a process (i.e., its stochastic behaviour) is highly relevant information which few existing algorithms make available in their automatically discovered models. This can be addressed by automatically discovered stochastic process models.

We introduce a framework for automatic discovery of stochastic process models, given a control-flow model and an event log. The framework introduces an estimator which takes a Petri net model and an event log as input, and outputs a Generalized Stochastic Petri net. We apply the framework, adding six new weight estimators, and a method for their evaluation. The algorithms have been implemented in the open-source process mining framework ProM. Using stochastic conformance measures, the resulting models have comparable conformance to existing approaches and are shown to be calculated more efficiently.

Keywords: Stochastic Petri nets · Process mining · Stochastic process mining · Stochastic process discovery

1 Introduction

The world abounds in information systems, generating data about the processes they mediate, execute, or observe. Using this data to compute and analyze process models is the concern of process mining [3], within the field of Business Process Management (BPM). BPM studies the impact and improvement of processes in organizations. Automatic process discovery is one aspect of process mining concerned with finding a formal process model computationally from an input event log.

To understand a process, we often want to know how likely an event is. If we travel to work, a journey where our train reliably arrives on time is different from one where the train sometimes breaks down, is sometimes replaced by a bus, or is often so crowded that it's quicker to ride a bike. A highly contagious disease with rare side effects differs importantly from one difficult to transmit but with severe side effects, even if observable symptoms are similar. Detecting fraud in

© Springer Nature Switzerland AG 2021
S. Leemans and H. Leopold (Eds.): ICPM 2020 Workshops, LNBIP 406, pp. 260–272, 2021.
https://doi.org/10.1007/978-3-030-72693-5_20

financial transactions depends on recognizing certain client actions happening more frequently than usual. Existing process mining techniques already recognize this: where noise or probability is considered in creating control flows (e.g. [19, 30]), they acknowledge the importance of likelihood in process modeling. Better stochastic representations and stochastic-aware techniques have been flagged as a key research challenge for process mining [2].

Process discovery techniques have become quite sophisticated at determining causal relationships between activities from event logs, and representing that in process models. There are far fewer techniques for discovering relative probabilities (discussed in Sect. 5). We introduce a framework in Sect. 3 which leverages this by allowing transformation of models with only control flows into stochastic process models. This extends an existing stochastic process discovery technique by Rogge-Solti et al. (RSD) [25,26], in two ways. Firstly, it generalizes one estimation algorithm to a general class of *weight estimators*. Secondly, it specializes the possible outputs from general probability distributions to Generalized Stochastic Petri Nets (GSPNs) [4]. The framework does not prescribe whether the estimation calculation is deterministic, uses stochastic simulation, or other techniques, and our introduced estimators include both deterministic and non-deterministic types.

We describe our approach as a form of Stochastic Process Discovery, as it takes an event log input and produces a GSPN output. In decoupling weight estimation from control flow discovery, the technique also shares some features with process model enhancement for time and probability [3, p. 290]. Unlike enhancement techniques, estimators can potentially change control flows when producing a stochastic process model. Stochastic process models have a corresponding, emerging, set of stochastic process conformance measures [16,20,21]. Consequently, the algorithms and models presented here are evaluated, in Sect. 4, as stochastic process discovery algorithms, using stochastic process conformance measures. Evaluation, which also includes performance, is against real-life event logs, multiple control flow discovery algorithms, and RSD [25].

In the next section, we introduce existing concepts. In Sect. 3, we describe the weight estimation framework and instantiate it by introducing novel estimators. In Sect. 4, the results of using the estimators on real-world event logs are presented. Related work is reviewed in Sect. 5, and Sect. 6 concludes the paper.

2   Preliminaries

Petri nets and Generalized Stochastic Petri Nets are well-established formalisms for modelling processes and a number of good overviews exist [4,8]. We use notations from the process mining literature [3,21].

A *Petri net* is a tuple $PN = (P, T, F, M_0)$, where P is a finite set of places, T is a finite set of transitions, and $F : (P \times T) \to (T \times P)$ is a flow relation. A *marking* is a multiset of places $\subseteq P$ that indicate a state of the Petri net, with M_0 the initial marking. A transition is enabled if every incoming place contains a token. A transition fires by changing the marking of the net to consume incoming tokens and producing tokens for its outgoing transitions. For a node $n \in P \cup T$, we define $\bullet n = \{y \mid (y,x) \in F\}$ and $n\bullet = \{y \mid (x,y) \in F\}$.

A *Generalized Stochastic Petri Net (GSPN)* is a tuple $(P, T, F, M_0, W, T_i, T_t)$ such that $T_i \subseteq T$, $T_t \subseteq T$ and $T_i \cap T_t = \emptyset$. Weight function $W : T \to \mathbb{R}^+$ assigns each transition a weight. T_i is a set of immediate transitions. If multiple transitions $T_e \subseteq T_i$ are enabled in a particular marking, the probability of a transition $t \in T_i$ firing is given by $\frac{W(t)}{\Sigma_{t' \in T_e} W(t')}$. T_t is a set of timed transitions. Immediate transitions take priority over timed transitions. A timed transition, if enabled, fires according to an exponentially distributed wait time. Given a set of enabled timed transitions $T_e \subseteq T_t$, a particular transition t fires first with probability $\frac{W(t)}{\Sigma_{t' \in T_e} W(t')}$ [4].

Event Logs. A process consists of activities from the set $\mathcal{A}$. A trace is a non-empty sequence of activities, and an event log L is a finite multiset of traces observing the underlying process. Partial function $\lambda : T \to \mathcal{A}$ designates labels for Petri net transitions that represent log activities. The number of traces in a log L is denoted with $|L|$, while the number of events is denoted with $||L||$.

Control Flow Process Discovery. A process discovery algorithm for Petri Nets is then defined by $cfd : L \to (P, T, F, M_0)$.

Sequence Operations. A finite sequence over $\mathcal{A}$ of length n is a mapping $\sigma \in \{1..n\} \to \mathcal{A}$ and denoted by $\sigma = \langle a_1, a_2, ..., a_n \rangle$ where $\forall_i a_i = \sigma(i)$. Concatenation operator $+$ appends one sequence to another such that $\langle a_1, ..., a_n \rangle + \langle b_1, ..., b_m \rangle = \langle a_1, ...a_n, b_1, ..., b_m \rangle$. The tail function is then $tail(\langle a \rangle + \sigma) = \sigma$.

Subsequence. Function ct returns the number of times a subsequence is present in a sequence:
$$ct(\varsigma, \sigma) = \begin{cases} 0 & \text{if } \sigma = \langle \rangle \\ 1 + ct(\varsigma, tail(\sigma)) & \text{if } \sigma = \varsigma + x \\ ct(\varsigma, tail(\sigma)) & \text{if } \sigma \neq \varsigma + x \end{cases}$$

Alignments. An alignment [1] represents paired paths between a log and a model. That is, a move is a tuple where (a, t) represents a synchronous move on activity a in a trace and a transition t in the model (with the same label: $\lambda(t) = a$), $(a, \perp)$ represents a log move, and $(\perp, t)$ represents a model move. For our purposes, we assume that a function γ is available taking a Petri net, a set of final markings and an event log, and that γ returns a sequence of move tuples that represent all moves necessary to align every trace in the log.

3   Stochastic Process Model Weight Estimation

In this section, we first introduce our framework to transform a Petri net into a GSPN using an event log. Then, we introduce six estimators using the framework, which we will illustrate using the running example shown in Fig. 2. Estimators are a large solution space with many potential algorithms. Our six estimators are chosen to emphasize broad applicability of inputs, computational tractability, using the implicit causal information in control flow models, and reapplying established process mining concepts.

3.1   A Framework for GSPN Discovery

The framework defines functions which together transform an event log into a GSPN, as shown in Fig. 1.

A stochastic process discovery algorithm for GSPNs ($mine\_spn$) is a function $mine\_spn : L \rightarrow (P, T, F, M_0, W, T_i, T_t)$. Our framework considers functions of the form $mine\_spn = est(cfd(L), L)$. Functions $est : L \times (P, T, F, M_0) \rightarrow (P, T, F, M_0, W, T_i, T_t)$ are termed $estimators$.

Functions $se : L \times (P, T, F, M_0) \rightarrow T \times \mathbb{R}^+$ are $simple\ weight\ estimators$ and use the control flow of the input Petri net intact in the output Petri net, such that for discovered control flow model $cfd(L) = (P_d, T_d, F_d, Md_0)$,

$$\exists_{pe \in est} pe = (P_d, T_d, F_d, Md_0, se(L, (P_d, T_d, F_d, Md_0)), T_d, \emptyset)$$

The estimators discussed next are of this simpler form.

Specific estimators may have further restrictions on their inputs, or provide guarantees on their outputs. For example, estimators discussed below do not distinguish transitions with duplicate labels. A challenge common to several estimators is treatment of silent transitions, as those transitions in a discovered model serve a structural role and do not directly represent an activity in the log. Assigning such a transition a weight of zero in a stochastic net is equivalent to deleting the transition, and all subsequent model paths. To avoid this impact, default values are assigned to silent transitions where the calculation would otherwise result in zero weights. In general, estimators make no distinction between silent transitions and transitions without a corresponding activity in the log. In the remainder of this section, we introduce several examples of estimators that instantiate this framework.

3.2   Frequency Estimator

The first estimator, w_{freq}, straightforwardly uses how often each transition t appeared in the event log L:

$$w_{freq}(L, t) = \max(1, \Sigma_{\sigma \in L}\ \text{ct}(\langle \lambda(t) \rangle, \sigma))$$

Silent transitions are assigned the arbitrary weight of 1, equivalent to a single observation in the log. The complexity of this estimator is linear in the number of events in the log. Figure 2c shows the results of this estimator on our running example, e.g. $w_{freq}(EL, b) = 15$.

3.3   Activity-Pair Frequency Estimators

An Activity-Pair Estimator uses the frequency of pairs of successor activities to better reflect the constraints of more general Petri nets. These are $edge$-$structured$ $estimators$, in that Petri net edges inform the weighting.

We first introduce some frequency definitions. The functions q_I and q_F capture how often an activity appears as the first/last in a trace. The function q_P

Fig. 1. Our framework for GSPN discovery.

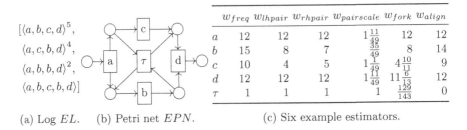

	w_{freq}	w_{lhpair}	w_{rhpair}	$w_{pairscale}$	w_{fork}	w_{align}
a	12	12	12	$1\frac{11}{49}$	12	12
b	15	8	7	$\frac{35}{49}$	8	14
c	10	4	5	$1\frac{1}{49}$	$4\frac{10}{11}$	9
d	12	12	12	$1\frac{11}{49}$	$11\frac{6}{13}$	12
τ	1	1	1	1	$\frac{129}{143}$	0

$[\langle a, b, c, d\rangle^5,$
$\langle a, c, b, d\rangle^4,$
$\langle a, b, b, d\rangle^2,$
$\langle a, b, c, b, d\rangle]$

(a) Log EL.   (b) Petri net EPN.   (c) Six example estimators.

Fig. 2. Running example of an event log and a Petri net, and the estimators.

captures the frequency of activity pairs in the log, that is, where the two given activities follow one another directly in the log:

$$q_I(L, t) = ||[\langle \lambda(t), \ldots\rangle \in L]||$$
$$q_F(L, t) = ||[\langle \ldots, \lambda(t)\rangle \in L]||$$
$$q_P(L, s, t) = \Sigma_{\sigma \in L} \, \text{ct}(\langle \lambda(s), \lambda(t)\rangle, \sigma)$$

There are both left-handed and right-handed variants of the Activity-Pair estimator, depending on whether weights are informed by successor or predecessor transitions, defined as:

$$w_{lhpair}(L, t) = \max(1, q_I(L, t) + q_F(L, t) + \sum_{s \in \bullet(\bullet t)} q_P(L, s, t))$$

$$w_{rhpair}(L, t) = \max(1, q_I(L, t) + q_F(L, t) + \sum_{s \in (t\bullet)\bullet} q_P(L, t, s))$$

There are no restrictions on input Petri nets and they can be calculated in time $O(||L|||F|)$, that is, the number of events times the number of model edges.

When using activity pair frequency data, two important types of path through the model are neglected for any given trace: paths from the initial place to the first transition, and the paths from the last transition to the final place. Traces of length one are also invisible from this perspective. To account for this, how often an activity appears as the initial or final activity in a trace is also included in the weight estimation. Note that not all activity pairs occurring in the log are used to calculate the resulting transition weights. For instance, where a given Petri net represents two transitions a and b as concurrent, the frequency of $\langle a, b\rangle$ will not be used. In our running example (see Fig. 2c), $w_{lhpair}(EL, c) = 4$ and $w_{rhpair}(EL, c) = 5$.

3.4   Mean-Scaled Activity-Pair Frequency Estimator

The previous estimators depend on the size of the log. Two logs with the same traces in the same ratios will result in two models with two distinct sets of weights, which challenges human analysis. Though comparison and comprehensibility of stochastic process models appears not to have been directly addressed in the literature, it is consistent with research that finds "small variations between models can lead to significant differences in their comprehensibility" [24] and the usability principle of minimizing user memory load. The mean-scaled activity-pair estimator $w_{pairscale}$ mitigates this effect by scaling weights by average transition frequency ($\frac{||L||}{|T|}$) in the log L:

$$pairscale(L,T,t) = \frac{q_I(L,t) + q_F(L,t) + \sum_{s \in (t\bullet)\bullet} q_P(L,t,s)}{\frac{||L||}{|T|}}$$

$$w_{pairscale}(L,(P,T,F,M_0),t) = \begin{cases} pairscale(L,T,t) & \text{if } pairscale(t) \neq 0 \\ 1 & \text{otherwise} \end{cases}$$

One effect of defaulting after scaling is that silent or unrepresented transitions are weighted more heavily, that is, the same as an activity of mean-frequency, rather than the equivalent of an activity occurring once in the log. In our running example of Fig. 2c, $||L|| = 49$, $|T| = 5$ and the numerator of $pairscale$ is equal to w_{rhpair} for a, b, c and d. Then, for instance $w_{pairscale}$ of c is $\frac{10}{\frac{49}{5}} = 1\frac{1}{49}$.

3.5   Fork Distribution Estimator

The Fork Distribution Estimator w_{fork} uses a two-stage approach: it first assigns weights to each place in a Petri net using activity-pair frequencies. Second, it distributes those weights to transitions according to the activity frequency in the event log.

$$pw(L,p) = \begin{cases} |L| & \text{if } p \in M_0 \\ \Sigma_{s \in \bullet p} \Sigma_{t \in p \bullet} q_P(L,s,t) & \text{otherwise} \end{cases}$$

$$placeWeights(L,p) = \max(1, pw(L,p))$$

$$w_{fork}(L,(P,T,F,M_0),t) = \Sigma_{p \in \bullet t} placeWeights(L,p) \frac{w_{freq}(t)}{\Sigma_{t'}^{p\bullet} w_{freq}(t')}$$

This estimator only applies to Petri nets which have at least one place without incoming edges, such as workflow nets [3, p. 81]. This is an edge-structured estimator informed by the structure of the input net. The complexity is $O(||L|| |F|)$. The w_{fork} estimator shares similarities with the Alpha algorithm [3, p. 167], in that it treats a place as defining a neighbourhood of related activities represented as transitions. In our example (Fig. 2), let p_1 be the top-right place and p_2 the bottom-right place. Then, $pw(EL,p_1) = q_P(c,d) + q_P(\tau,d) = 5$, $pw(EL,p_2) = q_P(\tau,d) + q_P(b,d) = 7$, $placeWeights(EL,p_1) = 5$, $placeWeights(EL,p_2) = 7$ and w_{fork} of $d = 5\frac{12}{12} + 7\frac{12}{13} = 11\frac{6}{13}$.

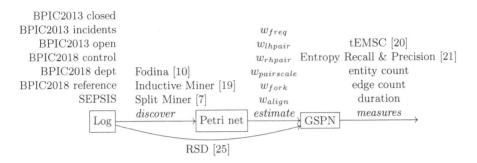

Fig. 3. Set-up of the evaluation.

3.6 Alignment Estimator

The estimator w_{align} applies alignments [1] to estimate weights. To this end, it counts the number of times a transition t appears either as a model move or as a synchronous move in the alignments:

$$w_{align}(L, PN, M_F, t) = |[(x, t) \in \gamma(PN, M_F, L)]|$$

This algorithm only applies to Petri nets with at least one final marking. The time complexity is $O(|T| \, |\gamma|)$ plus the time to compute γ. The alignment estimator has similarities with RSD [25], which fits duration distributions to aligned logs. In our example of Fig. 2, the last trace of log EL does not fit the model EPN, as b is executed a second time and c is executed. Thus, alignments will (based on a cost function, or if that does not discriminate the options an arbitrary choice) include a log move on either b or a log move on c. If the alignments choose a b for a log move, then $w_{align}(EL, EPN, M_F, b) = 14$ and $w_{align}(EL, EPN, M_F, \tau) = 0$. Alignments are not always deterministic, and consequently neither is w_{align}.

4 Implementation and Evaluation

4.1 Evaluation Design

The six estimators introduced in Sect. 3 were implemented in the ProM framework [13][1]. For our evaluation, a discovery algorithm was applied to an event log. Where necessary, the result was converted to a Petri net. Each estimator was invoked on the resulting Petri net, resulting in a GSPN. Finally, the conformance of the resulting GSPN was measured against the original log. For comparison, an existing stochastic discovery algorithm by Rogge-Solti et al. [25] (RSD) was also applied to the log. This direct discovery algorithm also outputs GSPNs, and the same conformance measures were applied. The implementation of this

---

[1] Source code is accessible via https://github.com/adamburkegh/spd_we.

plugin in ProM 6.9 uses the Inductive Miner internally as an initial control flow discovery step, which has been updated from the gradient-descent procedure described in [25]. Algorithms, reference event logs and conformance measures are summarized as Fig. 3.

Measures include (1) Truncated Earth Movers' Distance (tEMSC) [20], a measure expressing the cost of transforming the distribution of activity traces from one stochastic language into another. We use a minimum probability mass parameter setting of 0.8 for feasibility. (2) Entropy Precision and Recall [21] are stochastic conformance measures based on the entropy of equivalent automata constructed from a given log or model. (3) Petri net entity count (places and transitions) and (4) edge count are used as structural simplicity measures, ensuring that conformance quality has not been achieved by sacrificing model simplicity and comprehensibility. Entity and arc counts have existing uses in process model evaluation [14,17], and were preferred here over behavioural simplicity measures [16], though these measures also have limitations, including specificity to Petri nets, and insensitivity to the stochastic perspective of GSPNs. The duration of a discovery process was also captured, and direct discovery times are compared with combined runtimes for discovery and estimation.

The experiments were run on a Windows 10 machine with 2.3 GHz CPU and 50 GB of memory allocated to each process on JDK 1.8.0_222. All logs are publicly available at https://data.4tu.nl/. The full results for these experiments are available in an accompanying technical report [11].

4.2   Results and Discussion

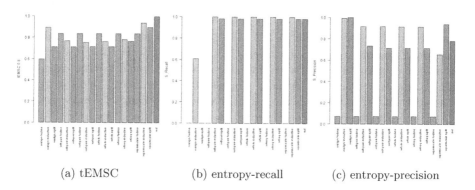

(a) tEMSC   (b) entropy-recall   (c) entropy-precision

Fig. 4. Results on BPIC 2018 control log categorized by {*estimator*}-{*control flow algorithm*}, plus RSD.

The estimators produced different, relevant, stochastic models when applied to a range of real-life logs. As seen in Figs. 4 and 5, stochastic conformance for these models was comparable, but not uniformly better, than existing techniques, and was highly dependent on the discovery algorithm and log.

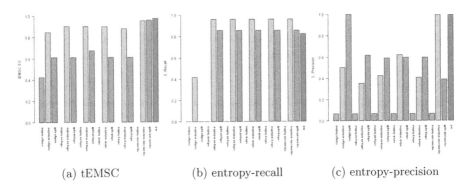

(a) tEMSC   (b) entropy-recall   (c) entropy-precision

Fig. 5. Results on BPIC 2018 reference log.

The estimators combined well with the Inductive Miner and Split Miner control discovery algorithms. Frequency-based estimators combined poorly with the Fodina discovery algorithm for some logs. This is at least partly due to Petri net representational bias in the presented framework. Fodina outputs a causal net, which was converted to a Petri net. The resulting Petri net includes a large number of silent transitions, often intermediating between transitions corresponding to activity pairs in the log. This can be seen distinctly in results for BPIC 2018 reference log in Fig. 5, where w_{align} produces a stochastically relevant model on the output of a Fodina input, but no other estimator does. For Split Miner and Inductive Miner, though they use other representations internally, the Petri net model produced used fewer silent transitions and were less impacted by this property.

For the BPIC 2013 closed and incidents logs, Fodina returned a model without an initial place, to which w_{fork}, w_{align}, tEMSC and Entropy-Recall and Entropy-Precision conformance measures do not apply. For some algorithm-estimator combinations, these conformance measures could not be calculated due to soundness, time or memory constraints. Nevertheless, in these results it is clear that tEMSC 0.8 is more sensitive to the stochastic perspective produced by estimators than the Entropy Precision and Recall measures. Where RSD [25] produced a model on which measures could be calculated, the resulting models often conformed well to the logs, but not consistently better than the estimator-produced models. There were a number of event logs where RSD returned no model within the constraints of time (12 h timeout) and machine memory, or where conformance measures were unable to be calculated within time (5 h timeout) and memory constraints.

The run time of the estimators, which took never more than 10 s, was always comparable or better than RSD, orders of magnitude better in some cases, as shown in Fig. 6. In the future, we aim to extend these experiments with larger logs containing more traces, events, and activities. However, even though our estimators returned results for each model and log combination quickly, the conformance measures were the limiting factors in these experiments in terms

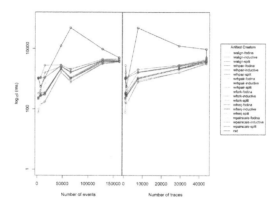

Fig. 6. Run times for control flow discovery and weight estimation by event and trace count. 12 h time out for RSD [25] on sepsis log is excluded.

of time and memory, which indicates that future research should be directed towards more efficient stochastic conformance checking techniques.

In summary, our new estimators, even the alignment-based w_{align}, are able to handle real-life event logs and outputs from existing discovery techniques much faster than existing approaches. Depending on the applied discovery technique, they can also achieve higher stochastic quality, providing alternatives to the existing RSD discovery technique when analyzing control flow and stochastic perspectives.

5   Related Work

Significant work exists on performance analysis using process mining and Stochastic Petri Nets (SPNs) with pre-existing normative models. This includes improving parameters from an input SPN [22,26,29], from models in UML [9], and industrial case studies [9,26]. These and other applications can benefit directly from automatic discovery of stochastic models.

RSD [25] is a technique, with publicly available implementation, for discovering Generally Distributed Transition Stochastic Petri Nets (GDT_SPNs), with some high level descriptions of techniques and algorithms preceding it [6,15,18]. RSD first discovers a control flow model in the form of a Petri net, then performs a fitness calculation, and attempts to repair the model if fitness is low. An alignment and replay calculation then informs the production of an output GDT_SPN. The distinction between control flow discovery and stochastic perspectives is extended by our proposed framework to many possible weight estimators. The post-control flow discovery steps in RSD are a *weight estimator*, but not a *simple estimator*, in our terminology.

In [27,28], queues are discovered in stochastic process mining using two formalisms, Process Trees [28] and Queue-Enabling Colored Stochastic Petri

Nets [27]. The Process Tree approach is informed by statistics theory and uses both Bayesian and Markov-Chain Monte-Carlo fitting.

Hidden Markov Models (HMMs) have seen some applications to stochastic process discovery [5,12]. For instance, [12] constructs HMMs for resource usage using a variant of the Alpha algorithm [3, p. 167], an early process mining algorithm with known weaknesses on real-world event data. [5] uses event log data to prune unlikely paths from a HMM process model in the context of a semi-automated stochastic process discovery procedure.

Declarative process models describe a process in terms of constraints on behaviour. This contrasts with control-flow based process models, such as Petri nets used in our framework, which describe permitted behaviour. Techniques for automatic process discovery of probabilistic declarative models also exist [23]. Transforming the significant differences between the forms of control-flow and declarative models, and evaluating the result for stochastic conformance, put rigorous comparison beyond the scope of this paper.

6   Conclusion

The likelihood of an event is important information in understanding many real-world processes. Automatically discovered stochastic process models may then help analyze and improve organizations. In this paper we presented a framework for discovery of Generalized Stochastic Petri Nets (GSPNs) from logs. The framework leverages existing control flow discovery algorithms, and introduces *estimators* which transform discovered Petri nets into GSPNs. We introduced six estimators; their implementation is publicly available, and evaluated against real-life logs using multiple stochastic conformance measures. The evaluation used three existing flow discovery algorithms, and an existing stochastic discovery technique, finding models of comparable quality, across a broader range of logs, in a generally shorter time.

The estimators presented here are not exhaustive, and we look forward to future research on novel, improved estimators. The estimator framework also implies the possibility of "direct stochastic discovery" algorithms which do not use a separate control flow algorithm, but produce a control flow model as a side-effect of a stochastic one. A simplicity measure sensitive to both structural representation and stochastic information in a process model would be a useful evaluation tool for work in this area, and is an avenue of future research.

Acknowledgement. Computational resources used included those provided by the eResearch Office at QUT.

References

1. van der Aalst, W.M.P., Adriansyah, A., van Dongen, B.: Replaying history on process models for conformance checking and performance analysis. DMKD **2**(2), 182–192 (2012)

2. van der Aalst, W.: Academic view: development of the process mining discipline. In: Reinkemeyer, L. (eds.) Process Mining in Action, pp. 181–196. Springer, Cham (2020). https://doi.org/10.1007/978-3-030-40172-6_21

3. van der Aalst, W.: Process Mining: Data Science in Action, 2nd edn. Springer, Heidelberg (2016). https://doi.org/10.1007/978-3-662-49851-4

4. Marsan, M.A., et al.: The effect of execution policies on the semantics and analysis of stochastic Petri nets. TSE **15**(7), 832–846 (1989)

5. Alharbi, A.M.: Unsupervised abstraction for reducing the complexity of healthcare process models. Ph.D. thesis, University of Leeds, July 2019

6. Anastasiou, N., Knottenbelt, W.: Deriving coloured generalised stochastic Petri net performance models from high-precision location tracking data. In: PE 2013, pp. 375–386 (2013)

7. Augusto, A., et al.: Split miner: automated discovery of accurate and simple business process models from event logs. KIS **59**, 251–284 (2019). https://doi.org/10.1007/s10115-018-1214-x

8. Bause, F., Kritzinger, P.S.: Stochastic Petri Nets: An Introduction to the Theory. Vieweg+Teubner Verlag (2002)

9. Bernardi, S., et al.: A systematic approach for performance evaluation using process mining: the POSIDONIA operations case study. In: QUDOS 2016, pp. 24–29 (2016)

10. vanden Broucke, S.K.L.M., et al.: Fodina: a robust and flexible heuristic process discovery technique. DSS **100**, 109–118 (2017)

11. Burke, A., et al.: Report on stochastic process discovery by weight estimation experimental results. Technical report, September 2020. https://eprints.qut.edu.au/204662/

12. Carrera, B., Jung, J.-Y.: Constructing probabilistic process models based on hidden Markov models for resource allocation. In: Fournier, F., Mendling, J. (eds.) BPM 2014. LNBIP, vol. 202, pp. 477–488. Springer, Cham (2015). https://doi.org/10.1007/978-3-319-15895-2_41

13. van Dongen, B.F., de Medeiros, A.K.A., Verbeek, H.M.W., Weijters, A.J.M.M., van der Aalst, W.M.P.: The ProM framework: a new era in process mining tool support. In: Ciardo, G., Darondeau, P. (eds.) ICATPN 2005. LNCS, vol. 3536, pp. 444–454. Springer, Heidelberg (2005). https://doi.org/10.1007/11494744_25

14. Gruhn, V., Laue, R.: Adopting the cognitive complexity measure for business process models. In: CI 2006, pp. 236–241 (2006)

15. Hu, H., Xie, J., Hu, H.: A novel approach for mining stochastic process model from workflow logs. JCIS **7**(9), 3113–3126 (2011)

16. Kalenkova, A., Polyvyanyy, A., La Rosa, M.: A framework for estimating simplicity of automatically discovered process models based on structural and behavioral characteristics. In: Fahland, D., Ghidini, C., Becker, J., Dumas, M. (eds.) BPM 2020. LNCS, vol. 12168, pp. 129–146. Springer, Cham (2020). https://doi.org/10.1007/978-3-030-58666-9_8

17. Kluza, K., Nalepa, G.J., Lisiecki, J.: Square complexity metrics for business process models. In: Mach-Król, M., Pełech-Pilichowski, T. (eds.) Advances in Business ICT. AISC, vol. 257, pp. 89–107. Springer, Cham (2014). https://doi.org/10.1007/978-3-319-03677-9_6

18. Leclercq, E., et al.: Identification of timed stochastic Petri net models with normal distributions of firing periods. IFAC **13**(4), 948–953 (2009)

19. Leemans, S.J.J., Fahland, D., van der Aalst, W.M.P.: Discovering block-structured process models from event logs - a constructive approach. In: Colom, J.-M., Desel, J. (eds.) PETRI NETS 2013. LNCS, vol. 7927, pp. 311–329. Springer, Heidelberg (2013). https://doi.org/10.1007/978-3-642-38697-8_17

20. Leemans, S.J.J., Syring, A.F., van der Aalst, W.M.P.: Earth movers' stochastic conformance checking. In: Hildebrandt, T., van Dongen, B.F., Röglinger, M., Mendling, J. (eds.) BPM 2019. LNBIP, vol. 360, pp. 127–143. Springer, Cham (2019). https://doi.org/10.1007/978-3-030-26643-1_8

21. Leemans, S.J.J., Polyvyanyy, A.: Stochastic-aware conformance checking: an entropy-based approach. In: Dustdar, S., Yu, E., Salinesi, C., Rieu, D., Pant, V. (eds.) CAiSE 2020. LNCS, vol. 12127, pp. 217–233. Springer, Cham (2020). https://doi.org/10.1007/978-3-030-49435-3_14

22. Chuang, L.I.N., Yang, Q.U., Fengyuan, R.E.N., Marinescu, D.C.: Performance equivalent analysis of workflow systems based on stochastic Petri net models. In: Han, Y., Tai, S., Wikarski, D. (eds.) EDCIS 2002. LNCS, vol. 2480, pp. 64–79. Springer, Heidelberg (2002). https://doi.org/10.1007/3-540-45785-2_5

23. Maggi, F.M., Montali, M., Peñaloza, R.: Probabilistic conformance checking based on declarative process models. In: Herbaut, N., La Rosa, M. (eds.) CAiSE 2020. LNBIP, vol. 386, pp. 86–99. Springer, Cham (2020). https://doi.org/10.1007/978-3-030-58135-0_8

24. Mendling, J., Reijers, H.A., Cardoso, J.: What makes process models understandable? In: Alonso, G., Dadam, P., Rosemann, M. (eds.) BPM 2007. LNCS, vol. 4714, pp. 48–63. Springer, Heidelberg (2007). https://doi.org/10.1007/978-3-540-75183-0_4

25. Rogge-Solti, A., van der Aalst, W.M.P., Weske, M.: Discovering stochastic Petri nets with arbitrary delay distributions from event logs. In: Lohmann, N., Song, M., Wohed, P. (eds.) BPM 2013. LNBIP, vol. 171, pp. 15–27. Springer, Cham (2014). https://doi.org/10.1007/978-3-319-06257-0_2

26. Rogge-Solti, A., et al.: Prediction of business process durations using non-Markovian stochastic Petri nets. IS **54**, 1–14 (2015)

27. Senderovich, A., et al.: Data-driven performance analysis of scheduled processes. In: Motahari-Nezhad, H.R., Recker, J., Weidlich, M. (eds.) BPM 2015. LNCS, vol. 9253, pp. 35–52. Springer, Cham (2015). https://doi.org/10.1007/978-3-319-23063-4_3

28. Senderovich, A., Leemans, S.J.J., Harel, S., Gal, A., Mandelbaum, A., van der Aalst, W.M.P.: Discovering queues from event logs with varying levels of information. In: Reichert, M., Reijers, H.A. (eds.) BPM 2015. LNBIP, vol. 256, pp. 154–166. Springer, Cham (2016). https://doi.org/10.1007/978-3-319-42887-1_13

29. Tsironis, L.C., et al.: Fuzzy performance evaluation of workflow stochastic Petri nets by means of block reduction. ToS **40**(2), 352–362 (2010)

30. Weijters, A.J.M.M., Ribeiro, J.T.S.: Flexible heuristics miner (FHM). In: CIDM 2011, pp. 310–317 (2011)

Graph-Based Process Mining

Amin Jalali[✉] [ID]

Department of Computer and Systems Sciences, Stockholm University,
Stockholm, Sweden
`aj@dsv.su.se`

Abstract. Process mining is an area of research that supports discovering information about business processes from their execution event logs. One of the challenges in process mining is to deal with the increasing amount of event logs and the interconnected nature of events in organizations. This issue limits the organizations to apply process mining on a large scale. Therefore, this paper introduces and formalizes a new approach to store and retrieve event logs into/from graph databases. It defines an algorithm to compute Directly Follows Graph (DFG) inside the graph database, which shifts the heavy computation parts of process mining into the graph database. Calculating DFG in graph databases enables leveraging the graph databases' horizontal and vertical scaling capabilities to apply process mining on a large scale. We implemented this approach in Neo4j and evaluated its performance compared with some current techniques using a real log file. The result shows the possibility of using a graph database for doing process mining in organizations, and it shows the pros and cons of using this approach in practice.

Keywords: Process mining · Graph database · Big data · Neo4j

1 Introduction

Business Process Management (BPM) is a research area that aims to enable organizations to narrow the gap between business goals and information technology support [21]. Business process evaluation is a key support in narrowing down this gap. There are two evaluation techniques to analyze business processes, a.k.a., model-based analysis, and data-based analysis [17]. While model-based analysis deals with analyzing business process models, the data-based analysis mostly focuses on analyzing business processes based on their execution event logs.

Process Mining is a discipline in the BPM area that enables data-based analysis for business processes in organizations [18]. It allows analysts not only to evaluate the business processes but also to perform process discovery, compliance checking, and process enhancement based on the execution result, a.k.a., event logs. As the volume of logs increases, new opportunities and challenges also appear. The large volume of logs enables the discovery of more information about business processes; while also raises some challenges, such as feasibility, performance, and data management.

© Springer Nature Switzerland AG 2021
S. Leemans and H. Leopold (Eds.): ICPM 2020 Workshops, LNBIP 406, pp. 273–285, 2021.
https://doi.org/10.1007/978-3-030-72693-5_21

The large volume of data is a challenge to perform process mining in organizations. There are different approaches to deal with this problem. This paper
proposes and formalizes a new approach to store and retrieve event logs in graph
databases to do process mining on a large volume of data. It also defines an algorithm to compute Directly Follows Graph (DFG) inside the graph database. As
a result, it enables i) removing the requirement to move data into analysts'
computer, and ii) scaling the DFG computation vertically and horizontally.

The approach is implemented in Neo4j, and its performance is evaluated
in comparison with some current techniques based on a real log file. The result
shows the feasibility of this approach in discovering process models when the data
is much bigger than the computational memory. It also shows better performance
when dicing data into small chunks.

The remainder of this paper is organized as follows. Section 2 gives a short
background on process mining and graph database. Section 3 introduces the
graph-based process mining approach, and Sect. 4 elaborates on the implementation of the approach in Neo4j. Section 5 reports the evaluation results. Section 6
discuss alternative approaches and related works, and finally, Sect. 7 concludes
the paper and introduces future research.

2   Background

2.1   Process Mining

Process Mining is a research area that supports business process data-based
analysis [18]. Process discovery is a sort of process mining technique that enables
identifying process models from event logs automatically. There are different
sorts of perspectives that can be discovered from event logs. Control-flow, which
describes the flow of activities that happened in a business process, is one of
the most important ones. Directly-Follows Graphs (DFGs) is a simple notation
widely used and considered a de-facto standard for commercial process mining
tools [19].

Figure 1 shows an overview of how a process model can be discovered from
event logs using DFG graphs. The process discovery starts by loading a log file
that stores business process execution results, a.k.a., log files. Each log contains

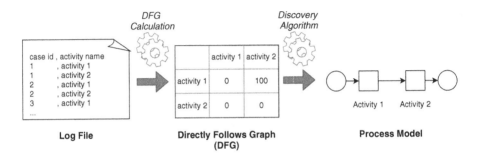

Fig. 1. Steps in a process discovery algorithm

a set of traces representing different cases that are performed in the business process. Each trace contains a set of events representing the execution result of activities in the business process. Thus, a log file shall contain information about traces and events at a minimum. Note that the events should be stored according to the execution order with this basic setup, unless we have information about execution time. It is usual to have more information like the execution time and the resource who has done the activity in the log file.

The next step is calculating the Directly Follows Graph (DFG). This graph shows the frequency of direct relations between activities that are captured in the log file. The result can be considered as a square matrix with the activity names as the index for rows and columns. Let's consider the cell with the index of *activity 1* for the row and *activity 2* for the column (see Fig. 1). The value of the cell shows the number of times that the *activity 2* happened after *activity 1*. Although the calculation of DFG comes back to alpha miner, which was introduced around 20 years ago, it is still the backbone for many process mining algorithms and tools [20]. There are different variations of DFG that store more information, but the basic idea is the same.

The last step is to infer the process model from DFG matrix based on rules that are specified by a process discovery algorithm. This step usually does not take much time since the computation is performed on top of DFG.

2.2 Graph Database

Graph databases are Database Management Systems (DBMS) that support creating, storing, retrieving, and managing graph database models. Graph database models are defined as the data structure where schema and instances are modeled as graphs, and the operation on graphs are graph-oriented [2]. The idea is not new, and it comes back to the late eighties when the object-oriented models were also introduced [2]. However, it recently got much attention in both research and industry due to its ability to handle the huge amount of data and networks. It enables leveraging parallel computing capabilities to analyze massive graphs. As a result, a new discipline is emerged in research, called Parallel Graph Analytics [15].

There are different sorts of graph databases with different features. For example, Neo4j is a graph DBMS that supports both vertical and horizontal scaling, meaning that not only the hardware of the system that runs the DBMS can be scaled out, but the number of physical nodes that run the DBMS as a network can be increased. These features enable having a considerable performance at runtime.

3 Approach

This paper proposes a new approach to store event logs and retrieve a DFG using a graph database. In this way, the scalability capabilities in graph databases can be used in favor of applying process mining. The aim is to introduce an alternative approach to enable discovering process models from large event logs.

Thus, the formal definitions of event repository in graph form are introduced. Then, the soundness property of such a repository log is defined. Finally, an algorithm to discover DFG is introduced.

Note that the formal definition is simplified by limiting the set of attributes to hold information about activities. In practice, the definition of attributes can be extended to store all information about the data perspective.

3.1   Definitions

Definition 1 (Event Repository). *An event repository is a tuple $G = (N = L \cup T \cup E \cup A, R)$, where:*

- *N is the superset of L, T, E, and A subsets which are pairwise disjoint, where:*
    - *L represents the set of logs,*
    - *T represents the set of traces,*
    - *E represents the set of events,*
    - *A represents the set of attributes, representing activities, where:*
    - *$L \cap T \cap E \cap A = \emptyset$.*
- *$R = L \times T \cup T \times E \cup E \times E \cup E \times A$ is the set of relations connecting:*
    - *logs to traces, i.e., $L \times T$*
    - *traces to events, i.e., $T \times E$,*
    - *events to events, i.e., $E \times E$,*
    - *events to attributes, i.e., $E \times A$, where:*
    - *$N \cap R = \emptyset$*

Let's also define two operators on the graph's nodes as:

- *$\bullet n$ represents the operator that retrieves the set of nodes from which there are relations to node n, i.e., $\bullet n = \{\forall e \in N | (e, n) \in R\}$.*
    - *This operator enables retrieving incoming nodes for a given node, e.g., retrieving the set of events that occurred for an activity.*
- *$n \bullet$ represents the operator that retrieves the set of nodes to which there are relations from node n, i.e., $n \bullet = \{\forall e \in N | (n, e) \in R\}$.*
    - *This operator enables retrieving outcoming coming nodes for a given node, e.g., retrieving the set of events that occurred for a trace.*

Note that the relations among logs, traces, events, and attributes are adopted from the eXtensible Event Stream (XES) standard [1]. The information is stored in attributes like XES standard which states: "Information on any component (log, trace, or event) is stored in attribute components" [1]. This is the reason why the activities are represented as attributes in this work. Note that we limit attributes to represent activities only in this work for making formalization simple for the sake of presentation. In practice, the attributes can have types to represent different properties. For example, they can be used to store different data properties of an event, e.g., who has performed it, what data it generates, etc. The usage of attributes in practice can also be extended to hold case id properties for traces and metadata information for the log node. Despite it is good to have the case id as an attribute, we kept the formalization simple by ignoring that as traces represent cases in this structure. Note that you need to know the case id to create such a structure, which is needed in the ETL process.

Definition 2 (Soundness). *An event repository* $G = (N = L \cup T \cup E \cup A, R)$, *where* N, L, T, E, A, R, *represent the set of Nodes, Logs, Traces, Events, Attributes, Relations respectively, is sound iff:*

- $\forall t \in T, |\bullet t| = 1$, *meaning that a trace must belong to 1 and only 1 log.*
- $\forall e \in E, |\bullet e \cap T| = 1$, *meaning that an event must belong to 1 and only 1 trace.*
- $\forall e \in E, |\bullet e \cap E| <= 1$, *meaning that an event can only have at most 1 input flow from another event.*
- $\forall e \in E, |e \bullet \cap E| <= 1$, *meaning that an event can only have at most 1 output flow to another event.*
- $\forall e \in E, |e \bullet \cap A| = 1$, *meaning that an event must be related to 1 and only 1 attribute.*

Note that the soundness is a property of event repository and shall not be mistaken by the soundness property of a modeling notation like Petri nets. It is worth mentioning that this formalization can be extended to enable several types of sequences among event logs. To calculate DFG, we need to count the number of direct relations among events for each activity pairs. Algorithm 1 defines how the DFG for a given sound event repository can be calculated.

Algorithm 1: Algorithm for calculating dfg

```
1  Algorithm dfgcalculator(G = (N = L ∪ T ∪ E ∪ A, R))
2  │   Ψ ← ∅;
3  │   foreach two attributes a, b ∈ A do
4  │   │   c ← 0;
5  │   │   foreach e ∈ •a, e′ ∈ •b do
6  │   │   │   if (e, e′) ∈ R then
7  │   │   │   │   c ← c + 1;
8  │   │   Ψ ← Ψ ∪ {(a, b, c)};
9  return Ψ;
```

3.2 Example

This section elaborates on the definitions through an example.

Figure 2 shows an example of a sound event repository graph. The set of nodes for Log, Trace, Event, and Attribute are colored as green, red, white, and

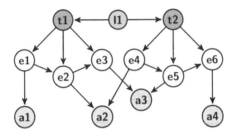

Fig. 2. An example of a sound event repository graph

yellow, respectively. This repository includes one log file, called *l1*, which has two traces, i.e., *t1* and *t2*. *t1* has three events that occurred in this order $e1 \rightarrow e2 \rightarrow e3$. *t2* also has three events that occurred in this order $e4 \rightarrow e5 \rightarrow e6$.

As it can be seen, each event is related to one activity, e.g., *e1* is the execution of activity *a1*. To get the list of events that happened for an activity *a1*, we can use •*a1* operator, which returns $\{e1\}$. For some activities, there might be more than one event, e.g., •*a2* returns $\{e2, e4\}$. Applying Algorithm 1 on this event repository will return the DFG. The DFG calculation is described as below:

– for each pair of activities, the algorithm will calculate the frequency. We show the calculation for one pair example, i.e., $a2, a3$:
  – •*a2* retreives $\{e2, e4\}$
  – •*a3* retreives $\{e3, e5\}$
  – $c = \sum_{\forall e \in \bullet a2, e' \in \bullet a3} |(e, e') \in R| = \sum_{\forall e \in \{e2, e4\}, e' \in \{e3, e5\}} |(e, e') \in R|$
  $= |\{(e2, e3), (e4, e5)\}| = 2$

If we calculate the frequencies for all pairs of activities, the result will be like Table 1.

Table 1. DFG calculation for the sample event repository graph

	a1	a2	a3	a4
a1	0	1	0	0
a2	0	0	2	0
a3	0	0	0	1
a4	0	0	0	0

4   Implementation

The approach presented in this paper is implemented using the Neo4j, which was chosen because it supports i) storing graphs and doing graph operations,

ii) both vertical and horizontal scaling, iii) querying the graph using Cypher, iv) containerizing the database, which allows controlling the computational CPU and memory.

We implemented a data-aware version of the approach. The main differences with the formalization are:

- Attributes store activity names and other attributes that might be associated with an event like resource id, case id, etc. To comply with PM4Py, we stored the log, case, and activity name by 'log_concept_name', 'case_concept_name', and 'concept_name' respectively.
- events have timestamps to enable dicing information based on time. Note that the timestamp cannot be defined as an attribute with its own key since we will end up with many extra nodes due to many timestamps that exist for each event. Thus, they are kept as an attribute of Event class, following the same practice to deal with times in data warehousing [14].

The calculation of DFG is implemented using a Cypher query as below:

```
match
(a1:Attribute {key:'concept_name'})<--(:Event)-[n]->(:Event)
   -->(a2:Attribute {key:'concept_name'})
return
a1.val as dfg_from, a2.val as dfg_to, count(n) as dfg_freq
```

The match clause in the query identifies all patterns in sub-graphs that match the expression. This expression selects two attributes *a1* and *a2* with the type of *concept_name*, which indicates that they are activities' names. Then, it selects all incoming events to those attributes where there is a direct relationship between those two events. The return clause retrieves all combinations of attributes in addition to the number of total direct relations between their events, which is the calculation that we formalized in Algorithm 1.

To limit the number of events based on their timestamp, we can easily add a where clause to the cypher query to limit the timestamp. For other attributes, the associated attribute node can be filtered.

5   Evaluation

This section reports the evaluation result of the approach, which is presented in this paper[1]. To evaluate the approach, we calculated DFG for a real public log file [6] using Process Mining for Python (PM4Py) library [3]. This dataset [6] is selected because it is published openly, which makes the experiment repeatable. It is also the biggest log file that we could find in the BPI challenges, which can help us to evaluate the performance.

---

[1] The data, code and instructions can be found at https://github.com/neo4pm/supporting_materials/tree/master/papers/Graph-based%20process%20mining.

To evaluate the performance, we need to control the resources that are available for performing process mining. Thus, we decided to containerize the experiments and run them with Docker. Docker is a Platform as a Service (PaaS) product that enables creating, running, and managing containers. It also enables the control of the resources that are available for each container, such as RAM and CPU.

Among different process mining tools, we chose PM4Py [3], because i) it is open-source; ii) the DFG calculation step and discovery step can be separated easily, and iii) it can easily be encapsulated in a container. The separation of DFG calculation and discovery step in this library also enables reusing all discovery algorithms along using our approach, which makes our approach very reusable.

We designed two experiments to evaluate our approach. In *Experiment 1*, we loaded the whole log file into both containers running neo4j and PM4Py, so we kept the number of event logs constant. We calculated DFG several times by changing the RAM and CPU, so we defined the computational resources as a variable. In *Experiment 2*, we kept RAM and CPU constant for both containers, and we calculated DFG by dicing the data. The dicing is done based on a time constraint, and we added more days in an accumulative way to increase the number of events. We ran the experiments for each container separately to make sure that the assigned resources are free and available (Table 2).

Table 2. Evaluation setting

	Constant	Variable
Experiment 1	Events in the log (9 million events)	CPU & RAM
Experiment 2	CPU & RAM	Events in the log

5.1   Experiment 1

To simulate the situation where the computational memory is less than the log size, we started by assigning 512 megabytes of ram to each container. We added the same amount of RAM in each experiment round until we reached 4 gigabytes. We also changed the CPU starting from half of a CPU (0.5), by adding the same amount at each round until we reached 4.0.

Figure 3 shows the execution result for both containers, where the x, y and z axes refer to the available memory (RAM) (in megabytes), DFG calculation time (in seconds), and available CPU quotes, respectively. The experiment related to neo4j and PM4Py containers is plotted in red and blue, respectively. As can be seen, PM4Py could not compute DFG when the memory was less than the size of the log, i.e., around 1.5 gigabytes, while neo4j could calculate DFG in that setting. This shows that the graph database can compute DFG when computational memory is less than the log size, which is an enabler when applying process mining on a very large volume of data.

As it can be seen in the figure, the increasing amount of memory reduced the time that neo4j computed the DFG, while it has very little effect on PM4Py. This is no surprise for in-memory calculation since if the log fits the memory, then the performance will not be increased much by adding more memory. It is also visible that assigning more CPU does not affect the performance of either of these approaches.

It should also be mentioned that despite increasing memory can reduce the DFG calculation time for neo4j significantly; it cannot be faster than PM4Py when calculating the DFG on the complete log file. The reason can be that graph databases shall process metadata, which adds more computation than in-memory calculation approaches. Thus, for small log files that can fit the computer's memory, the in-memory approach can be better if the security and access control are not necessary.

5.2   Experiment 2

Event logs usually contain different variations that exist in the enactment of business processes [4]. These variations make process mining challenging because discovering the process based on the whole event log usually produces the so-called spaghetti models, which usually cannot be comprehended by humans, so they have very little value. Thus, analysts need to filter data to produce a meaningful model, which is a common practice in applying process mining [4,11]. Therefore, we designed this experiment to compare our approach and PM4Py when calculating DFG on a filtered subset of data without scaling the infrastructure.

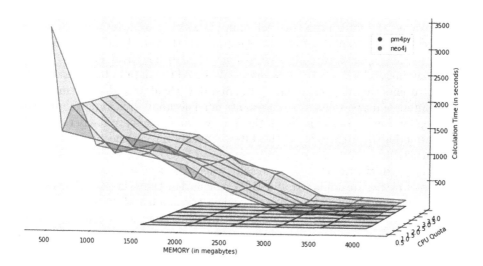

Fig. 3. Evaluating DFG calculation time by scaling resources

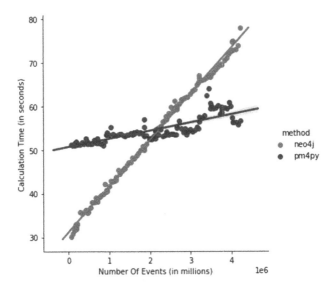

Fig. 4. Evaluating DFG calculation time by dicing the log

To evaluate this scenario, we kept the resources (RAM and CPU) constant for both containers, but we changed the condition for filtering the data. The condition is set based on the dates in which the event occurred. We started by filtering events for a day range, and we calculated DFG for the filtered data. Then, we expanded the filter range by including events that occurred the day after, and we calculated DFG again. We repeated the process for 30 days. As we expanded the filter range by including events that occurred on more days, we increased the number of events. This means that we kept the number of events in the log as a variable. We assigned 14 Gb for RAM and 4 CPU for each container, which was run separately. We diced the data in both settings by filtering events that happened during the first day; then, we added one more day to the filter condition to increase the events in an accumulative way. We repeated this step for almost four months. In this way, we could compare the performance by considering how the size of the filtered events affects the performance of calculating DFG.

Figure 4 shows the evaluation result, where the x and y axes refer to the number of events (in millions) and DFG calculation time (in seconds). As can be seen, our approach performed better when the number of events is less than 2 million. Note that this is still a very big sub-log to analyze for process mining, so this shows that our approach can improve the performance of process mining when dealing with sub-sets of the log. However, PM4Py performed better when the number of events exceeded 2 million. This is no surprise since PM4Py loaded logs into memory first, so increasing the size will have less effect on its performance. Indeed, the difference is only related to filtering the log and retrieving the biggest chunk of data in each iteration.

6   Related Work and Discussion

The related work can be divided into two categories: those related to scalability and those using graph databases.

6.1   Scalability

The scalability issue in process mining is a big concern for applying the techniques on a large volume of data. Thus, different researchers investigated this problem through different techniques.

Hernández, S. et al. computed intermediate DFG and other matrixes through the MapReduce technique over a Hadoop cluster [10]. The evaluation of their approach shows a similar trend for a performance like what we presented in Fig. 4. The performance cannot be compared precisely due to different setup and resources. This is the closest approach to ours.

MapReduce has been used by other researchers for the aim of process mining, e.g., [9,16]. As discussed by [10], MapReduce has been used to support only event correlation discovery in [16], and it is used to discover process models using Alpha Miner and the Flexible Heuristics Miner in [9].

6.2   Graph Database

There are different attempts to use graph databases with process mining.

Esser S. and Fahland D. used the graph database to query multi-dimensional aspects from event logs. This is one important use case that has been introduced by a graph database, i.e., adding more features to the data [7]. They have used Neo4j as the graph database and used Cypher to query the logs. The approach uses a graph database as a log repository to store data without any predefined structure, which is quite different from the topic of this paper. In this regard, the approach is similar to [5], where a relational database is used to store the data. The main difference is that [7] demonstrates that the graph database has more capability to add more features to data, which is a very important topic in any machine learning related approach in general.

Joishi J. and Sureka A. also used a graph database for storing non-structured event logs [12,13]. They also demonstrated that Actor-activity matrix could be calculated using Cypher. However, the approach is context-dependent since the logs are not standardized like our approach. Also, the approach cannot be used with other process discovery algorithms since it does not shift and separate the computation of DFG to a graph database.

Parallel to this work, we realized that Esser S. and Fahland D. [8] extended their approach [7] to discover different perspectives from events which are stored in neo4j. They also introduced an approach to discover DFG from their repository. The approach is similar in creating the event repository, yet this paper also focuses on evaluating the performance and scalability to some extent. This study also confirms the benefits of using a graph database for process mining, which can extend the application of process mining in practice.

7　Conclusion

This paper introduced and formalized a new approach to support process mining using graph databases. The approach defines how log files shall be stored in a graph database, and it also defines how Directly Follows Graphs (DFG) can be calculated in the graph database. The approach is evaluated in comparison with PM4Py by applying it to a real log file. The evaluation result shows that the approach supports mining processes when the event log is bigger than computational memory. It also shows that it is scalable, and the performance is better when dicing the event log in a small chunk.

Graph databases can bring more benefits to process mining than what we have presented in this paper. They are useful to support complex analysis, which requires taking the interconnected nature of data into account. Thus, they can enable more advanced analysis by incorporating data relations while applying different process mining techniques. As future work, we aim to extend the formalization to represent the data-aware event repository. It is also interesting to compare this approach with process discovery approaches that can be implemented in Apache Spark. We also intend to develop a new library to support the use of a graph database for process mining for practitioners and researchers.

References

1. IEEE standard for extensible event stream (XES) for achieving interoperability in event logs and event streams. IEEE Std 1849–2016, pp. 1–50 (2016)
2. Angles, R., Gutierrez, C.: Survey of graph database models. ACM Comput. Surv. (CSUR) **40**(1), 1–39 (2008)
3. Berti, A., van Zelst, S., van der Aalst, W.: Process Mining for Python (PM4Py): bridging the gap between process-and data science, pp. 13–16 (2019)
4. Bolt, A., De Leoni, M., van der Aalst, W., Gorissen, P.: Exploiting process cubes, analytic workflows and process mining for business process reporting: a case study in education. In: SIMPDA, pp. 33–47 (2015)
5. De Murillas, E., Reijers, H., van der Aalst, W.: Connecting databases with process mining: a meta model and toolset (2016)
6. Dees, M., van Dongen, B.: BPI challenge 2016: clicks not logged in (2016)
7. Esser, S., Fahland, D.: Storing and querying multi-dimensional process event logs using graph databases. In: Di Francescomarino, C., Dijkman, R., Zdun, U. (eds.) BPM 2019. LNBIP, vol. 362, pp. 632–644. Springer, Cham (2019). https://doi.org/10.1007/978-3-030-37453-2_51
8. Esser, S., Fahland, D.: Multi-dimensional event data in graph databases. arXiv preprint arXiv:2005.14552 (2020)
9. Evermann, J.: Scalable process discovery using map-reduce. IEEE Trans. Serv. Comput. **9**(3), 469–481 (2014)
10. Hernández, S., Ezpeleta, J., van Zelst, S., van der Aalst, W.: Assessing process discovery scalability in data intensive environments. In: Big Data Computing (BDC), pp. 99–104. IEEE (2015)
11. Jalali, A.: Exploring different aspects of users behaviours in the Dutch autonomous administrative authority through process cubes. Business Process Intelligence (BPI) Challenge (2016)

12. Joishi, J., Sureka, A.: Vishleshan: performance comparison and programming process mining algorithms in graph-oriented and relational database query languages. In: International Database Engineering & Applications Symposium, pp. 192–197 (2015)
13. Joishi, J., Sureka, A.: Graph or relational databases: a speed comparison for process mining algorithm. arXiv preprint arXiv:1701.00072 (2016)
14. Kimball, R., Ross, M.: The Data Warehouse Toolkit: The Complete Guide to Dimensional Modeling. Wiley, Hoboken (2011)
15. Lenharth, A., Nguyen, D., Pingali, K.: Parallel graph analytics. Commun. ACM **59**(5), 78–87 (2016)
16. Reguieg, H., Toumani, F., Motahari-Nezhad, H.R., Benatallah, B.: Using Mapreduce to scale events correlation discovery for business processes mining. In: Barros, A., Gal, A., Kindler, E. (eds.) BPM 2012. LNCS, vol. 7481, pp. 279–284. Springer, Heidelberg (2012). https://doi.org/10.1007/978-3-642-32885-5_22
17. van der Aalst, W.: Business process management: a comprehensive survey. Int. Sch. Res. Not. **2013**, 37 (2013). ISRN Software Engineering
18. van der Aalst, W.: Process Mining: Data Science in Action. Springer, Heidelberg (2016). https://doi.org/10.1007/978-3-662-49851-4
19. van der Aalst, W.: A practitioner's guide to process mining: limitations of the directly-follows graph (2019)
20. van der Aalst, W.: Academic view: development of the process mining discipline. In: Process Mining in Action: Principles, Use Cases and Outlook (2020)
21. Weske, M.: Business Process Management: Concepts, Languages, Architectures. Springer, Heidelberg (2019). https://doi.org/10.1007/978-3-662-59432-2

3rd International Workshop on Process-Oriented Data Science for Healthcare (PODS4H)

Third International Workshop on Process-Oriented Data Science for Healthcare (PODS4H)

The world's most valuable resource is no longer oil, but data. The ultimate goal of data science techniques is not to collect more data, but to extract knowledge and insights from existing data in various forms. To analyze and improve processes, event data is the main source of information. In recent years, a new discipline has emerged which combines traditional process analysis and data-centric analysis: Process-Oriented Data Science (PODS). The interdisciplinary nature of this new research area has resulted in its application to analyze processes in different domains such as education, finance, and especially healthcare.

The International Workshop on Process-Oriented Data Science for Healthcare 2020 (PODS4H20) aimed at providing a high-quality forum for interdisciplinary researchers and practitioners (both data/process analysts and a medical audience) to exchange research findings and ideas on healthcare process analysis techniques and practices. PODS4H research includes a wide range of topics from process mining techniques adapted for healthcare processes to practical issues on implementing PODS methodologies in healthcare centers' analysis units. For more information, visit pods4h.com.

Despite being a virtual workshop this year, the third edition of the workshop attracted 13 submissions, from which 7 Regular Papers were selected for presentation. The papers included a wide range of topics such as process mining, disease trajectories, simulation, data standards, and clinical pathways. The conference also included 2 Forum Papers, cases, and a discussion panel.

This edition of the workshop included two awards, the Best Paper Award and the Best Student Paper Award. The PODS4H20 Best Paper Award was given to "Deriving a sophisticated clinical pathway based on patient conditions from electronic health record data" by Jungeun Lim, Kidong Kim, Minsu Cho, Hyunyoung Baek, Seok Kim, Hee Hwang, Sooyoung Yoo, and Minseok Song. The PODS4H20 Best Student Paper Award was given to "Process Mining on FHIR – An Open Standards-Based Process Analytics Suite for Healthcare" by Emmanuel Helm, Oliver Krauss, Anna Lin, Andreas Pointner, Andreas Schuler, and Josef Küng. The awards included a voucher for professional Process Scientist Training, provided by the Celonis Academic Alliance.

The workshop was an initiative of the Process-Oriented Data Science for Healthcare Alliance (PODS4H Alliance). The goal of this international alliance is to promote the research, development, education, and understanding of process-oriented data science in healthcare. For more information about the organization's activities and members, visit pods4h.com/alliance.

The organizers would like to thank all the Program Committee members for their valuable work in reviewing the papers, and the ICPM 2020 organizing committee for supporting this successful event.

Organization

Workshop Chairs

Jorge Munoz-Gama   Pontificia Universidad Católica de Chile
Carlos Fernández Llatas   Universitat Politècnica de València
Niels Martin   Hasselt University
Owen Johnson   University of Leeds
Marcos Sepúlveda   Pontificia Universidad Católica de Chile
Emmanuel Helm   University of Applied Sciences Upper Austria

Program Committee

Wil van der Aalst   RWTH Aachen University
Davide Aloini   Università di Pisa
Robert Andrews   Queensland University of Technology
Andrea Burattin   Technical University of Denmark
Daniel Capurro, M. D.   University of Melbourne
Claudio di Ciccio   Sapienza University of Rome
Marco Comuzzi   Ulsan National Institute of Science
     and Technology

Benjamin Dalmas   École des Mines de Saint-Étienne
Carlos Fernández Llatas   Universitat Politècnica de València
Renè de la Fuente, M. D.   Pontificia Universidad Católica de Chile
Roberto Gatta   Università degli Studi di Brescia
Emmanuel Helm   University of Applied Sciences Upper Austria
Zhengxing Huang   Zhejiang University
Owen Johnson   University of Leeds
Felix Mannhardt   Eindhoven University of Technology
Ronny Mans   Philips
Niels Martin   Hasselt University
Mar Marcos   Universitat Jaume I
Renata Medeiros de Carvalho   Eindhoven University of Technology
Jorge Munoz-Gama   Pontificia Universidad Católica de Chile
Simon Poon   University of Sydney
Luise Pufahl   Technische Universität Berlin
Ricardo Quintano   Philips Research
Hajo Reijers   Utrecht University
David Riaño   Universitat Rovira i Virgili
Stefanie Rinderle-Ma   Technical University of Munich

A Process Mining Approach to Statistical Analysis: Application to a Real-World Advanced Melanoma Dataset

Erica Tavazzi[1,2]([✉]) [ID], Camille L. Gerard[2] [ID], Olivier Michielin[2,3],
Alexandre Wicky[2], Roberto Gatta[2] [ID], and Michel A. Cuendet[2,3,4] [ID]

[1] Department of Information Engineering, University of Padova, Padova, Italy
erica.tavazzi@unipd.it
[2] Precision Oncology Center, Lausanne University Hospital (CHUV),
Lausanne, Switzerland
{erica.tavazzi,camille.gerard,olivier.michielin,alexandre.wicky,
roberto.gatta,michel.cuendet}@chuv.ch
[3] Molecular Modeling Group, Swiss Institute of Bioinformatics,
Lausanne, Switzerland
[4] Department of Physiology and Biophysics, Weill Cornell Medicine, New York, USA

Abstract. Thanks to its ability to offer a time-oriented perspective on the clinical events that define the patient's path of care, Process Mining (PM) is assuming an emerging role in clinical data analytics. PM's ability to exploit time-series data and to build processes without any *a priori* knowledge suggests interesting synergies with the most common statistical analyses in healthcare, in particular survival analysis. In this work we demonstrate contributions of our process-oriented approach in analyzing a real-world retrospective dataset of patients treated for advanced melanoma at the Lausanne University Hospital. Addressing the clinical questions raised by our oncologists, we integrated PM in almost all the steps of a common statistical analysis. We show: (1) how PM can be leveraged to improve the quality of the data (data cleaning/pre-processing), (2) how PM can provide efficient data visualizations that support and/or suggest clinical hypotheses, also allowing to check the consistency between real and expected processes (descriptive statistics), and (3) how PM can assist in querying or re-expressing the data in terms of pre-defined reference workflows for testing survival differences among sub-cohorts (statistical inference). We exploit a rich set of PM tools for querying the event logs, inspecting the processes using statistical hypothesis testing, and performing conformance checking analyses to identify patterns in patient clinical paths and study the effects of different treatment sequences in our cohort.

Keywords: Process mining · Oncology · Melanoma · Statistical analysis

© The Author(s) 2021
S. Leemans and H. Leopold (Eds.): ICPM 2020 Workshops, LNBIP 406, pp. 291–304, 2021.
https://doi.org/10.1007/978-3-030-72693-5_22

1   Introduction

Process Mining (PM) is a family of process analysis methods that aim at discovering, monitoring and improving the efficiency of real processes by extracting knowledge from the Event Logs (EL) recorded by an information system. Analytic algorithms are applied to ELs with the main goals of: (i) mining the data in order to represent the process able to produce them (*Process Discovery*, PD), (ii) measuring to which extent a given process can represent an input EL or how much an EL complies with a given process (*Conformance Checking*, CC), and (iii) improving process efficiency, by allowing problem diagnosis and delay prediction, recommending process redesigns or supporting decision making (*Process Enhancement*) [2].

In PM for Healthcare (PM4HC), processes are meant as a graph of activities which can be performed with the aim of diagnosing, treating and/or preventing diseases to improve the patients' health status. The activities can be clinical and non-clinical and may represent different behaviours according to the specific organization [12]. Often, such processes are highly dynamic, complex, increasingly multidisciplinary [8]. Notably, the complexity increased recently due to the advent of personalized approaches to care, in which treatments are tailored to the specific profile of the patient and disease, such that the diversity of therapeutic pathways exploded compared to traditional standardized care guidelines.

Pragmatically, PM4HC has shown interesting applications in many domains, and in Oncology in particular, PM4HC was successfully applied to identify the most common patterns of care for many kinds of tumors, even though the purpose remained exploratory. Rectal cancer [7], gynecological cancer [11], and melanoma [13] were investigated both in terms of PD and CC, even if in most cases the focus was more on CC, while the application of PD remained descriptive of the general trend [9]. From this perspective, there were only few cases where the PM4HC analysis was used for statistical inference, *i.e.* to concretely develop predictive models assessing the role of covariates in determining disease evolution or patient clinical pathway. While the idea of applying a combination of PM and statistics for a complete statistical analysis is not entirely new [4,10], it is not a very common approach and still requires to be consolidated, in particular to integrate survival analysis, which plays a forefront role in Oncology.

In this work, we focus on exploring the contributions of PM when performing statistical analyses in Oncology. As an application, we examined a real-world cohort of advanced melanoma patients treated at the Lausanne University Hospital (CHUV); here we show how PM can guide and/or assist researchers in all the classical steps of statistical analysis, that is, data preprocessing, descriptive statistics, and inferential statistics. Figure 1 summarizes these steps.

In the preprocessing step, we approached the data inspecting their structure, their information content, and their quality: after identifying the clinical milestones of interest (like diagnosis, treatments, survival outcome), data were first shaped as EL. We then employed the visualization tools provided by PM to detect data inconsistencies due to input errors or missing values. This allowed

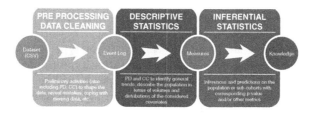

Fig. 1. Workflow of the classical steps of a statistical analysis, here implemented exploiting a process-oriented approach.

us to go back to the data sources, recheck and correct the recorded information, thus recursively improving the data quality.

In the descriptive analysis step, we first employed the EL time-oriented structure to inspect cardinality and order of the administered pharmacological treatments. Then, we implemented both unsupervised and supervised methods to capture the flow of the patients' pathways over data-driven graphs (PD approach) or user-defined graphs (CC approach), respectively. In this part of the analysis, the graphical output provided by PM allows a fast access to the design and/or interpretation of the models, and an immediate assessment of the treatments in terms of type, order and timing of consecutive administrations.

Finally, in the inferential statistics step, we build upon the processes constructed in the previous step to quickly select sub-cohorts of patients characterized by similar patterns of care and/or clinical attributes. The cohorts were then compared in terms of time-to-event outcome and overall survival (OS), using Kaplan-Meier analysis and log-rank test.

2   Materials and Methods

2.1   Material

In this work, we analyzed the data of a cohort of patients treated at the CHUV and diagnosed with advanced melanoma.

Melanoma is an aggressive cancer that arises from melanocytes (pigment cells). Cutaneous melanoma is the most common type. However, it exists also uveal and mucosal melanomas, which occur in the eye and in the mucosa (such as the mouth or the vulva), respectively. The primary risk factor of cutaneous melanoma is ultraviolet light exposure. As outdoor activities are a way of life in Switzerland, the melanoma incidence is high in the country [3]. The extent of the disease progression is described by a staging system, ranging from I to IV: Stage IV indicates metastatization of melanoma cells to distant organs. Surgery is the most common and resolutive approach for the lowest stages, but when the disease is more extensive, systemic treatments such as Immunotherapy are required, with Radiotherapy also used as palliative or local treatment.

The study cohort includes 184 patients diagnosed with advanced melanoma between March 18th, 2008 and November 17th, 2019, with follow-up up to 2019,

December 30th.[1] Data were sourced from the electronic healthcare records available at CHUV and curated by trained oncologists.

Data includes: sex, date of birth, primary tumor type, stage and diagnosis date, advanced tumor diagnosis date and mutation type (among BRAF-V600, BRAF-nonV600, NRAS, wild type (wt)), pharmacological treatments, and survival information (date of death or last follow-up). In this study, only the medications administered after the stage IV diagnosis were considered.

2.2   Methods

We implemented the classical statistical analysis pipeline shown in Fig. 1 by employing PM4HC techniques to achieve the goals of each step. To perform the analyses, we used pMineR, an open source R library implementing PM4HC functionalities [5]. By handling data in the form of EL, it allows, among its features, to implement PD and CC analyses.

We started with the raw data set, which we first assumed to be *clean* from mistakes. First, we cast the data in the form of EL, by selecting the main clinical milestones of interest for the analysis and defining the rules to cope with missing values. Then, we implemented a PD algorithm based on First Order Markov Models (FOMMs) [5], to provide a fast and easy-to-understand representation of the subsequent events. This representation allowed us to identify visually some unexpected links between clinical events (*e.g.* due to mistakes in some dates). With the help of a physician, we iteratively reviewed the data and rerun the PD algorithm in order to increasingly approach the expected graph and thus refine the data quality.

To describe the general statistics of the population and quantify the flux of patients though different patterns of cares (the second step in Fig. 1), we exploited both PD and CC techniques. The unsupervised PD analysis is based on the same FOMM model as described above. The supervised CC approach is based on a pre-defined representation of the different treatment lines implemented with the Pseudo-Workflow formalism (PWF) available in the software tool. Once performed PD and CC, the patients were grouped according to their paths through the graphs using the selection language provided by the tool. Then Kaplan-Meier survival curves and log-rank tests were used to quantify statistical differences between the groups, considering as end-points time-to-event in PD and OS in CC.

Process Discovery. In PD, one of the most diffused process representation exploits the directly-follows graphs (DFGs): in this graphical representation, directed edges link all the couples of nodes representing subsequent activities in the EL. Even if DFGs have some well-known limitations [1], they are very

---

[1] This study was approved by the Research Ethical Committee of Canton de Vaud (CER-VD) and includes only patients who did not oppose usage of their data, and was conducted according to the Swiss Federal Act on Research involving Human Beings.

intuitive and can be helpful to share with clinicians a first representation of the data. In the pMineR implementation, DFGs correspond to FOMMs.

Conformance Checking. CC was performed by using the PWF, designing a diagram that describes the expected flow of events in terms of diagnoses, treatment lines, and survival events. Graphically, this results in a set of nodes, representing the *status* that the subjects can assume, and a set of conditions (*triggers*) which fire transitions between status [6]. This representation allows to count which triggers/status are activated while automatically running down the events of each subjects, thus capturing the population behaviours through the diagram.

3   Results

3.1   Data Preprocessing

Event Log. For each patient, we built the EL with the following events, each associated with a time stamp:

- *Primary Stage*: the primary diagnosis, with melanoma type, tumor stage at the diagnosis, and somatic mutation harboured by the tumor as attributes;
- *Stage IV*: the diagnosis of stage IV;
- *T-Begin*: the begin of a line of treatment, with the type of the given drug(s) as attribute;
- *T-End*: the end of a line of treatment, with the type of the given drug(s) as attribute;
- *Dead, Censored*: the survival information, consisting in the dead of the patient or in the last follow-up date, respectively.

The collected treatments belong to the following categories:

- *Immunotherapy (IO)*: anti-CTLA4, anti-PD1, anti-CTLA4 + anti-PD1 (in combination), or other IO;
- *Chemotherapy (Chemo)*;
- *Targeted therapy*: tyrosine kinase inhibitors (TKI), other targeted therapy (TT).

In this study, only the treatments after stage IV diagnosis were considered.

Missing Data. In time-oriented analyses, missing information can consist either in unrecorded events or in missing dates associated to the events themselves. In order to preserve the clinical information we kept only complete treatments lines: the EL of patients with an incomplete line were thus truncated to the last available certain information (stage IV diagnosis or end of a previous line), artificially introducing a *Censored* event before the line with missing information.

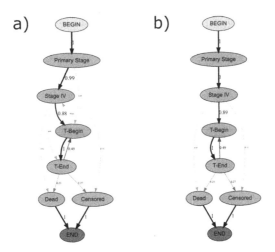

Fig. 2. First Order Markov Models obtained on all the events constituting the EL: a) before cleaning the information of a subject with an error in the dates, b) after data cleaning.

Data Cleaning. To detect mistakes in the data, we adopted an iterative approach: a FOMM process was discovered and visually analyzed to detect inconsistencies on unexpected edges. Then, the data were updated and the the procedure repeated until no more mistakes were found.

To give a practical example of detection, we report in Fig. 2a) the FOMM resulting from an intermediate version of the dataset, where unexpected edges emerge because the beginning of the first line of treatment was erroneously dated before the stage IV diagnosis for one patient in the source data. In Fig. 2b) we can observe the FOMM after correction of the inaccurately collected information. This updated graph presents, conversely, only relations fully compliant with the nature (and the collection design) of the data.

With this approach we revealed some previously uncaught mistakes in the original data, such as inconsistency in data representation (*e.g.* dd/mm/yy vs dd/mm/yyyy), or temporal event inversion (*e.g.* cancer treatment begin before a tumor diagnosis).

3.2 Descriptive Statistics

A first descriptive statistics was performed by querying the input EL, consisting of 1196 records: this allowed us to explore in the first instance cardinality and order of the administered treatments. Then, we delved into the data by using the FOMM, to obtain an agnostic data representation, and a PWF diagram, to verify the consistency of the process with respect to the expected behaviour.

Event Log Querying. By analysing the EL it was possible to perform some first descriptive investigations. We focused, specifically, on the treatments administered to the patients. Considering the events of all the patients, regardless of the position in the path of care, we extracted a total of 322 administered treatments. Table 1 reports, for each treatment category, its absolute and relative frequency of occurrence, and its duration in terms of median and inter-quartile range (25%–75%).

Out of 163 patients that received at least one recorded line of treatment, we identified 49 distinct patterns of treatment sequence. The most frequent ones are reported in Table 2.

Table 1. Occurrences and duration (in days) of the administered treatments collected in the data. The inter-quartile ranges (IQR) are computed at 25% and 75%.

Drug category	Occurrences (n = 322)	(%)	Median (IQR) duration [days]
TKI	76	(23.6)	122 (76.5–228.0)
anti-CTLA4 + anti-PD1	70	(21.7)	46.5 (0.0–167.8)
anti-PD1	66	(20.5)	84.0 (33.0–253.2)
anti-CTLA4	66	(20.5)	61.5 (31.0–63.0)
Chemo	29	(9.0)	44.0 (22.0–67.0)
Other IO	13	(4.0)	92.0 (22.0–203.0)
TT	2	(0.6)	461.5 (300.7–622.2)

Table 2. Most frequent patterns of treatment recorded in the data. The relative frequency of occurrence is computed over the total number of patients with at least one recorded treatment.

First line	Second line	Occurrence (n = 163)	(%)
anti-CTLA4 + anti-PD1	–	36	(22.1)
anti-PD1	–	22	(13.5)
anti-CTLA4	–	11	(6.7)
anti-CTLA4 + anti-PD1	TKI	11	(6.7)
Chemo	anti-CTLA4	9	(5.5)
anti-CTLA4	anti-PD1	8	(4.9)
TKI	anti-CTLA4	6	(3.7)
anti-CTLA4	TKI	5	(3.1)
TKI	–	3	(1.8)

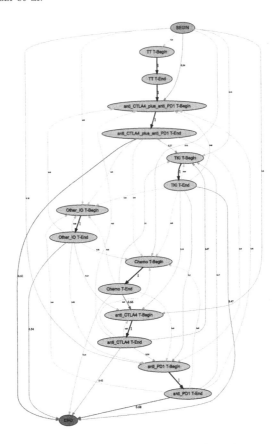

Fig. 3. First Order Markov Models obtained on the treatments.

Process Discovery on Treatment Sequences. Figure 3 shows the FOMM obtained from the clean EL considering only the administered treatments (ignoring diagnosis and survival events). Such a process allows to inspect the temporal causality of the treatments, highlighting the most frequent connections over all the population. It also provides a first overview of the position of the treatments in the paths.

Conformance Checking for Treatment Sequences. We designed a PWF able to capture the chronological order of the events: at the top, we represented the events related to the staging, and then the different treatment lines. In order to be able to define treatments paths at different levels of granularity we added a further status for each treatment line, that is, *IO* (immunotherapy). This is doable thanks to the possibility in the PWF formalism to define simultaneous activation of multiple status. Finally, we introduced two additional status to catch the survival outcomes, namely *Dead* and *Censored*, that can be activated without constraints on the previous status, as soon as a survival event is read

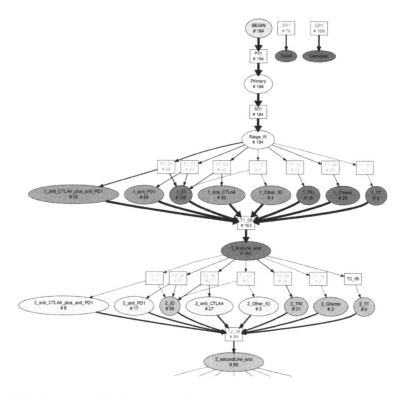

Fig. 4. Conformance Checking model (limited to the first two lines of treatments) reporting the status activated by the patients' processes over the used-defined PWF.

in the EL. The activation of the survival status terminates the inspection of the flow of events for that patient.

Figure 4 reports the result of the run on our cohort. Nodes and boxes report the number of times that a status/trigger was reached/fired. Due to space constraints, we limited the plot to the first two lines of treatment, even if the PWF included all the 7 lines of treatments available in the data.

By inspecting the graph, it is possible to follow the population's paths and read the corresponding number of subjects that run specific patterns. For instance, we can observe that all the patients included in the dataset (and thus with a BEGIN event) had a Stage IV diagnosis (expected by design), that the most frequent first line of treatment was the combination of anti-CTLA4 and anti-PD1 with a total of 56 occurrences, or that only 163 over 184 patients had a first line recorded, followed in 89 cases by a second line.

The survival nodes (*Dead* and *Censored*) are graphically separated from the others in order to limit the number of edges in the graph. However they can be reached from any point in the graph, and the available query tool can inspect at what precise point they were activated.

3.3   Inferential Statistics

By exploiting the EL, the FOMM and the PWF diagrams of the previous analyses, we could easily select cohorts characterized by specific patterns of interest and perform survival analyses. While the FOMM strongly reflects (and is limited to) the events and the information present in the EL, the PWF represents an abstraction where the user has the opportunity to provide additional knowledge in the definition of the PWF structure itself. This enhanced semantic expressiveness is one of the main reasons why PWF was previously used in structuring Clinical Guidelines [5]. Descriptive statistics can help in suggesting hypotheses: in our case, the previous PWF and FOMM diagrams allowed to easily identify and query cohorts for statistical inference analyses. We report below two examples of the investigations we performed.

First, we inspected the relationship between type of somatic tumor mutation and time between primary and Stage IV diagnosis. Here, we consider the following mutation status: BRAF V600 mutated, BRAF non-V600 mutated, NRAS mutated, and wt. For this study, we limited the cohort to cutaneous melanoma patients, exploiting filtering tool to easily query the EL attributes.

We implemented a survival analysis by first using the FOMM structure of Fig. 2 to query the path of interest (between the nodes Primary Stage and Stage IV) and obtain the time between the two events. Then, the Kaplan-Meier estimator is computed, with patients stratified by mutation status, as shown in Fig. 5a). Even if a difference between the BRAF v600 mutated and the NRAS mutated sub-cohorts seems to emerge, the log-rank test computed between all the survival distributions pairs report no significant differences (all p-values were >0.05) for any combinations.

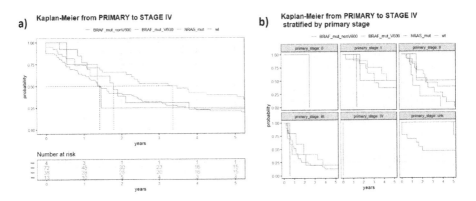

Fig. 5. Time-to-event analysis based on a mined FOMM: time from primary to stage IV diagnosis, stratified by: a) mutation, b) mutation and type of primary.

To demonstrate the potential of the analysis – even if in this case limited by the sample cardinality – we performed a further stratification of the data, distinguishing patients by their primary stage. Also here, pMineR facilitates this step, by allowing direct selection on the patient attributes. Figure 5b) reports the plot of the corresponding Kaplan-Meier estimator. Even if, as expected, no statistically significant clinical evidence emerges from this analysis, mainly due to the low number of subjects per category, it is interesting to observe how rapidly this approach allows to enrich the analysis' level of detail.

The second survival analysis exploits the PWF defined in Fig. 4. We queried the data in order to identify any differences in terms of OS based on the following patterns of interest: (1) only IO (any BRAF status), (2) IO → TKI, (3) TKI → IO, (4) only TKI. In defining the rules, we grouped together consecutive lines belonging to the same category. Patterns interspersed with TT or Chemo treatments were excluded. Upon the suggestions of clinicians, in case of sequences with multiple treatment lines, only the first occurring pattern was considered. The resulting OS survival curves are shown in Fig. 6. Table 3 reports the frequency of occurrence of each pattern, the median OS time (in years), and the percentage of patients alive at 1.5 and 3 years (CI at 95%), respectively. Statistical significance of OS differences was assessed with the log-rank test, which turned out to be significant for IO vs IO → TKI (p-value < 0.0001) and IO vs TKI → IO (p-value: 0.012). The difference between IO and IO → TKI is expected because patients who receive TKI after IO are those who did not respond to IO. Knowing that the benefits of TKI are usually only temporary, it is not surprising that these patients have shorter OS. The difference between IO and TKI → IO is interesting, as it may be related to recent biological findings showing that acquired resistance to TKI may hinder IO efficacy.

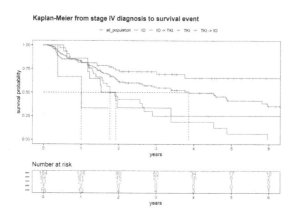

Fig. 6. Overall survival analysis based on a CC graph: time from stage IV diagnosis to death, stratified by treatment pattern.

Table 3. OS for the main treatment patterns of interest.

Treatment path	Frequency	Median OS [years]	1.5-year OS % (95% CI)	3-year OS % (95% CI)
all	100 %	3.87	72.7 (66.1–80.1)	54.9 (47.1–64.1)
IO	45.7 %	NA	76.9 (68.0–86.9)	69.4 (59.1–81.5)
IO → TKI	17.9 %	1.77	63 (48.3–82.1)	18.6 (7.7–45.2)
TKI → IO	8.7 %	1.92	57.4 (36.6–90.1)	25.1 (9.7–65.3)
TKI	1.6 %	1.00	0	0

4   Discussion and Conclusion

PM4HC is expected to have an increasingly relevant role in the analysis of health-care data, in particular in Oncology. Process-oriented representations, together with tools able to interrogate the data in terms of temporal patterns identified through paths in a workflow, are efficient ways to easily generate clinically-relevant hypotheses and measure statistical significance, in particular in survival analysis.

In this preliminary work, we demonstrated the added value of a process-oriented approach when performing three classical steps of data analysis: pre-processing, descriptive statistics, and inferential statistics. The main remarkable points emerging from this experience are: (a) query languages for EL, PD and CC are efficient tools for data cleaning and preprocessing, by quickly identifying previously unrecognized mistakes; (b) graphical representations can promote dialogue between clinicians and data scientists, suggesting alternative perspectives and possible research questions; (c) PD gives a relevant contribute in representing the data in an agnostic way; on the other hand CC (with formalisms such as PWF) allows implementing multi-scale data abstractions and identifying patterns or inconsistencies of the data in pre-defined workflows; (d) the process representations, both in PD and CC, effectively support survival analysis techniques, allowing rapid definition of sub-cohorts of interest and providing immediate statistical measures of differences between various paths of the graph.

Noticeably, each step of this study was performed in close cooperation between clinicians and PM scientists, in the effort of creating a multidisciplinary team with shared PM skills. The final goal will be to give full autonomy to physicians to perform PM analyses themselves.

In the future, PM4HC has great potential to be developed further in synergy with classical statistical tools to analyze healthcare-related data. In particular, the fast-growing amount of real-world clinical data produced in modern hospitals, each patient's therapeutic journey being by nature a temporal process, represents a formidable opportunity for PM4HC to contribute to the advent of precision medicine.

References

1. van der Aalst, W.: A practitioner's guide to process mining: limitations of the directly-follows graph. Procedia Comput. Sci. **164**, 321–328 (2019)
2. van der Aalst, W., et al.: Process mining manifesto. In: Daniel, F., Barkaoui, K., Dustdar, S. (eds.) BPM 2011. LNBIP, vol. 99, pp. 169–194. Springer, Heidelberg (2012). https://doi.org/10.1007/978-3-642-28108-2_19
3. Bulliard, J., Panizzon, R., Levi, F.: Melanoma prevention in Switzerland: where do we stand? Revue medicale suisse **2**(63), 1122–1125 (2006)
4. Cowey, C.L., Liu, F.X., Boyd, M., Aguilar, K.M., Krepler, C.: Real-world treatment patterns and clinical outcomes among patients with advanced melanoma: a retrospective, community oncology-based cohort study (A STROBE-compliant article). Medicine (Baltimore) **98**(28), e16328 (2019)
5. Gatta, R., et al.: pMineR: an innovative R library for performing process mining in medicine. In: ten Teije, A., Popow, C., Holmes, J.H., Sacchi, L. (eds.) AIME 2017. LNCS (LNAI), vol. 10259, pp. 351–355. Springer, Cham (2017). https://doi.org/10.1007/978-3-319-59758-4_42
6. Gatta, R., Vallati, M., Lenkowicz, J., et al.: Generating and comparing knowledge graphs of medical processes using pMineR. In: Proceedings of the Knowledge Capture Conference. K-CAP 2017, Association for Computing Machinery, New York (2017)
7. Geleijnse, G., Aklecha, H., et al.: Using process mining to evaluate colon cancer guideline adherence with cancer registry data: a case study. In: AMIA (2018)
8. Homayounfar, P.: Process mining challenges in hospital information systems. In: 2012 Federated Conference on Computer Science and Information Systems, pp. 1135–1140. IEEE (2012)
9. Kurniati, A.P., Johnson, O., Hogg, D., Hall, G.: Process mining in oncology: a literature review. In: 2016 6th International Conference on Information Communication and Management, pp. 291–297. IEEE (2016)
10. Lenkowicz, J., Gatta, R., et al.: Assessing the conformity to clinical guidelines in oncology: an example for the multidisciplinary management of locally advanced colorectal cancer treatment. Manage. Decis. **56**(10), 2172–2186 (2018)
11. Mans, R., Schonenberg, H., Song, M., Aalst, W.V., Bakker, P.: Application of process mining in healthcare - a case study in a Dutch hospital. In: BIOSTEC (2008)
12. Mans, R.S., van der Aalst, W.M.P., Vanwersch, R.J.B.: Process Mining in Healthcare. SBPM. Springer, Cham (2015). https://doi.org/10.1007/978-3-319-16071-9
13. Rinner, C., Helm, E., Dunkl, R., et al.: Process mining and conformance checking of long running processes in the context of melanoma surveillance. Int. J. Environ. Res. Public Health **15**(12), 2809 (2018)

Process Mining of Disease Trajectories in MIMIC-III: A Case Study

Guntur Kusuma[1,3]($\boxtimes$) (ID), Angelina Kurniati[2] (ID), Ciarán D. McInerney[1] (ID),
Marlous Hall[4] (ID), Chris P. Gale[4] (ID), and Owen Johnson[1] (ID)

[1] School of Computing, University of Leeds, Leeds LS2 9JT, UK
scgpk@leeds.ac.uk
[2] School of Computing, Telkom University, Bandung 40257, Indonesia
[3] School of Applied Science, Telkom University, Bandung 40257, Indonesia
[4] Leeds Institute of Cardiovascular and Metabolic Medicine, University of Leeds, Leeds
LS2 9JT, UK

Abstract. A temporal disease trajectory describes the sequence of diseases that
a patient has experienced over time. Electronic health records (EHRs) that contain coded disease diagnoses can be mined to find common and unusual disease
trajectories that have the potential to generate clinically valuable insights into the
relationship between diseases. Disease trajectories are typically identified by a
sequence of timestamped diagnostic codes very similar to the event logs of timestamped activities used in process mining, and we believe disease trajectory models can be produced using process mining tools and techniques. We explored this
through a case study using sequences of timestamped diagnostic codes from the
publicly available MIMIC-III database of de-identified EHR data. In this paper,
we present an approach that recognised the unique nature of disease trajectory
models based on sequenced pairs of diagnostic codes tested for directionality. To
promote reuse, we developed a set of event log transformations that mine disease trajectories from an EHR using standard process mining tools. Our method
was able to produce effective and clinically relevant disease trajectory models
from MIMIC-III, and the method demonstrates the feasibility of applying process
mining to disease trajectory modelling.

Keywords: Disease trajectories · Process mining · Electronic Health Records

1 Introduction

There is a small but growing body of literature exploring the generation of disease
trajectories using electronic health records (EHR) [1, 2]. The rich collection of patient
data in the EHR is a valuable source to get an extensive trail of disease diagnoses over
time [3]. Mining the trails of disease diagnoses and the temporal information may help
to identify patterns in disease trajectories of clinical value. A better understanding of
patterns of disease may advance precision medicine to improve care at an individual level
[4] and improve medical understanding of common disease progression at the population

S. Leemans and H. Leopold (Eds.): ICPM 2020 Workshops, LNBIP 406, pp. 305–316, 2021.
https://doi.org/10.1007/978-3-030-72693-5_23

level [5, 6]. A study by Jensen et al. [7] had identified the disease trajectories of a large cohort by combining a data-driven and statistical approach. However, their trajectories were built based on overlapping pairs of diagnostic codes suggesting the presence of longer trajectories without confirming if such trajectories are available in the data. Based on this, we propose an improvement by incorporating process mining as a toolset and method for mining end-to-end disease trajectories.

Process mining utilises a set of tools to discover process models using data from an organisation's information system. Extracted data are transformed into an event log, a collection of activities and its corresponding timestamps, sometimes supplemented with additional attributes. There is now a large body of literature applying process mining to the domain of healthcare, typically focussed on discovery of actual care processes [8], conformance to guidelines and enhancement to improve the quality of healthcare services [9], the safety of the patients, and better management of resources [10, 11].

Jensen et al. [7] defined a disease trajectory as the patient's orderly series of diagnoses. The definition is comparable to the concept of a trace in process mining where a trace is the sequence of activities for an individual case [12]. We hypothesise that it should be feasible to apply process mining to discover a disease trajectory model [2]. To the best of our knowledge, this is the first time process mining has been used to identify disease trajectories from a real world EHR.

In this paper, we present a novel disease trajectory mining method using process mining techniques applied to the MIMIC-III open access EHR database. We identified the sequence of diagnoses (trace) based on the temporal aspect of the patients' admissions, broke down each trace into pairs of diagnoses, statistically analysed the pair's correlation and represented the identified disease trajectories using a directly-followed graph produced by standard process mining visualisation tools [12]. The research questions are as follow: *Q1-Can disease trajectories be identified using a process-mining approach? Q2-What are the most followed trajectories and what exceptional trajectories are followed? Q3-Are there differences in trajectories followed by different patient groups (by sex, by age group, by mortality status)?* And, *Q4-What are the longest and shortest average time transition trajectories?*

2 Background

Process mining provides a set of techniques and tools to uncover the real behaviour of processes from a range of perspectives including, but not limited to [12]: control-flow, performance, conformance, and organisational. There are three types of process mining: first, process discovery to generate process models from event log data, second, process conformance to check either a process model conforms to an event log or vice versa and third, process enhancement to improve a process model using the information of the actual process recorded in the event log [12].

In healthcare, process mining techniques may help the clinicians answer questions associated to each characteristic of the healthcare processes (e.g. primary care, secondary care, tertiary care, etc.) [8]. The rich information in the EHR is the source of answer to the four types of data science questions: *"what happened?"*, *"why did it happen?"*, *"what will happen?"*, and *"what is the best that may happen?"*. In this study, we followed the most widely used methodology, the PM^2 framework, which describes six process mining stages and defines the set of activities to complete each stage.

The diagnostic codes available within electronic health records result from diagnostic decisions made by clinical specialists after considering the patient's health problem [13]. Jutel [14] described the diagnosis as a process of assessing and making a formal judgement based on a specific physical symptom that takes place at a particular time involving both patient and doctor. Once the disease is determined it is recorded in the EHR using standard diagnostic codes such as the World Health Organisation's International Classification of Diseases (ICD) [15].

3   Method

The goal of this case study was to identify patients' disease trajectories using a process-mining approach. We conducted a retrospective cohort study of patients who were admitted to critical care using the MIMIC-III database as our data source [16]. The MIMIC-III database contains a detailed record of patients' clinical care that has been de-identified to respect the sensitive nature of the data. It is available online to researchers (https://mimic.physionet.org) under an open access policy. We obtained access through two mandatory steps: a training program in human research subject protections and a data user agreement. The Process Mining Project Methodology (PM^2) was followed in this study as the methodology allows us to have multiple research questions that require iterations of analyses [17].

3.1   Data Source for the Case Study

MIMIC-III provides a database of de-identified electronic health records containing the medical history from 2001 to 2012 of 46,520 critical care patients extracted from the EHR of the Beth Israel Deaconess Medical Centre in Boston, USA [16]. The database includes data on patient demographics, laboratory tests, diagnostic codes (in ICD-9 coding standard), medications, bedside monitoring, clinicians' notes and reports, and death records (linked to Social Security Death Index for outpatient death). As part of the anonymisation process, the timestamps used in the MIMIC-III dataset have been intentionally shifted into the future (between 2100 and 2200) by a random offset generated for each patient. This means that the sequence of disease codes and the time intervals between disease codes has been preserved for individual patients but no comparisons between patients are possible. This does not affect disease trajectory mining, but does limit other process-mining approaches such as the identification of bottlenecks. Our group has experience of applying process mining to MIMIC-III and in earlier work have published a data quality assessment on the suitability of the various MIMIC-III data components that are compatible with process mining [18].

3.2 PM^2 for Disease Trajectory Mining

In this section, we identify those sections of the PM^2 that we have adapted for disease trajectory mining. For a full understanding of the PM^2 method see [17].

In Stage 1 (Planning), our research questions were identified from a literature review and confirmed by a project team composed of a clinician, and epidemiologist and process mining and data science researchers.

In Stage 2 (Extraction), we defined the scope by determining the granularity level of data, the time period, and attributes of interest. The MIMIC-III database contains admissions of adult patients aged 16 years old or older [16] who were admitted to the hospital between 1 June 2001 and 10 October 2012. Only patients with at least two admissions were selected to capture the progression of the disease. Patients were followed up for mortality status until the last available discharge as the last censoring date and time for those who died within the hospital. The censoring date for patients who died outside of the hospital is the date recorded in the social security master death index in the MIMIC-III database. We used the first 3-digit ICD-9 codes to indicate diagnoses, [19] but excluded codes known not to be related to development of diseases, e.g. administration codes. Event data were extracted from the ADMISSIONS, PATIENTS, and DIAGNOSES_ICD tables in MIMIC-III database as the input for creating an event log (Table 1). The time of admission was used as the activity timestamp and the diagnostic code as the activity name. The patients were grouped according to their age in bands of 5 years. The attribute of age group was calculated from the patient's age at first admission.

In Stage 3 (Data Processing), we created the event log as defined in the PM^2 by creating the views, then filtering and enriching them. The case identifier for each event was taken from the patient identifier (`subject_id`), the diagnostic code was used as the event name (`diagnosis_code`), and the admission time as the timestamp (`admit-time`). The event log was filtered by removing recurring diagnostic codes (retaining the first occurrence), then reapplying the exclusion of patients with only one diagnostic code. The sequences of diagnostic codes for each patient in the event log informed a set of ordered pairs of diagnostic codes, D1→D2, where the diagnostic code D1 preceded the diagnostic code D2. For example, a patient's event log, D1→D2→D3 , informed two ordered pairs of diagnostic codes, D1→D2and D2→D3 . We excluded ordered pairs that occurred only once. To measure the strength of association between the ordered pairs, we compared the probability of diagnosis D2 occurring among patients who did and did not have a D1 diagnosis previously in the event log. This relative risk (RR) [20] indicated whether the D2 diagnosis was more incident in the group with a D1 diagnosis (RR > 1), less incident in the group with a D1 diagnosis (RR < 1), or equivalent (RR = 1). The RR is calculated as

$$RR = \frac{(a/(a+b))}{(c/(c+d))} \tag{1}$$

where a is the number of patients having D1 and D2, b is the number of patients having D1 but not D2, c is the number of patients without having D1 but having D2, and d is the number of patients neither having D1 nor D2.

Following Jensen et al. [7], only pairs with RR > 1 were carried forward for further processing. For a given pair of diagnoses D1 and D2, it was possible for both D1→D2and D2→D1trajectories to satisfy the RR > 1 threshold. Our goal was to identify disease trajectories that were acyclic, so we carried forward the dominant directionality of a given pair of diagnostic codes, only. We applied one-tailed binomial tests [21] to define

Table 1. Source of the required data from MIMIC-III database

Variables	Table source in MIMIC-III	Field name
Case identifier	PATIENTS	subject_id
Event	DIAGNOSES_ICD	hadm_id, icd9_code, seq_num
Activity name	DIAGNOSES_ICD	icd9_code (first 3 digits)
	ADMISSIONS	hospital_expire_flag
	PATIENTS	expire_flag (translated into 1:Dead, 0:End of data)
Time stamps	ADMISSIONS	admittime, dischtime, deathtime
	PATIENTS	dod, dod_hosp, dod_ssn,
Sex	PATIENTS	gender
Age*	PATIENTS	dob
	ADMISSIONS	admittime
Age group**	PATIENTS	dob
	ADMISSIONS	admittime

* the age calculation using PATIENT's dob and ADMISSIONS's admittime.
** the variable was added to group the patients' age.

the dominant directionality of pairs, i.e. D1→D2 or D2→D1. Using a significance level of $\alpha = 0.5$, only ordered pairs of diagnostic codes with one statistically significant direction were carried forward to define the final pairlog.

The final pairlog was transformed back into an event log and recurring diagnoses in each trace were merged to avoid loops. The event log was then enriched by adding attributes of age at admission, sex, age group and the mortality status. These attributes were not used to define the disease trajectory models, but allowed post-hoc analyses to determine differences between disease trajectories according to each attribute. The enriched event log was then loaded into ProM, an open-source process mining tool (https://promtools.org). A START and END event was added to every case in the event log to provide common start and end points of traces. The final event log then converted into the XES format. Common traces were grouped in trace variants using the Explore Event Log (Trace Variants/Searchable/Sortable) feature in ProM [22].

In Stage 4 (Mining and Analysis) we used ProM to analysed the event log to identify unique trace variants, performed process discovery, visualised the discovered model and performed conformance checking. For process analysis, we calculated descriptive summary statistics of the disease trajectories that were identified, including stratification by patient groups. The event log was visualised using the Explore Event Log (Trace variants/Searchable/Sortable). The Interactive Data-aware Heuristics Miner (iDHM) [23] plug-in was used to discover the disease process models.

The quality of the discovered models were evaluated using replay fitness, precision and generalisation [24]. Replay fitness is a measure of how many traces from the log can be reproduced in the process model, with penalties for skips and insertions. Precision is a measure of how 'lean' the model is at representing traces from the log. Lower values indicate superfluous structure in the model. Generalisation is a measure of generalisability as indicated by the redundancy of nodes in the model; The more redundant the nodes, the more variety of possible traces that can be represented. The value of each measure represents by a number between 0–1. Discovery and conformance checking used plugins in ProM. The Replay a Log on Petri Net for Conformance Analysis plug-in for measuring the fitness [25], Align-ETConformance plug-in [26] for the precision, and the Measure Precision/Generalization plugin for measuring the generalisation. Other tools used in this study were PostgreSQL as the database management system of MIMIC-III, and Python through Jupyter Notebook [27].

4 Results

An event log was extracted from an EHR to identify disease trajectories, pairs of diagnoses were identified and analysed for correlation measurement and tested for directionality. The discovery algorithm is applied to produce the disease trajectory model and represented using the directly-followed graph.

In Stage 1 (Planning), we aimed to mine the disease trajectory agnostically without any specific selection of diagnosis and time window. Following the literature review in Sect. 2, we defined the main research question as: *(Q1) Can disease trajectories be identified using a process-mining approach?* Further questions added which were motivated by the frequently posed question for process mining in healthcare [28]: *(Q2) What are the most followed trajectories and what exceptional trajectories are followed?(Q3) Are there differences in trajectories followed by different patient groups (by sex, by age group, by mortality status)? (Q4) What are the longest and shortest average time transition trajectories?*

In Stage 2 (Extraction), Of the 58,976 unique admissions in MIMIC-III from 46,520 patients, there were 6,984 unique ICD-9 diagnostic codes used for 651,000 diagnoses. From this dataset, we excluded 172,685 (26.5%) diagnostic codes that are medically known to be codes related to external factors not directly related to the development of diseases [5], including pregnancy (ICD-9 3-digit codes 630–679, 760–779), general symptoms and signs not related to a disease (780–799), external cause (800–999, E800-E999), and administration (V01-V89). We further excluded 436,483 (67%) secondary diagnostic codes and focused on the 41,832 primary diagnostic codes whilst there will be valuable opportunity in exploring the secondary diagnostic codes.

In Stage 3 (Data Analysis), we composed the selected variables in a way that follows the minimum requirements of event log (see Fig. 1a). The traces of each patients are illustrated in Fig. 1b. We removed 2,692 (16.2%) recurrent diagnoses, retained the first occurrence, excluded patients with only one admission, and subsequently excluded patients who were less than 16 years old at their first ever admission. A total of 4,911 patients remained in the event log consisting of 11,725 diagnostic codes. Figure 1 shows the transformation of event logs into a log of ordered pairs of diagnostic codes (pairlog)(see Fig. 1c). The resulting pairlog contained 6,814 ordered pairs of diagnostic codes. Only 3,781 pairs remained after filtering for RR > 1 and the binomial tests for directionality suggested there were 826 ordered pairs of diagnostic codes with a statistically significant dominant direction. The resulting data contained 796 traces where each trace represents a patient's disease trajectory.

subject_id	diagnostic_code	timestamp
21	410	11/09/2134 12:17
21	038	30/01/2135 20:50
124	433	24/06/2160 21:25
124	441	17/12/2161 03:39
124	440	21/05/2165 21:02
124	569	31/12/2165 18:55

(a) The extracted event log

#21: 410→038

#124: 433→441→440→569

(b) The trace of diagnosis

subject_id	Antecedent	Subsequent	Time1	Time2
21	410	038	11/09/2134 12:17	30/01/2135 20:50
124	433	441	24/06/2160 21:25	17/12/2161 03:39
124	441	440	17/12/2161 03:39	21/05/2165 21:02
124	440	569	21/05/2165 21:02	31/12/2165 18:55

(c) The *pairlog*

Fig. 1. Illustration of the transformation steps of event log for pairwise analysis. (a) The extracted event log from MIMIC-III; (b) the illustration of traces of diagnoses for each patient; (c) the transformed event log into pairlog.

In the last step of filtering, we transformed the pairlog back to an event log and enriched with age at admission, sex, age group and the mortality status. We then loaded the enriched event log into ProM, artificial 'START' and 'END' events were added and then analysed the trace variants using the Explore Event Log feature. Among the 796 traces, we further removed twenty traces that were unique to a single, individual patients as part of good anonymisation practice. Finally, the 776 common traces found in the event log were grouped into 81 trace variants.

In Stage 4 (Mining and Analysis), there were eighty one unique trace variants informed the processing discovery algorithms to answer the Q1. The conformance of the discovered disease trajectory model demonstrated fitness = 0.93, precision = 0.94, and generalisation = 0.92. Further evaluation was done by 5-folds cross-validation where the original event log was randomly divided into five groups of sub-event log equally. One sub-event log was used as the validation data and the remaining four sub-event logs as training data. The cross-validation process was done five times to allow each sub-event log used once as the validation data. The average value from the cross-validation are expected to be lower than the conformance, resulting fitness = 0.92 (SD: 0.006), precision = 0.82 (SD: 0.06), and generalisation = 0.88 (SD: 0.02). This suggests that the discovered trajectory model (Fig. 2) is robust to sampling, allows the traces seen in the event log, is precise enough to not allow behaviour unrelated to what was seen in the event log, and general enough to reproduce future behaviour of the trajectories.

In respond to the Q2, among 776 patients there are 81 distinct trajectories (Table 2). The most-followed trajectory (n = 80; 10.3%) was acute myocardial infarction to

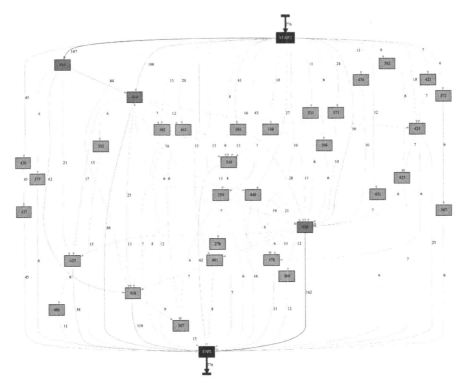

Fig. 2. The directly-follow graph representation of Disease Trajectory Model of Critical Care patients in MIMIC-III with the minimum case frequency = 6.

Table 2. The three most-common and least-common trace variants.

Traces (%)	Trace Variant	Median (months)	Dead (%)	Male (%)
80 (10.31%)	START→410→414→END	6.5	75	70
62 (7.99%)	START→410→428→END	3.9	72.58	54.84
45 (5.80%)	START→430→437→END	3.9	4.44	35.56
...	...	...	...	...
2 (0.26%)	START→410→427→486→END	28.3	100	50
2 (0.26%)	START→507→491→482→END	43.6	50	100
2 (0.26%)	START→518→250→038→END	14.6	100	0

ICD-9 Codes translation: 038 = Septicaemia, 250 = Diabetes mellitus, 410 = Acute myocardial infarction, 414 = Ischemic heart disease, 427 = Cardiac dysrhythmias, 428 = Heart failure, 430 = Subarachnoid haemorrhage, 437 = Other and ill-defined cerebrovascular disease, 482 = Other bacterial pneumonia, 486 = Pneumonia, organism unspecified, 491 = Chronic bronchitis, 507 = Pneumonitis due to solids and liquids, 518 = Other diseases of lung.

ischemic heart disease, which is consistent with the published literature [7, 29, 30]. Septicaemia occurred most frequently (n = 212; 27.3%), both as a precedent (n = 50; 6.4%) and subsequent (n = 162; 20.9%), with mortality in the end (n = 143; 66.9%).

This supported previous findings that it is associated with morbidity and mortality [16, 31]. There are three exceptional trajectories of two patients each (0.26%) (Table 2).

The third question was (Q3) Are there differences in trajectories followed by different patient group? We answered the question by comparing trajectories by sex (male, female) and age band (18–34 years, 35–64 years, and > 64 years). The male cohort consisted of 447 patients with the median duration of follow-up 6.98 months (IQR 1.6–28.2) where 252 cases (56.3%) ending in death. The most-common trajectory was acute myocardial infarction followed by other forms of chronic ischemic heart disease (56 cases, 12.5%) with median interval 6.5 months (IQR 1.5–35.3). In the female cohort, there were 329 patients with the median duration of follow-up 7 months (IQR 2–24.4) where 176 cases (54.4%) ending in death. The most-common trajectory was subarachnoid haemorrhage followed by other and ill-defined cerebrovascular disease (29 cases, 8.8%) with median interval 3.4 months (IQR 2.3–7.5). The most-followed trajectory in a group of 18 to 34-year-old cohort was diabetes followed by hypertensive chronic kidney disease (3 cases) with median interval 55.8 months (IQR 33–56.5). For the group of 35 to 64 years, there were 44 cases (14.5%) with acute myocardial infarction followed by ischemic heart disease, with median interval 7.8 months (IQR 1.9–39.7). Among 329 cases in this age group, there were 133 cases (40.4%) ending in death. Patients in >64 years, there were 293 (68.1%) deaths while the most-common trajectory was acute myocardial infarction followed by heart failure, with median interval 4.7 months (IQR 1.5–21.8).

The fourth question was (Q4) What are the longest and shortest average time transition trajectories? The longest disease progression at 63 months was *Ischemic heart disease* to *Diverticula of intestine* while the shortest progression was *Gastrointestinal hemorrhage* to *Liver abscess and sequelae of chronic liver disease* with average time transition is less than a month (0.98) (Table 3).

Table 3. The three longest and shortest average time interval trajectories in MIMIC-III.

Antecedent	Subsequent	Mean*	Median (IQR)**
A. *The three longest average time interval trajectories (descending)*			
Chronic ischemic heart disease	Diverticula of intestine	63	75.9 (54–84.8)
Chronic ischemic heart disease	Occlusion of cerebral arteries	52.7	51.2 (40.4–52.6)
Chronic ischemic heart disease	Heart failure	46	41.5 (4.6–89.7)
B. *The three shortest average time interval trajectories (ascending)*			
Gastrointestinal hemorrhage	Liver abscess and sequelae of chronic liver disease	0.98	0.81 (0.6–1.3)
Other diseases of endocardium	Other diseases of pericardium	1	0.8 (0.6–1.13)
Chronic bronchitis	Other bacterial pneumonia	2.2	2.2 (1.6–2.7)

*Mean is in months. **Median is in months (IQR); IQR = interquartile range.

5 Discussion

We present a case study of 776 patient admissions associated with 81 different disease transitions to demonstrate the feasibility of using a process-mining approach to reveal disease trajectories using a hospital electronic health record database. We show that the PM^2 framework is suitable for mining disease trajectories and is complemented by the addition of descriptive summary statistics in Stage-3 (Data Processing). Our approach applies a number of transformations to the data, which were adapted from published disease trajectory methods for constructing selected pairs of diagnoses with strong correlation, followed by testing the pairs' directionality to form the trajectories.

Process mining offers techniques to discover disease trajectories and measure the quality of the algorithm to discover the trajectory model. In this work we presented replay fitness, precision, generalisation and cross-validation to validate the model. The process-mining approach opens opportunities to cross-reference discovered disease trajectories with other critical care event data by defining workflows that can actioned using widely-available software. By conducting conformance checking, we have the indicators to show if the discovered model has a good quality. We note that the earlier study by Jensen et al. [7], did measure the robustness of their discovered disease trajectory model with one indicator that is similar to the replay fitness in process mining. This approach is useful to validate that the final model conforms closely to the data.

A particular benefit of the process-mining approach to constructing disease trajectories is that it may provide summaries of cases, events and time interval between occurrences of disease. For example, our method identified the trajectory of *acute kidney injury* (AKI) (584) followed by *septicaemia* (038) with an average interval of 16.22 months. This finding supports the conclusion of [32] where sepsis was a frequent consequence after AKI in intensive care setting. Also, the process-mining approach could provide an estimation of sepsis development after AKI as suggested in [33]. Our method also incorporates additional case attributes that easily facilitate outputs to be stratified by specific characteristics, e.g. sex, age group, and mortality status. For example, although the data were not pre-stratified for females, process mining tools made it easy to query the event log to reveal a dominant trajectory in females – *subarachnoid haemorrhage* (430) followed by *other and ill-defined cerebrovascular disease* (437) – that agrees with previous research [34].

6 Conclusion

In this paper, we have presented the mining of disease trajectories using a process-mining approach. The mining used the MIMIC-III dataset which is comparable to many databases from EHR systems in use at hospitals across the world. Our study included the use of PM^2 framework to mine a representative disease trajectory model from an EHR and addressed quality dimension standards. This study opens opportunities for future works in implementation of the technique using population sized EHR data. We believe the association of pairs of diagnoses might be improved by null hypothesis significance testing of relative risk rather than magnitude-based testing. Future work might assess the sensitivity of the method to the choice of process discovery algorithm used to mine the disease trajectory model.

Acknowledgements. The research was supported by the National Institute for Health Research (NIHR) Yorkshire and Humber Patient Safety Translational Research Centre (NIHR YH PSTRC) and the Indonesia Endowment Fund for Education (LPDP).

References

1. Allam, A., et al.: Patient Similarity Analysis with Longitudinal Health Data. arXiv preprint arXiv:2005.06630 (2020)
2. Kusuma, G., et al.: Process mining of disease trajectories: a feasibility study. In: 13th International Conference on Health Informatics, pp. 705–712 (2020). https://doi.org/10.5220/000 9166607050712
3. Weber, G.M., Mandl, K.D., Kohane, I.S.: Finding the missing link for big biomedical data. JAMA **311**, 2479–2480 (2014). https://doi.org/10.1001/jama.2014.4228
4. Jensen, P.B., et al.: Mining electronic health records: towards better research applications and clinical care. **13**, 395–405 (2012). https://doi.org/10.1038/nrg3208
5. Hanauer, D.A., Ramakrishnan, N.: Modeling temporal relationships in large scale clinical associations. J. Am. Med. Inform. Assoc. **20**, 332–341 (2013). https://doi.org/10.1136/ami ajnl-2012-001117
6. Rothman, K.J., Greenland, S.: Causation and causal inference in epidemiology. Am. J. Public Health **95**, S144-150 (2005). https://doi.org/10.2105/AJPH.2004.059204
7. Jensen, A.B., et al.: Temporal disease trajectories condensed from population-wide registry data covering 6.2 million patients. Nature Comm. **5**, 1–10 (2014). https://doi.org/10.1038/nco mms5022
8. Mans, R.S., et al.: Process Mining in Healthcare Evaluating and Exploiting Operational Healthcare Processes. Springer, Cham (2015). https://doi.org/10.1007/978-3-319-16071-9
9. Partington, A., et al.: Process mining for clinical processes: a comparative analysis of four australian hospitals. ACM Trans. Manag. Inform. Syst. Article **5**, 1–18 (2015). https://doi.org/10.1145/2629446
10. Rojas, E., et al.: Process mining in healthcare: a literature review. J. Biomed. Inform. **61**, 224–236 (2016). https://doi.org/10.1016/j.jbi.2016.04.007
11. Fernandez-Llatas, C., et al.: Process mining methodology for health process tracking using real-time indoor location systems. Sensors **15**, 29821–29840 (2015). https://doi.org/10.3390/s151229769
12. van der Aalst, W.M.P.: Process Mining: Data Science in Action. Springer, Heidelberg (2016). https://doi.org/10.1007/978-3-662-49851-4
13. Committee on Diagnostic Error in Health Care; Board on Health Care Services; Institute of Medicine; The National Academies of Sciences, Engineering, and Medicine. https://www.ncbi.nlm.nih.gov/books/NBK338596/
14. Jutel, A.: Sociology of diagnosis: a preliminary review. Sociol. Health Illn. **31**, 278–299 (2009). https://doi.org/10.1111/j.1467-9566.2008.01152.x
15. World Health Organization. https://www.who.int/classifications/icd/en/
16. Johnson, A.E.W., et al.: MIMIC-III, a freely accessible critical care database. Sci. Data **3**, 160035 (2016). https://doi.org/10.1038/sdata.2016.35
17. van Eck, M.L., Lu, X., Leemans, S.J.J., van der Aalst, W.M.P.: PM^2: a process mining project methodology. In: Zdravkovic, J., Kirikova, M., Johannesson, P. (eds.) CAiSE 2015. LNCS, vol. 9097, pp. 297–313. Springer, Cham (2015). https://doi.org/10.1007/978-3-319-19069-3_19

18. Kurniati, A.P., et al.: The assessment of data quality issues for process mining in healthcare using MIMIC-III, a publicly available e-health record database. Health Inf. J. **25**, 1878–1893 (2017). https://doi.org/10.1177/1460458218810760

19. National Center for Health Statistics. https://www.cdc.gov/nchs/data/icd/icd9cm_guidel ines_2011.pdf

20. StatPearls Publishing. https://www.ncbi.nlm.nih.gov/books/NBK430824/

21. Kang, S.-H., Ahn, C.W.: Tests for the homogeneity of two binomial proportions in extremely unbalanced 2 x 2 contingency tables. Stat. Med. **27**, 2524–2535 (2008). https://doi.org/10.1002/sim.3055

22. Mannhardt, F.: Tools & Software—ProM—Event Log Explorer (2018)

23. Mannhardt, F., et al.: Heuristic mining revamped: An interactive, data-Aware, and conformance-Aware miner. In: BPM 2017, pp. 1–5. CEUR-WS.org (2017)

24. Buijs, J.C.A.M., van Dongen, B.F., van der Aalst, W.M.P.: On the role of fitness, precision, generalization and simplicity in process discovery. In: Meersman, R., et al. (eds.) OTM 2012. LNCS, vol. 7565, pp. 305–322. Springer, Heidelberg (2012). https://doi.org/10.1007/978-3-642-33606-5_19

25. Adriansyah, A.: Replay a Log on Petri Net for Conformance Analysis-plugin.pdf. (2012)

26. Adriansyah, A., et al.: Measuring precision of modeled behavior. IseB **13**, 37–67 (2015). https://doi.org/10.1007/s10257-014-0234-

27. Kluyver, T., et al.: Jupyter Notebooks—a publishing format for reproducible computational workflows (2016)

28. Mans, R.S., van der Aalst, W.M.P., Vanwersch, R.J.B., Moleman, A.J.: Process mining in healthcare: data challenges when answering frequently posed questions. In: Lenz, R., Miksch, S., Peleg, M., Reichert, M., Riaño, D., ten Teije, A. (eds.) KR4HC/ProHealth -2012. LNCS (LNAI), vol. 7738, pp. 140–153. Springer, Heidelberg (2013). https://doi.org/10.1007/978-3-642-36438-9_10

29. Asaria, P., et al.: Acute myocardial infarction hospital admissions and deaths in England: a national follow-back and follow-forward record-linkage study. Lancet Public Health **2**, e191–e201 (2017). https://doi.org/10.1016/S2468-2667(17)30032-4

30. Hall, M., et al.: Multimorbidity and survival for patients with acute myocardial infarction in England and Wales: Latent class analysis of a nationwide population-based cohort. PLoS Med. **15**. 52 (2018). https://doi.org/10.1371/journal.pmed.1002501

31. Sakr, Y., et al.: Sepsis in intensive care unit patients: worldwide data from the intensive care over nations audit. Open Forum Infect. Dis, **5**, ofy313–ofy318 (2018). https://doi.org/10.1093/ofid/ofy313

32. Mehta, R.L., et al.: Sepsis as a cause and consequence of acute kidney injury: program to improve care in acute renal disease. Intensive Care Med. **37**, 241–248 (2011). https://doi.org/10.1007/s00134-010-2089-9

33. Peerapornratana, S., et al.: Acute kidney injury from sepsis: current concepts, epidemiology, pathophysiology, prevention and treatment. Kidney Int. **96**, 1083–1099 (2019). https://doi.org/10.1016/j.kint.2019.05.026

34. Eden, S.V., et al.: Gender and ethnic differences in subarachnoid hemorrhage. Neurology **71**, 731–735 (2008). https://doi.org/10.1212/01.wnl.0000319690.82357.44

The Need for Interactive Data-Driven Process Simulation in Healthcare: A Case Study

Gerhardus van Hulzen[1]([✉]) [iD], Niels Martin[1,2,3] [iD], and Benoît Depaire[1] [iD]

[1] Research group Business Informatics, Hasselt University, 3500 Hasselt, Belgium
{gerard.vanhulzen,niels.martin,benoit.depaire}@uhasselt.be
[2] Research Foundation Flanders (FWO), 1000 Brussels, Belgium
[3] Data Analytics Laboratory, Vrije Universiteit Brussel, 1050 Brussels, Belgium

Abstract. In healthcare, more and more process execution information is stored in Hospital Information Systems. This data, in conjunction with data-driven process simulation, can be used, e.g. to support hospital management with Capacity Management decisions. However, real-life event logs in healthcare often suffer from data quality issues, affecting the reliability of simulation results. In this work, we illustrate the effects of disregarding data quality issues on simulation outcomes and the importance of domain knowledge using a case study at the radiology department of a hospital. Current literature on data-driven process simulation acknowledges the need for domain expertise but does not provide a framework for conceptualising the involvement of domain experts. Therefore, we propose a novel conceptual framework which interactively involves experts during data-driven simulation model development.

Keywords: Data-driven process simulation · Data quality · Domain knowledge · Interactive modelling · Healthcare processes

1 Introduction

Worldwide, healthcare systems are under constant pressure. Increasing population numbers, lifestyle factors, ageing populations, and new technologies are the main drivers for increasing healthcare expenses. Simultaneously, healthcare budgets are under pressure due to national budget deficits and savings [14]. Healthcare managers have to improve their care processes to maintain high-quality care for all patients. One key aspect of ensuring this is efficient Capacity Management (CM), which is used to determine the suitable levels of resources, such as equipment, facilities, and staff size [28].

To support hospital management during CM decisions, *Business Process Simulation (BPS)* can be used to determine suitable resource levels objectively. BPS uses a (computer) model to imitate the process. This allows to evaluate the effect of various process modifications without actually implementing them into, nor

© Springer Nature Switzerland AG 2021
S. Leemans and H. Leopold (Eds.): ICPM 2020 Workshops, LNBIP 406, pp. 317–329, 2021.
https://doi.org/10.1007/978-3-030-72693-5_24

disrupting, the real process [21]. For instance, the effect on throughput rates and patient waiting times of installing an additional X-ray scanner can be simulated to determine suitable equipment levels.

Conducting a simulation study is often time-consuming and builds upon subjective inputs, such as interviews and observations. The emerging field of data-driven process simulation in Process Mining (PM) can overcome some of the limitations of "traditional" simulation model development by using data from Information Systems. Data-driven process simulation refers to the automated discovery of a simulation model from process execution data, i.e. an event log [9]. A key challenge in this field is data quality, given its strong impact on the reliability of the simulation results [31]. Because data quality issues are often encountered in healthcare event logs, it is imperative to assess these issues and correct them if needed. This will require domain knowledge. Current literature on data-driven simulation does not provide a clear framework to involve domain experts in model development.

This paper demonstrates the need for interactive data-driven process simulation in healthcare by assessing the impact of data quality issues on simulation results. To this end, a case study at the radiology department of a hospital is considered. In addition, we propose a novel conceptual framework which structures the integration of domain knowledge in the interactive development of data-driven simulation models.

The remainder of this paper is structured as follows. Section 2 gives an overview of the related work. The context of the case study is presented in Sect. 3. The experimental design, results, and discussion are presented in Sect. 4. Section 5 introduces our proposed framework for interactive data-driven process simulation. The paper ends with a conclusion in Sect. 6.

2 Related Work

This work relates to three key domains: (i) simulation for CM decisions in healthcare, (ii) data-driven process simulation, and (iii) data quality in process mining. The following paragraphs give a brief overview of these domains.

Simulation for Capacity Management Decisions in Healthcare. *Capacity Management decisions* in healthcare are concerned with determining the suitable levels of resources, such as staff size, equipment, and facilities [28]. In literature, simulation has been used to determine the required number of beds in general surgery [30]; the number of nurses, doctors, and buffer beds in an Emergency Department (ED) [7]; and the number of computed tomography (CT) scanners in a radiology department [27]. Within the radiology department, the context of our case study, Vieira et al. [32] gave an overview of Operations Research (OR) techniques – which includes simulation – for optimising resource levels and scheduling. For further reference on CM and the use of simulation in healthcare, the reader is referred to one of the existing review papers [26,28,33].

Data-Driven Process Simulation. *Data-driven process simulation* aims to "discover" BPS models from event logs automatically [9]. While existing PM research can support the discovery of individual BPS model components [19] – e.g. control-flow discovery, decision mining, or organisational mining – less work has been devoted to integrating all these components into a single, simulation-ready model. Rozinat et al. [24] made a first attempt by discovering Coloured Petri Nets (CPNs) to describe the control flow. In addition, gateway routing logic and resource pools were also included. Later, the authors extended their method with activity execution times and case inter-arrival times [25]. Khodyrev and Popova [16] described a similar approach. However, the resource perspective was not included, assuming no resource constraints [16]. Gawin and Marcinkowski [13] provided support for activity durations, control-flow, resources, gateway routing logic, resource schedules, and inter-arrival times. However, the latter two were not automatically derived from data and had to be defined by domain experts [13]. *ClearPath* [15] provides a methodology for discovering and simulating Care Pathways (CPs). Their approach follows an agile, iterative method which facilitates the interaction between the modeller and domain expert, but the obtained process models still have to be manually recreated in their simulation tool *NETIMIS* [15]. *Simod* was the first tool to automatically integrate all components into a single, simulation-ready model to support BPS [6]. In addition, *Simod* is also capable of measuring the accuracy of the derived model and improve it using hyperparameter optimisation [6].

Data Quality in Process Mining. Real-life event logs tend to suffer from *data quality issues*, especially when they originate from flexible environments with substantial manual recording, such as healthcare [5,23]. These issues include missing events and incorrect timestamps, where the latter is often caused by batched registrations by healthcare staff [18,31]. Given the potential impact of event log quality issues on the reliability of PM outcomes, research attention on this topic is increasing. Research efforts are centred around three key topics. Firstly, several frameworks are developed which define event log quality issues [5,29,31]. For instance, Bose et al. [5] define 27 event logs quality issues and group them in four broad classes (i.e. missing, incorrect, imprecise, and irrelevant data). Secondly, research is performed on data quality assessment, targeting the systematic identification of event log quality issues. In this respect, the R-package *DaQAPO* [20], the log query language *QUELI* [1], and the *CP-DQF* [12] for Electronic Health Records (EHRs) provide tools and frameworks to operationalise data quality assessment. They are based on the event log quality issues defined in Vanbrabant et al. [31], Suriadi et al. [29], and Bose et al. [5], respectively. Thirdly, heuristics have been developed which tackle specific data quality issues, e.g. adding missing events [10], imputing missing case identifiers [3], and handling event ordering issues [11].

3   Background: Capacity Management at the Radiology Department

To illustrate the impact of data quality issues in the context of data-driven simulation, a real-life case study is used. This section introduces the case study.

3.1   General Context

The case study relates to a project at the radiology department of a hospital. Hospital management is preparing plans to build new facilities and is requesting input from each department regarding the required capacity. For the radiology department, this relates to the number of examination rooms – i.e. scanners – and the size of the waiting rooms – i.e. the number of seats – for each examination room. The radiology department wants to approach this Capacity Management problem in a data-driven way.

To support this data-driven analysis, process execution data is obtained from the Radiology Information Systems (RIS). This system supports the entire process flow, of which a simplified representation is shown in Fig. 1. The process starts when a patient arrives at the registration desk, after which (s)he is registered. Afterwards, the patient will wait in the waiting room until (s)he is called into the examination room. A nurse helps the patient onto the scanning table and correctly positions the scanner. Next, the image is created. In case the patient needs an additional scan of the same type, e.g. an X-ray scan of both shoulder and neck, this image can be made without leaving the room. After all required scans have been made, the patient can leave the examination room, and the nurse will post-process the images. If the patient still requires additional scans – of a different kind than the previous (e.g. also a CT scan) – (s)he will go to the waiting room of the other examination room. After all scans have been made, the patient can leave the radiology department and return home. Note that the interpretation of the scans by a radiologist is out of scope as it does not impact the required scanner and waiting room capacity.

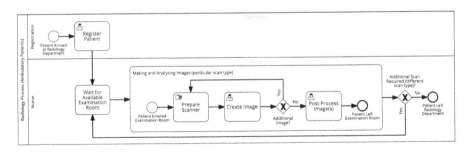

Fig. 1. Simplified process flow of (ambulatory) patients at the radiology department.

To solve the CM problem in this process, *Discrete-Event Simulation (DES)* is used due to the stochastic nature of the process. DES uses simulation to compare policy alternatives before implementing them in practice [33]. Arena v15 [2] was used to simulate the model.

In a DES model, *entities* are dynamic objects which move through the process and trigger the execution of activities [19]. In this case study, entities are patients visiting the radiology department. Four patient types are distinguished:

(i) *ambulatory patients (A)* which are outpatients, (ii) *day hospital patients (D)* which are admitted to the hospital for at most one day, (iii) *hospitalised patients (H)* which are inpatients, and (iv) *emergency patients (S)* which are transferred from the Emergency Department (ED).

The process flow depicted in Fig. 1 actually gives an overview of ambulatory patients. Nevertheless, the flow of the other patient types is, in essence, the same. Only the way patients arrive and where they wait are different. Hospitalised and day hospital patients will wait in their room until they are called in. Emergency patients will wait at the ED.

Depending on the type of scan, a different scanner – and thus a different examination room – is used. In this case study, there are six different types of scans of interest: *angiogram (ANGIO)*, *computed tomography (CT)*, *echocardiogram (ECHO)*, *mammogram (MAMMO)*, *magnetic resonance (MR)*, and *X-ray (RX)*. CT, ECHO, MAMMO, and MR all require separate rooms. ANGIO and RX are performed in RX rooms.

3.2   Data Description

To support the development of the DES model, two years of data from the RIS – from March 2017 until March 2019 – was available. The dataset includes various key timestamps for each patient visit, such as time of registration, and start and end time of scanning. Other attributes, such as the scan type (e.g. ECHO, RX, etc.) and patient type (e.g. ambulatory, emergency, etc.), were also recorded for each patient visit.

The dataset contains 404,750 individual patient visits. The proportions per patient type were 60%, 23%, 15%, and 2% for ambulant, hospitalised, emergency, and day hospital patients, respectively. In total, 464,053 scans were recorded, indicating that the majority of patients only needed one scan. Most scans were RX, i.e. 45%. ECHO represented 19%, followed by MR, 16%, 14% CT, and 5% MAMMO. A very small proportion, less than 0.001%, were ANGIO.

In the process, the activity "Create Image" (cf. Fig. 1) has the most considerable impact on waiting times and throughput rates because it generally takes longer than all other activities. Both start and end timestamps are available of this activity and are recorded when the nurse starts and stops the scanning device, respectively. We initially expected that this activity would not suffer much from quality issues because it is recorded automatically. However, this appeared not to be the case.

Table 1 gives an overview of the scan duration times per scan type. According to the data, some scans took over several years to complete. A few observations even had a negative duration, caused by the end timestamp being recorded before the start timestamp. Given its impact on capacity requirements, the scenario analysis will focus on the effect of scanning time data with data quality issues on simulation outcomes.

Table 1. Scan execution times (in mins).

Scan Type	Min	Max	Mean	Median	SD	IQR
ANGIO	0.00	323,258	14,372.11	26.05	57,336.29	87.83
CT	−726.53	30,605	6.73	1.97	196.86	2.23
ECHO	−79.00	116,685	71.36	23.38	636.37	28.48
MAMMO	−6.48	40,780	16.41	2.98	531.00	1.35
MR	0.00	946,449	161.22	11.48	9,679.90	6.90
RX	−1,031.63	2,109,457	22.69	0.55	5,111.25	1.20

4   Scenario Analysis: The Impact of Data Quality Issues

4.1   Experimental Design

To illustrate the impact of data quality issues w.r.t scanning times, we consider two scenarios:

- **Scenario 1 – Direct sampling:** In this scenario, actual observed data is sampled. This is useful when no theoretical distribution, such as the Gaussian, exponential, or gamma distribution, fits the data well. However, the disadvantage is that only the observed values can be used, which is problematic for smaller datasets [17].
- **Scenario 2 – Distribution fitting:** In this scenario, a distribution is fitted to the observed data. We used the distribution with the *least worst* fit because not a single distribution fitted the data well. With this approach, we follow the state-of-the-art of data-driven BPS techniques.

For each scenario, three alternative data filtering approaches are compared:

- **Alternative 1 – Validated filtering (VF):** In this alternative, which is the baseline, we used filtered data validated by domain experts. For scenario 2, we had to use empirical distributions for this alternative as none of the theoretical distribution provided a good fit. In the other two alternatives, we always used theoretical distributions.
- **Alternative 2 – No filtering (NF):** Here, we used the unfiltered data directly. Only observations less than zero were filtered out because the simulation model cannot handle negative activity durations.
- **Alternative 3 – Context-agnostic filtering (CAF):** Even without any domain knowledge, one would immediately notice that the maximum values in Table 1 are unrealistic. Therefore, this alternative uses filtered data to exclude anomalies. We adopted the commonly used *box plot rule* to detect anomalies in the absence of domain knowledge. Any observation smaller than $Q_1 - 1.5IQR$ or larger than $Q_3 + 1.5IQR$ is removed [8]. If the lower limit was less than zero, zero was used instead.

The length of the simulation run was set at two years for each alternative in each scenario. Initial experimentation showed that outliers in Alternative 2 caused severe queue accumulation, which resulted in i.a. extreme waiting times. Therefore, we integrated a weekly "reset", which removed all patients from queues and ongoing scans. We will refer to this reset as "flushing" and kept track of the weekly number of flushed patients.

To compare the alternatives, we focused on patient throughput and waiting times. Moreover, we looked at the flush count mentioned above. To measure the true effect of the different distributions used in each alternative, *common random number streams (CRNs)* are used. Consequently, the same random numbers are sampled across all alternatives. To compare the difference between alternatives, we used the non-parametric *Wilcoxon-Mann-Whitney (WMW)* test. Instead of using the original observations, ranks are used to compare the difference between two samples. This has the advantage that no underlying distribution is assumed [22]. To control the *false discovery rate (FDR)* of the multiple testing problem, we used the *Benjamini–Yekutieli* procedure [4] to adjust the p-values.

4.2 Results

Throughput Times. The *throughput time* measures the elapsed time between the patient's arrival and departure. Because a patient could require multiple scans, the *average throughput time per examination* is considered by dividing the throughput time of a patient by the number of scans. Patients who were "flushed" did not complete all scans and are therefore excluded from this measure.

As shown in Table 2, the throughput times for NF are much higher than VF, e.g. in Scenario 2, the average throughput time per examination for hospitalised patients is almost 100 times longer. The differences between CAF and VF are also statistically significant, albeit much smaller. For day hospital patients, representing 0.5% of the observations for this measure, the differences between VF and CAF were not statistically significant. Nevertheless, important differences in mean throughput times are observed due to larger outlier values for CAF.

Waiting Times. The *waiting time* is the time a patient spends in a queue before undergoing a scan. Table 3 shows comparable differences as the throughput times. Again, large differences between VF and NF are observed, e.g. the average waiting time for hospitalised patients is more than 150 times longer in NF than VF for Scenario 2. For day hospital patients, only the difference between VF and CAF in Scenario 1 is not significant, even though the absolute difference between the means is, again, rather large, indicating the presence of outliers.

Flush Counts. The more patients are flushed at the end of a week, the more this indicates that queues have accumulated throughout that week. Especially in the NF alternative, many patients have to be flushed to "reset" the process at the end of a week, in some cases even more than a thousand patients in total. The differences between VF and CAF are much smaller, i.e. on average less than one patient more was flushed in CAF. However, it should be noted that

Table 2. Throughput times per examination (in min) per patient type (A, D, H, S) and alternative (VF, NF, CAF).

ADHS	Model 1	Model 2	Mean Model 1	Mean Model 2	Adj. p-value	Significance
			Scenario 1			
A	VF	NF	25.8493	556.622	<0.0001	****
A	VF	CAF	25.8493	45.8470	<0.0001	****
D	VF	NF	25.0905	129.1714	0.0077	**
D	VF	CAF	25.0905	42.2703	0.6525	ns
H	VF	NF	29.6365	606.8702	<0.0001	****
H	VF	CAF	29.6365	34.1812	<0.0001	****
S	VF	NF	13.4890	92.7292	<0.0001	****
S	VF	CAF	13.4890	16.3655	<0.0001	****
			Scenario 2			
A	VF	NF	25.8622	938.1933	<0.0001	****
A	VF	CAF	25.8622	48.8646	<0.0001	****
D	VF	NF	24.6727	363.3950	<0.0001	****
D	VF	CAF	24.6727	118.0489	0.1479	ns
H	VF	NF	29.6764	2,740.0265	<0.0001	****
H	VF	CAF	29.6764	253.3411	<0.0001	****
S	VF	NF	13.5586	76.9503	<0.0001	****
S	VF	CAF	13.5586	17.5182	<0.0001	****

****: p-value < 0.0001, ***: p-value < 0.001, **: p-value < 0.01, *: p-value < 0.05, ns: not signif.

Table 3. Waiting times (in min) per patient type (A, D, H, S) and alternative (VF, NF, CAF).

ADHS	Model 1	Model 2	Mean Model 1	Mean Model 2	Adj. p-value	Significance
			Scenario 1			
A	VF	NF	8.7470	536.3071	<0.0001	****
A	VF	CAF	8.7470	26.4227	<0.0001	****
D	VF	NF	10.8402	112.1685	<0.0001	****
D	VF	CAF	10.8402	27.4978	1.0000	ns
H	VF	NF	17.3111	592.3354	<0.0001	****
H	VF	CAF	17.3111	20.7817	<0.0001	****
S	VF	NF	1.7663	76.4688	<0.0001	****
S	VF	CAF	1.7663	3.6703	<0.0001	****
			Scenario 2			
A	VF	NF	8.7252	912.6071	<0.0001	****
A	VF	CAF	8.7252	31.3267	<0.0001	****
D	VF	NF	10.3187	337.0036	<0.0001	****
D	VF	CAF	10.3187	99.9275	0.0140	*
H	VF	NF	17.2705	2,737.3727	<0.0001	****
H	VF	CAF	17.2705	243.9300	0.0140	*
S	VF	NF	1.7545	47.0165	<0.0001	****
S	VF	CAF	1.7545	5.3972	<0.0001	****

****: p-value < 0.0001, ***: p-value < 0.001, **: p-value < 0.01, *: p-value < 0.05, ns: not signif.

sometimes the maximum number of flushed patients in CAF was much higher than in VF, e.g. for Scenario 2, VF flushed at most two hospitalised patients, whereas in CAF this was at most 46. For ambulatory patients, this was smaller, i.e. nine and seventeen, respectively.

4.3   Discussion

The results illustrate the need to consider data quality issues seriously. The unfiltered alternative – which completely neglects these issues – exhibits much higher throughput times, waiting times, and flush counts than the validated baseline. The difference between context-agnostic and validated filtering is smaller but still highly relevant. For instance, waiting times for hospitalised patients are up to eight times longer in CAF. However, for other performance metrics, such as flush counts, the differences between VF and CAF are smaller.

In this case study, the cut-off points for outliers in VF and CAF happened to be reasonably close to each other, except for echocardiograms. The domain experts indicated a maximum of 30 mins, whereas the box plot rule returned 84.64 min. However, this does not give any guarantee for other cases as context-agnostic filtering does not take into account the specificities of a particular domain in any way. Therefore, domain knowledge is always required to achieve accurate simulation results.

When comparing the differences between the two scenarios for each alternative (i.e. comparing the outcomes under direct sampling with their counterpart under distribution fitting), large differences are often observed between throughput and waiting times, even though the same input data was used. A possible explanation is that the theoretical distributions did not fit the data well. Therefore, we highlight the need to report goodness-of-fit (GoF) statistics in state-of-the-art data-driven BPS discovery algorithms and use direct sampling or empirical distributions in case no theoretical distribution fits the data well.

5   Interactive Data-Driven Process Simulation

As illustrated in the case study, data quality issues can have a profound impact on the reliability of simulation results. Moreover, domain knowledge plays a vital role in the development of a simulation model. Without domain knowledge, it is, e.g. challenging to determine whether particular observations are exceptional – but plausible – or data errors. Even though current literature on data-driven process simulation acknowledges the need for domain expertise for i.a. validation purposes, no framework conceptualises how this knowledge should be incorporated.

To enhance the integration of domain knowledge in the development of data-driven simulation models, we propose a novel conceptual framework which interactively involves experts during model building. This framework, which is visualised in Fig. 2, distinguishes three interaction cycles. In the *first cycle*, the initial model is constructed. For each required modelling task (e.g. entity arrival rate, activity durations, resource roles, etc.) – of which an overview is presented in Martin et al. [19] – the data requirements are verified. For instance, mining resource roles requires the presence of a resource attribute. If these requirements are not fulfilled, the domain expert is asked for additional input to perform this modelling task. Conversely, if the requirements are fulfilled, the quality of the data is assessed, and an applicable discovery algorithm is employed. Next, the results of the discovery algorithm and detected data quality issues are presented for a check by the domain expert. (S)he can then solve any data quality-related issues and tweak the discovery parameters until the results are satisfactory.

The *second cycle* integrates all discovered model components from the first cycle into a single, simulation-ready model. The entire model is simulated, and the domain expert checks the preliminary results. If the simulation outputs do not satisfactorily reflect reality, the model can be "calibrated" by altering the simulation parameters. An estimation of the impact of the altered parameter on simulation outcomes is delivered in real-time, so the expert does not have to wait until the entire simulation has been completed before receiving an indication whether the altered parameter results in the desired change.

The final and *third cycle* is concerned with the validation of the model. The calibrated model from the second cycle is simulated comprehensively and validated by the domain expert. In addition, a validation dataset – which was not used to discover the model – can be used as well. If the desired accuracy level is not achieved, the domain expert can modify the simulation parameters again. The final validated model can be used for the evaluation of various scenarios and further analyses.

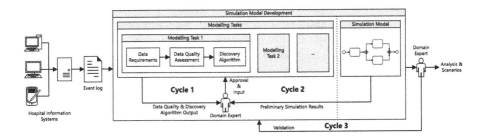

Fig. 2. Interactive data-driven process simulation framework.

6   Conclusion

Data-driven process simulation has great potential within a healthcare context, e.g. to support hospital management with Capacity Management decisions. However, real-life data extracted from Hospital Information Systems tend to suffer from data quality issues, which affects the reliability of simulation results. The presented case study at the radiology department of a hospital illustrates the impact of these issues, as well as the importance of domain knowledge.

Current literature on data-driven process simulation acknowledges the need for domain expertise but does not provide a framework to conceptualise the involvement of domain experts. Therefore, we propose a novel conceptual framework which interactively involves experts during data-driven simulation model building. In this framework, a distinction is made between three cycles: an initial development cycle, a calibration cycle, and a validation cycle.

Future work will focus on how the interaction between the domain expert and the framework will occur more specifically. Ultimately, our goal is to implement our framework into a tool to support the integration of domain knowledge into the development of data-driven process simulation models. In addition, this case study highlights the need for further research on identifying and remedying data quality issues in a healthcare context.

References

1. Andrews, R., Suriadi, S., Ouyang, C., Poppe, E.: Towards event log querying for data quality. In: Panetto, H., Debruyne, C., Proper, H.A., Ardagna, C.A., Roman, D., Meersman, R. (eds.) OTM 2018. LNCS, vol. 11229, pp. 116–134. Springer, Cham (2018). https://doi.org/10.1007/978-3-030-02610-3_7
2. Arena: Rockwell Automation, Inc. (2016). https://www.arenasimulation.com/
3. Bayomie, D., Awad, A., Ezat, E.: Correlating unlabeled events from cyclic business processes execution. In: Nurcan, S., Soffer, P., Bajec, M., Eder, J. (eds.) CAiSE 2016. LNCS, vol. 9694, pp. 274–289. Springer, Cham (2016). https://doi.org/10.1007/978-3-319-39696-5_17
4. Benjamini, Y., Yekutieli, D.: The control of the false discovery rate in multiple testing under dependency. Ann. Stat. **29**(4), 1165–1188 (2001). https://doi.org/10.1214/aos/1013699998
5. Bose, R.P.J.C., Mans, R.S., van der Aalst, W.M.P.: Wanna improve process mining results? It's high time we consider data quality issues seriously. In: Proceedings of the 2013 IEEE Symposium on Computational Intelligence and Data Mining, pp. 127–134 (2013). https://doi.org/10.1109/CIDM.2013.6597227
6. Camargo, M., Dumas, M., González-Rojas, O.: Automated discovery of business process simulation models from event logs. Decis. Support Syst. **134**, 113284 (2020). https://doi.org/10.1016/j.dss.2020.113284
7. Carmen, R., Defraeye, M., Van Nieuwenhuyse, I.: A decision support system for capacity planning in emergency departments. Int. J. Simul. Model **14**(2), 299–312 (2015). https://doi.org/10.2507/ijsimm14(2)10.308
8. Chandola, V., Banerjee, A., Kumar, V.: Anomaly detection: a survey. ACM Comput. Surv. **41**(3), 15:1–15:58 (2009). https://doi.org/10.1145/1541880.1541882
9. Depaire, B., Martin, N.: Data-driven process simulation. In: Sakr, S., Zomaya, A. (eds.) Encyclopedia of Big Data Technologies. Springer, Cham (2018). https://doi.org/10.1007/978-3-319-63962-8_102-1
10. Di Francescomarino, C., Ghidini, C., Tessaris, S., Sandoval, I.V.: Completing workflow traces using action languages. In: Zdravkovic, J., Kirikova, M., Johannesson, P. (eds.) CAiSE 2015. LNCS, vol. 9097, pp. 314–330. Springer, Cham (2015). https://doi.org/10.1007/978-3-319-19069-3_20
11. Dixit, P.M., et al.: Detection and interactive repair of event ordering imperfection in process logs. In: Krogstie, J., Reijers, H.A. (eds.) CAiSE 2018. LNCS, vol.

10816, pp. 274–290. Springer, Cham (2018). https://doi.org/10.1007/978-3-319-91563-0_17

12. Fox, F., Aggarwal, V.R., Whelton, H., Johnson, O.A.: A data quality framework for process mining of electronic health record data. In: Proceedings of the 2018 IEEE International Conference on Healthcare Informatics, pp. 12–21 (2018). https://doi.org/10.1109/ICHI.2018.00009

13. Gawin, B., Marcinkowski, B.: How close to reality is the "as-is" business process simulation model? Organizacija **48**(3), 155–175 (2015). https://doi.org/10.1515/orga-2015-0013

14. Hicks, C., McGovern, T., Prior, G., Smith, I.: Applying lean principles to the design of healthcare facilities. Int. J. Prod. Econ. **170**, 677–686 (2015). https://doi.org/10.1016/j.ijpe.2015.05.029

15. Johnson, O.A., Ba Dhafari, T., Kurniati, A., Fox, F., Rojas, E.: The ClearPath method for care pathway process mining and simulation. In: Daniel, F., Sheng, Q.Z., Motahari, H. (eds.) BPM 2018. LNBIP, vol. 342, pp. 239–250. Springer, Cham (2019). https://doi.org/10.1007/978-3-030-11641-5_19

16. Khodyrev, I., Popova, S.: Discrete modeling and simulation of business processes using event logs. In: Proceedings of the 14th International Conference on Computational Science. Procedia Comput. Sci. **29**, 322–331 (2014). https://doi.org/10.1016/j.procs.2014.05.029

17. Law, A.M.: Simulation Modeling and Analysis, 5th edn. McGraw-Hill Education, New York (2014)

18. Mans, R.S., van der Aalst, W.M.P., Vanwersch, R.J.B.: Process Mining in Healthcare: Evaluating and Exploiting Operational Healthcare Processes. Springer, Cham (2015). https://doi.org/10.1007/978-3-319-16071-9

19. Martin, N., Depaire, B., Caris, A.: The use of process mining in business process simulation model construction. Bus. Inf. Syst. Eng. **58**(1), 73–87 (2015). https://doi.org/10.1007/s12599-015-0410-4

20. Martin, N., Van Houdt, G., Janssenswillen, G.: Towards more structured data quality assessment in the process mining field: the DaQAPO package. In: Proceedings of the European R Users Meeting (2020)

21. Melão, N., Pidd, M.: Use of business process simulation: a survey of practitioners. J. Oper. Res. Soc. **54**(1), 2–10 (2003). https://doi.org/10.1057/palgrave.jors.2601477

22. Neuhäuser, M.: Wilcoxon-Mann-Whitney test. In: Lovric, M. (ed.) International Encyclopedia of Statistical Science, pp. 1656–1658. Springer, Berlin (2011). https://doi.org/10.1007/978-3-642-04898-2_615

23. Rebuge, Á., Ferreira, D.M.R.: Business process analysis in healthcare environments: a methodology based on process mining. Inf. Syst. **37**(2), 99–116 (2012). https://doi.org/10.1016/j.is.2011.01.003

24. Rozinat, A., Mans, R.S., van der Aalst, W.M.P.: Mining CPN models: discovering process models with data from event logs. In: Proceedings of the 7th Workshop and Tutorial on Practical Use of Coloured Petri Nets and the CPN Tools. DAIMI PB, vol. 579, pp. 57–76 (2006)

25. Rozinat, A., Mans, R.S., Song, M., van der Aalst, W.M.P.: Discovering simulation models. Inf. Syst. **34**(3), 305–327 (2009). https://doi.org/10.1016/j.is.2008.09.002

26. Salleh, S., Thokala, P., Brennan, A., Hughes, R., Booth, A.: Simulation modelling in healthcare: an umbrella review of systematic literature reviews. Pharmacoeconomics **35**(9), 937–949 (2017). https://doi.org/10.1007/s40273-017-0523-3

27. Shakoor, M.: Using discrete event simulation approach to reduce waiting times in computed tomography radiology department. Int. J. Ind. Manuf. Eng. **9**(1), 177–181 (2015). https://doi.org/10.5281/zenodo.1338044

28. Smith-Daniels, V.L., Schweikhart, S.B., Smith-Daniels, D.E.: Capacity management in health care services: review and future research directions. Decis. Sci. **19**(4), 889–919 (1988). https://doi.org/10.1111/j.1540-5915.1988.tb00310.x
29. Suriadi, S., Andrews, R., ter Hofstede, A.H.M., Wynn, M.T.: Event log imperfection patterns for process mining: towards a systematic approach to cleaning event logs. Inf. Syst. **64**, 132–150 (2017). https://doi.org/10.1016/j.is.2016.07.011
30. VanBerkel, P.T., Blake, J.T.: A comprehensive simulation for wait time reduction and capacity planning applied in general surgery. Health Care Manag. Sci. **10**(4), 373–385 (2007). https://doi.org/10.1007/s10729-007-9035-6
31. Vanbrabant, L., Martin, N., Ramaekers, K., Braekers, K.: Quality of input data in emergency department simulations: framework and assessment techniques. Simul. Model. Pract. Theory **91**, 83–101 (2019). https://doi.org/10.1016/j.simpat.2018.12.002
32. Vieira, B., Hans, E.W., van Vliet-Vroegindeweij, C., van de Kamer, J., van Harten, W.: Operations research for resource planning and-use in radiotherapy: a literature review. BMC Med. Inform. Decis. Mak. **16**(1) (2016). Article number: 149. https://doi.org/10.1186/s12911-016-0390-4
33. Zhang, X.: Application of discrete event simulation in health care: a systematic review. BMC Health Serv. Res. **18**(1), 687 (2018). https://doi.org/10.1186/s12913-018-3456-4

Process Mining on the Extended Event Log to Analyse the System Usage During Healthcare Processes (Case Study: The GP Tab Usage During Chemotherapy Treatments)

Angelina Prima Kurniati[1]([email]) [iD], Geoff Hall[2,3] [iD], David Hogg[4] [iD],
and Owen Johnson[3] [iD]

[1] School of Computing, Telkom University, Bandung 40257, Indonesia
angelina@telkomuniversity.ac.id
[2] School of Medicine, University of Leeds, Leeds LS2 9JT, UK
[3] Leeds Teaching Hospitals NHS Trust, Leeds LS9 7TF, UK
[4] School of Computing, University of Leeds, Leeds LS2 9JT, UK

Abstract. In healthcare, process mining has been used in many case studies to discover and analyse process models of patient treatments. Process mining is generally applied to analyse the event log of patient treatments as extracted from the Electronic Health Record (EHR). In this study, we proposed an approach to combine the event log of patient treatments with the clinical user access log of the hospital information system to analyse system usage during patient treatments. Our case study combined an event log of breast cancer patients receiving chemotherapy treatments in the Leeds Cancer Centre with the user access log in the hospital information system. The event log of patient records during chemotherapy was extracted from the EHR system. The clinical user access log was extracted from the Splunk web-based log management system in the hospital. Combining records from those two logs has been useful to provide information on system usage during patient treatment. Our experiment focused on the GPTab, a functionality that allows clinicians during consultations to check on patient records on their GP visits. We applied both statistical and clinical evaluations to ensure that the findings are statistically correct and clinically meaningful. We captured the phenomena of the decreasing number of patients on the subsequent cycles of chemotherapy and when GPTab has been used during the course of chemotherapy. This approach is potentially useful for general cases to analyse system usage during process execution and can be applied to investigate the effects of system changes to process executions.

Keywords: Process mining · Extended event log · Clinical user access log · Chemotherapy · Cancer treatment · EHR

1 Introduction

As a large group of diseases, cancer is very complex and can affect any part of the body [1]. There are at least 65 recognised types of cancer [2]. Breast cancer is the most

© Springer Nature Switzerland AG 2021
S. Leemans and H. Leopold (Eds.): ICPM 2020 Workshops, LNBIP 406, pp. 330–342, 2021.
https://doi.org/10.1007/978-3-030-72693-5_25

common cancer in women affecting about 12% of women in the world [3]. In the UK, breast cancer is one of the four most common cancer types, along with prostate cancer, lung cancer, and colorectal cancer [4]. Breast cancer [5] is diagnosed by physical exam, mammogram, ultrasound, MRI, blood chemistry studies, and biopsy of the affected area of the breast. Surgery is the primary treatment, which may be followed by chemotherapy or radiation therapy, or both [6]. A course of chemotherapy [7] is usually done in six cycles, where each cycle is given 21 days after the previous one. Some patients might not be able to get a cycle of chemotherapy due to some adverse events, including emergency admission and neutropenia.

Process mining is a process-oriented data science approach that uses event logs for discovering and analysing business process models [8]. An event log is a record of timestamped activities generated automatically by the information system. Process mining has been applied in healthcare processes [9] for quality improvement, patient safety, and resource optimisation in healthcare settings [10]. Our literature review of process mining in Oncology [11], the study of cancer, found the limited availability and accessibility of suitable datasets for process mining. Our earlier study explored a publicly available dataset for process mining in healthcare [12, 13]. In this study, we were fortunate in having access to explore the in-house developed PPM EHR system including the database, the software developers of the system, the training team, clinical staff and senior clinicians involved in the process.

Our case study is based on a de-identified extract from the Patient Pathway Manager (PPM) database of the PPM EHR system [14]. The patient dataset has been used in the previous study to define real-life clinical pathways during chemotherapy [15]. This paper presents a worked example to analyse General Practitioner (GP) Tab usage during chemotherapy treatment on breast cancer patients. GPTab is a menu that allows clinicians to access patient records in the GP system. The GPTab presents clinical information (diagnosis, allergies, medications, etc.) recorded in the registered Leeds GPs. Accessing GP Tab during consultations in chemotherapy cycles improves understanding of patient condition and support decision making for patient treatment. We described an approach to enhance a process model through an extension of the event log, by combining patient records with the user access log. This approach is potentially useful in many other cases to enhance process mining approaches with user access log describing real user accessing information systems.

2   Patient Pathways Manager (PPM) EHR System

The PPM EHR system is used in the Leeds Teaching Hospitals NHS Trust (LTHT), the largest provider of specialised services in England that manages six hospitals, including St James's University Hospital (SJUH) [16]. The SJUH hosts the Leeds Cancer Centre, one of Europe's large cancer centres [17]. The PPM system integrates data from multiple systems within the LTHT, including patient admissions, treatments (chemotherapy, surgery, and radiotherapy), pathology, investigations, Multidisciplinary Team (MDT) meetings, consultations, and outpatients.

The PPM database contains clinical information about all patients within the hospital, including cancer patients. We gained access to the PPM database through an IRAS application that allows direct access to a secure SQL database on a virtual machine. The data has been checked, cleaned, and aggregated before approval for access by the research team. The PPM database consists of clinical data of more than 3 million patients, of which more than 270,000 patients have at least one cancer-related diagnosis. The PPM EHR system is connected to patient records in other service providers, including General Practitioners (GPs), Mental Health, and Community services. Figure 1 shows a screenshot of GPTab screen in the PPM EHR system.

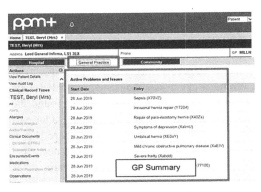

Fig. 1. Screenshot of the GPTab in the PPM EHR system, from the PPM support website [18].

The clinical user access log is recorded in PPM Splunk. The PPM Splunk is web-based application management that captures real-time user access to the PPM system, which is useful in analysing system usage for specific functionalities. Every time a user views data in the PPM EHR system, the system automatically recorded the activity in the PPM Splunk. In this study, the healthcare user access log was focused on the GPTab access log, as a representative of functionalities related to cancer treatment. GPTab is a functionality that can be used by clinicians to access patient records in the GP system, to support clinical decisions related to patient treatment.

3 Methodology

The general methodology is based on the Process Mining Project Methodology (PM2) [19] with a focus on the Mining and analysis step (Fig. 2).

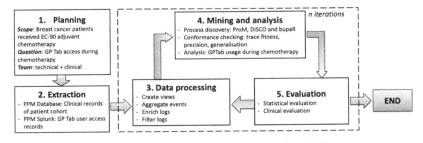

Fig. 2. The general methodology (based on PM2)

We did the stages in the methodology in at least two iterations: once with only the clinical records as the input, and once with a combination of the clinical records and the healthcare user access log. For simplicity and ease-of-understanding, this paper describes only the final iteration and summarise the findings in the intermediate iterations as part of the final iteration.

The Planning stage identified the scope, the team, and the research questions in the study. The scope of this study was to analyse GPTab usage during chemotherapy treatment of breast cancer patients in the PPM system. The research questions were:

Q1. What are the most followed paths and the exceptional paths?
Q2. How did clinicians use GPTab during the course of chemotherapy?

Our team consisted of process mining experts, clinical experts, representatives of the development and training teams of the PPM EHR system. We did at least one meeting in each stage of the study to discuss the plan, progress of the study, and validation of the findings. The discussion was done to ensure domain expert engagement during all stages of the study, as suggested in the ClearPath method [20].

The Extraction stage included the patient clinical records from the PPM database and the user access log from PPM Splunk. The patient clinical records are included if (1) the patient had at least one diagnosis of breast cancer (ICD-10 C50) and received epirubicin and cyclophosphamide (EC90) chemotherapy as adjuvant treatment and (2) the patient was first diagnosed with breast cancer between 2014 and 2018. The EC90 is one of the most commonly used regimens in Leeds Cancer Centre in the specified time period. The GPTab user access records from PPM Splunk are included if clinicians access GP records of patients in the cohort during their cancer treatment between 2014 (when GPTab was introduced) and 2018. Combining patient clinical records with user access records is useful to get additional data from user access log that is not recorded in the patient clinical records, in this study, adding GPTab access activity to the chemotherapy pathways. The extraction stage is illustrated in Fig. 3.

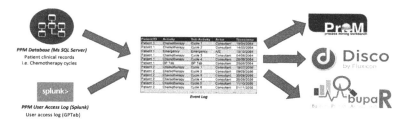

Fig. 3. The extraction stage, combining patient clinical records with user access records.

The Data Processing stage consisted of creating views, aggregating events, enriching logs, and filtering logs. The views were created by focusing on the chemotherapy cycles of breast cancer patients. Instead of aggregating events, we used the fine-grained event names, which are Cycle 1, Cycle 2, up to Cycle 6, representing the cycle number of chemotherapy. Log enrichment added information to the event log, in this case, the process duration for each patient that was calculated as the number of days from the first activity to the last one in the recorded treatment. We also included Emergency and Neutropenia events as suggested by clinical experts to be the two events potentially affecting chemotherapy progressions. We extracted the Emergency events as they were recorded in the Admission table with an Emergency Admission type. Neutropenia is a condition where a patient had a neutrophil count less than normal ($<1.5 \times 109$/L). More details about those two additional events had been described in our previous study using the same dataset [15]. An attribute-based log filtering was done by filtering in selected events to include only the chemotherapy cycle events of patient treatment. The patient records were transformed into an event log, which contains {case_id, activity, resource, timestamp}. The event log was loaded into ProM tools and R for analysis in the next stage.

The Mining and Analysis stage included process discovery, conformance checking [21], enhancement, and process analytics. *Process discovery* was done in the fine-grained level to model chemotherapy cycles of patients in the selected cohort. The adjuvant chemotherapy for breast cancer patients is commonly given in six cycles, sequentially from Cycle 1 to Cycle 6. The main tools for process discovery were ProM 6.8 [22], DISCO [23], and bupaR [24]. ProM is an academic platform that is widely used in process mining projects. DISCO is used in this study to get an early model easily, based on the fuzzy miner algorithm. BupaR is a library in R that was used in this study to support a more detailed statistical analysis. *Enhancement* was done by extending the event log of the patient records with the GPTab access log in the PPM Splunk. Figure 4 shows a screenshot containing detailed data on the date and time, page address, patient id and user id recording a time when a clinician had accessed the GP Tab page of a patient. There is also a bar chart visualising the number of records on a daily basis. The bar chart shows an obvious pattern of weekday- and weekend- usages.

Fig. 4. A query result in the PPM Splunk. Confidential information such as Patient ID and dates are blocked in black.

The Evaluation stage was done to diagnose, verify, and validate the results of the previous stages. In this study, the evaluation analysed all findings from the statistics and clinical perspectives. The statistical evaluation was done to verify and validate the result quantitatively, which was later confirmed to the clinical experts and the representative of the development team. The clinical evaluation was done to make sure that the findings reflected reality, supported and enhanced prior knowledge of the clinical experts about patient treatment.

4   Results and Discussion

4.1   The Extracted Data

We extracted Leeds patients diagnosed with breast cancer (C50) who received EC-90 as adjuvant chemotherapy, whose GP Tab was accessed by clinicians from 2014 to 2018. There were 738 patients included in this selection. Table 1 shows a list of the eight selected events for process discovery, which consists of six cycles of chemotherapy and two adverse events (emergency admission and neutropenia).

Table 1. Selected events for process discovery

Event name	Cycle						Emergency	Neutropenia
	1	2	3	4	5	6		
Patients (n)	738	725	699	487	402	380	380	412
Percentage	100%	99%	95%	66%	55%	52%	-	-
Med (days)	21	21	21	21	21	-	-	-

Table 1 shows that 738 patients received *Cycle 1* of chemotherapy, but the number decreases in the following cycles. The median duration from a Cycle to the next one is 21 days, which reflects the typical duration of treatment in reality. This finding has been discussed with clinical experts. It has been confirmed to reflect the reality where patients might find several conditions that prevent them from completing the course of chemotherapy. It is shown that among patients who started receiving *Cycle 1* of EC-90 as adjuvant chemotherapy, only around half of them (n = 380; 52%) completed *Cycle 6*. This condition needs to be explored more, to learn what were the possible conditions preventing patients from completing the treatment.

4.2   Discovered Process Models and the Conformance

We presented Table 2 to show the 15 most common trace variants out of 289 variants in total. Each of those 15 variants followed by at least seven patients.

Table 2 shows that the most common variant is a sequence of *Cycle 1* to *Cycle 6* (n = 120; 16.26%), followed by the second variant that is a sequence of *Cycle 1* to *Cycle 3* (n = 56; 7.59%). Our clinical experts confirmed that even though a complete sequence of *Cycle 1* to *Cycle 6* is expected, a lot of patients needed a consultation after *Cycle 3* to decide if the chemotherapy regimen can be continued. Patients might also change regimen after *Cycle 3* and therefore are not captured in this study.

Figure 5 shows a dotted chart of routine chemotherapy cycles of patients treatments of up to 7 years. The chart shows groups of patients who had not completed six cycles

Table 2. Top fifteen trace variants

Var	Trace variant	n	(%)
1	*Cycle 1 - Cycle 2 - Cycle 3 - Cycle 4 - Cycle 5 - Cycle 6*	120	16.26
2	*Cycle 1 - Cycle 2 - Cycle 3*	56	7.59
3	*Cycle 1 - Cycle 2 - Cycle 3 – Emergency*	37	5.01
4	*Cycle 1 - Cycle 2 - Cycle 3 - Cycle 4 - Cycle 5 - Cycle 6 - Emergency*	25	3.39
5	*Cycle 1 - Cycle 2 - Cycle 3 - Cycle 4*	14	1.90
6	*Cycle 1 - Cycle 2 - Cycle 3 - Cycle 4 - Cycle 5 - Neutropenic - Cycle 6*	11	1.49
7	*Cycle 1 - Neutropenic - Cycle 2 - Neutropenic - Cycle 3 - Neutropenic*	10	1.36
8	*Cycle 1 - Cycle 2 - Cycle 3 - Cycle 4 - Neutropenic - Cycle 5 - Cycle 6*	10	1.36
9	*Cycle 1 - Cycle 2 - Cycle 3 - Cycle 4 - Emergency*	9	1.22
10	*Cycle 1 - Cycle 2 - Cycle 3 – Emergency - Neutropenic*	9	1.22
11	*Cycle 1 - Cycle 2 - Cycle 3 – Neutropenic*	8	1.08
12	*Cycle 1 - Cycle 2 - Cycle 3 - Cycle 4 - Cycle 5*	8	1.08
13	*Cycle 1 - Cycle 2 - Cycle 3 - Neutropenic - Cycle 4 - Cycle 5 - Cycle 6 - Emergency*	8	1.08
14	*Cycle 1 - Cycle 2 - Cycle 3 - Neutropenic - Emergency*	8	1.08
15	*Cycle 1 - Cycle 2 - Cycle 3 - Neutropenic - Cycle 4 - Cycle 5 - Cycle 6*	7	0.95

of chemotherapy (the one-third top part of the chart), who completed six cycles of chemotherapy (the middle part), and who had more complicated courses of treatment (the bottom part). In total, 51% (n = 376) patients completed all six cycles, without any acute event (n = 158; 21%) or having at least one acute event including *Emergency Admission* or *Neutropenia* (n = 218; 30%). The patients who did not complete six cycles (n = 392; 49%), might had acute events (n = 207; 28%) or not completing for other reasons (n = 155; 21%). Based on our discussion with clinical experts, some of those reasons are missing appointments, disease complications, and personal reasons.

This dotted chart had been shown to the clinicians and all of them agreed that this visualisation helped them understanding the situation more clearly. There are only about a third of patients had the normal and 'happy' path of six cycles of chemotherapy, while the others had incomplete or overly complicated paths of treatment. Some example of patients were picked and discussed with clinical experts to see specific cases where patient conditions preventing them from completing the treatment. Those specific cases are not presented in this paper because presenting data of a small number of patient would breach ethical approvals.

Further analysis of the result was examining the cycles leading to an emergency admission or a neutropenic condition. Table 3 shows that most patients who had emergency admission got it after *Cycle 3* (n = 117; 16%), *Cycle 6* (n = 90; 12%), or *Cycle 1* (n = 81; 11%); while most patients who had *Neutropenic* got it after *Cycle 3* (n = 142; 19%), *Cycle 2* (n = 123; 17%), or *Cycle 1* (n = 94; 13%). Collectively, adverse events

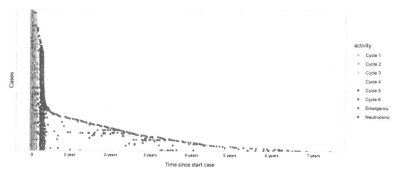

Fig. 5. Dotted chart showing adverse events during six chemotherapy cycles. The x-axis shows duration from the first activity to the last one. The y-axis shows patient id, sorted by durations.

(Emergency or Neutropenic) have mostly occurred after *Cycle 3*. Table 2 summarised the pattern of the cycles leading to an acute event and might have a one-to-many relation to trace variants presented in Table 2. For example, *Cycle 3* leading to a *Neutropenic* event in Table 3 (n = 142; 19%) is related to variants 7, 11, 13, 14, 15 and other infrequent variants in Table 2.

Table 3. The cycles leading to an acute event

Activity	Leads to emergency		Leads to neutropenic	
	N (%)	Med; Mean	N (%)	Med; Mean
Cycle 1	81(11)	8 d; 18.4 d	94(13)	19 d; 23.1 d
Cycle 2	52(7)	8 d; 43.9 d	123(17)	19 d; 20.6 d
Cycle 3	117(16)	28 d; 27.3 w	142(19)	18 d; 61.1 d
Cycle 4	64(9)	14d; 27.3 w	84(11)	19 d; 16 d
Cycle 5	22(3)	13.5 d; 19.2 w	70(9)	19 d; 33.5 d
Cycle 6	-	-	-	-

It is also important to note that the median and mean duration of acute events after a chemotherapy cycle are generally under 21 days, within the expected duration of a cycle to the next one. This means that patients experienced one or more acute events before the next cycle of chemotherapy, got treated, and continue to the next cycle of chemotherapy as planned. On the last row, Emergency and Neutropenic events after Cycle 6 are not presented because they are not part of this study.

4.3 The Enhanced Process Model

There were 339 out of 738 patients (46%) who had their GPTab accessed by clinicians. This percentage is higher than the percentage of all cancer patients who had their GPTab accessed by clinicians (46,547 out of 339,127 patients; 37%), which showed that clinicians made use of the patient records in the GPTab to support their decisions on the next treatment for their patients. Figure 6 shows the process model containing the flow from Cycle 1 to Cycle 6 of chemotherapy. During the course of chemotherapy, the GPTab might be accessed by clinicians. The most frequent sequence is that GPTab was accessed after Cycle 6 (n = 160; 47%), followed by GPTab access after Cycle 3 (n = 110, 32%) and GPTab access after Cycle 4 (n = 31; 9%).

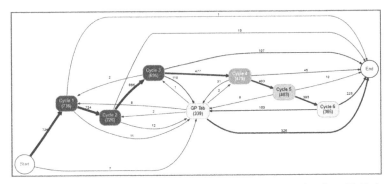

Fig. 6. Process model showing GPTab access during chemotherapy cycles (bupaR). It shows that GPTab was mostly accessed after Cycle 3, Cycle 6, or Cycle 4.

These results have been confirmed by the clinical experts to reflect reality. The clinicians are most likely need to check on patient records in GPTab after the sixth cycle to decide whether to discharge the patient, to follow on the next cycle of chemotherapy, or to suggest another treatment. Clinicians might need to check on patient records in GPTab after Cycle 3, to decide if the next cycles should be delivered as planned or not. Another finding was that GPTab click is mostly the last activity in the pathways, or at the end of treatment (n = 326; 96%). The enhanced process model revealed some important insights into how GPTab has been used during the treatment process.

4.4 Process Analytics

Process analytics was done to analyse GPTab usage chemotherapy. This was based on a discussion with a representative of the PPM development team who mentioned that the GPTab had been through some changes during the study period. We followed up this discussion by exploring the increasing pattern of GPTab usage over time. Figure 7 shows a bar chart of the number of GPTab clicks from July 2014 to December 2018.

Further exploration of the PPM Splunk records shows that in March 2018, the first version of GPTab (GPv1) has been replaced by the second version (GPv2). In September 2017 to February 2018 both versions were accessed by clinicians, and this has been confirmed as the transition period. The transition period from GPv1 to GPv2 can be captured in the monthly usage from 2017 to 2018, as shown in Fig. 8. This has not been seen in Fig. 7, which shows that the transition from the first version to the second one has been done smoothly.

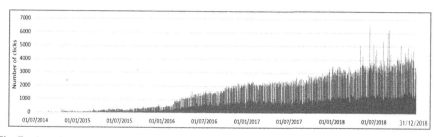

Fig. 7. GPTab clicks each day. It shows that the number of clicks generally increased over time, with steady fluctuations showing the pattern of weekday- and weekend- usages.

4.5   Statistical and Clinical Evaluation

The evaluation was done in both statistical and clinical aspects. Statistical evaluation was done throughout the stages by analysing the occurrence numbers and percentages of events in the process. This has been presented in the relevant steps in the previous sections of this paper.

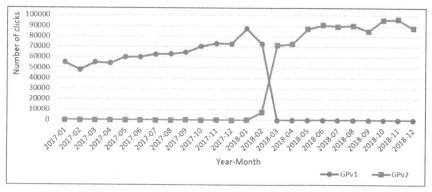

Fig. 8. Monthly usage of GPTab during 2017–2018. The blue dots are monthly usage of the first version (GPv1) and the orange dots are those of the second version (GPv2).

Clinical evaluation was done through discussion with clinical experts. In the Planning stage, clinical experts suggested the scope of the study. The GPTab functionality was chosen based on the availability of the related data to enhance process model of patient treatment. One important insight from the software training team was that for some new features introduced in the PPM software, there was a period when training was given to the clinicians to introduced the use of the new feature, such as GPTab. During the Extraction stage, clinical experts evaluated and suggested details the extraction step. One important suggestion in this stage was the specific type of chemotherapy for breast cancer selected in this study, which is EC90 for adjuvant treatment. In the Data processing stage, clinicians suggested focusing on the effect of the GPTab introduction to the chemotherapy cycles. The findings from the Mining and analysis stage have been discussed with clinical experts. Some of their comments had been presented in the relevant part in Sect. 4.1 to Sect. 4.4. The GPTab supported clinicians to decide on the next treatment suitable for their patients, such as to follow with the next cycle of chemotherapy, to change the regimen of chemotherapy, or to discharge the patient.

5 Conclusion

This paper described a process analytics approach by combining patient clinical records with user access log to analyse system usage during patient treatment. A case study presented in this paper was GPTab usage during chemotherapy treatment. Two research questions had been established and answered through a structured experiment following the PM2 stages. The first research question has been answered in the Mining and analysis stage, specifically in the process model (see Sect. 4.2). Additional analysis to support this answer has been presented in a trace variant list (Table 2) and a dotted chart (Fig. 5). The second research question has been answered by the enhanced process model (Fig. 6) which shows how GPTab has been used to support clinician to decide the next treatment for their patients. General comments of the findings throughout the stages are that process mining is potentially useful to improve clinical pathway analysis by providing visualisation of process models and additional results such as trace variance diagrams and dotted charts. Those visualisations supported discussions with the multi-disciplinary team.

Some limitations and potential improvements in this study are as follow. The first is to explore the aggregated events to see how chemotherapy has been given in the sequence from a referral, diagnosis, and a set of treatments. Second, the idea of combining user access records in PPM Splunk with the treatment records in the PPM database was good to analyse the effect of system functionality to the treatment process. Another possibility discussed was to analyse PPM Splunk separately to be compared to the discovered process model from the patient records. Since PPM Splunk recorded all actions done by clinicians during patient treatment, the treatment process itself should be reflected in the records. Third, the extraction and data processing in this study relied on the selection of the best set of events of the specific cohort of patients, based on the understanding of the data and problem domain. Further improvement might be to explore possible ways to select the best set of events based on the data attributes, with less dependence on clinical expert judgments.

Acknowledgment. This research was supported by ClearPath Connected Health Cities Project and was partly funded by the Indonesia Endowment Fund for Education (LPDP). Access to data used in this study is under the Health Research Authority (HRA) Approval Number 206843.

References

1. American Cancer Society. The History of Cancer (2011). www.cancer.net/patient/Advocacy and Policy/Treatment_Advances_Timeline.pdf. Accessed 09 Aug 2016
2. CRUK. "Your cancer type," Cancer Research UK (2014). https://www.cancerresearchuk.org/about-cancer/type/. Accessed 09 Aug 2016

3. National Cancer Institute. Breast Cancer - Patient Version (2018). https://www.cancer.gov/types/breast/patient/breast-treatment-pdq

4. Office of National Statistics. Cancer registration statistics, England (2016)

5. National Institute for Health and Care Excellence. "Advanced breast cancer overview - NICE Pathways," NICE Pathways (2016). https://pathways.nice.org.uk/pathways/advanced-breast-cancer

6. National Chemotherapy Advisory Group. "Chemotherapy Services in England: Ensuring quality and safety," Dep. Heal., no. August 2009

7. Royal Cornwall Hospitals NHS Trust. Clinical Guideline for the Assessment and Management of Chemotherapy Induced Diarrhoea, p. 11

8. van der Aalst, W.M.P.: Process Mining: Data Science in Action, 2nd edn. Springer-Verlag, Heidelberg (2016). https://doi.org/10.1007/978-3-662-49851-4

9. Rojas, E., Munoz-Gama, J.: Process mining in healthcare: a literature review. J. Biomed. Inform. **61**, 224–236 (2016)

10. Mans, R.S., van der Aalst, W.M.P., Vanwersch, R.J.B., Moleman, A.J.: Process mining in healthcare: data challenges when answering frequently posed questions. In: Lenz, R., Miksch, S., Peleg, M., Reichert, M., Riaño, D., ten Teije, A. (eds.) KR4HC/ProHealth -2012. LNCS (LNAI), vol. 7738, pp. 140–153. Springer, Heidelberg (2013). https://doi.org/10.1007/978-3-642-36438-9_10

11. Kurniati, A.P., Johnson, O., Hogg, D., Hall, G.: Process mining in oncology: a literature review. In: Proceedings of the 6th ICICM 2016, pp. 291–297. https://doi.org/10.1109/INFOCOMAN.2016.7784260

12. Kurniati, A.P., Rojas, E., Hogg, D., Johnson, O.: The assessment of data quality issues for process mining in healthcare using MIMIC-III, a publicly available e-health record database, no. 2 (2017)

13. Kurniati, A.P., Hall, G., Hogg, D., Johnson, O.: Process mining in oncology using the MIMIC-III dataset. IOP J. Phys. Conf. Ser. **971**(012008), 10 (2018)

14. Newsham, A., Johnston, C., Hall, G.: Development of an advanced database for clinical trials integrated with an electronic patient record system. Comput. Biol. Med. **41**(8), 575–586 (2011)

15. Baker, K., et al.: Process mining routinely collected electronic health records to define real-life clinical pathways during chemotherapy. Int. J. Med. Inform. **103**, 32–41 (2017). https://doi.org/10.1016/j.ijmedinf.2017.03.011

16. Leeds Teaching Hospitals NHS Trust. "Leeds Teaching Hospital" (2016). https://www.leedsth.nhs.uk/. Accessed 26 Jul 2016

17. Hazell, W.: Analysed: the biggest NHS providers of specialised services|News|Health Service Journal (2015). https://www.hsj.co.uk/home/analysed-the-biggest-nhs-providers-of-specialised-services/5091147.article. Accessed 30 July 2019

18. LTHT. "Leeds Care Records GP Tab," Leeds (2019). https://www.leedscarerecord.org/lcr/widget/whats-in/gp-tab.pdf

19. van Eck, M.L., Lu, X., Leemans, S.J.J., van der Aalst, W.M.P.: PM²: A process mining project methodology. In: Zdravkovic, J., Kirikova, M., Johannesson, P. (eds.) CAiSE 2015. LNCS, vol. 9097, pp. 297–313. Springer, Cham (2015). https://doi.org/10.1007/978-3-319-19069-3_19

20. Johnson, O.A., Ba Dhafari, T., Kurniati, A., Fox, F., Rojas, E.: The ClearPath method for care pathway process mining and simulation. In: Daniel, F., Sheng, Q.Z., Motahari, H. (eds.) BPM 2018. LNBIP, vol. 342, pp. 239–250. Springer, Cham (2019). https://doi.org/10.1007/978-3-030-11641-5_19

21. Buijs, J.C.A.M., van Dongen, B.F., van der Aalst, W.M.P.: On the role of fitness, precision, generalization and simplicity in process discovery. In: Meersman, R., et al. (eds.) OTM 2012.

LNCS, vol. 7565, pp. 305–322. Springer, Heidelberg (2012). https://doi.org/10.1007/978-3-642-33606-5_19

22. van der Aalst, W.M.P., Van Dongen, B.F., Gunther, C., Rozinat, A., Verbeek, H.M.W., Weijters, A.J.M.M.: ProM: the process mining toolkit. In: CEUR Workshop Proceedings, vol. 489 (2009)

23. Günther, C.W., Rozinat, A.: Disco: discover your processes. In: BPM 2012 Demonstration Track, vol. 940, pp. 40–44 (2012)

24. Janssenswillen, G.: bupaR: Business Process Analysis in R, R package version 0.4.2 (2019). https://cran.r-project.org/package=bupaR

Process Mining on FHIR - An Open Standards-Based Process Analytics Approach for Healthcare

Emmanuel Helm[1,3](✉) ⓘ, Oliver Krauss[1,3] ⓘ, Anna Lin[1] ⓘ,
Andreas Pointner[1] ⓘ, Andreas Schuler[2,3] ⓘ, and Josef Küng[3]

[1] Research Department Advanced Information Systems and Technology,
University of Applied Sciences Upper Austria, 4232 Hagenberg, Austria
`emmanuel.helm@fh-hagenberg.at`
[2] Department of Medical and Bioinformatics, University of Applied Sciences
Upper Austria, 4232 Hagenberg, Austria
`andreas.schuler@fh-hagenberg.at`
[3] Johannes Kepler University, 4040 Linz, Austria
`josef.kueng@jku.at`
`https://aist.fh-hagenberg.at`

Abstract. Process mining has become its own research discipline over the last years, providing ways to analyze business processes based on event logs. In healthcare, the characteristics of organizational and treatment processes, especially regarding heterogeneous data sources, make it hard to apply process mining techniques. This work presents an approach to utilize established standards for accessing the audit trails of healthcare information systems and provides automated mapping to an event log format suitable for process mining. It also presents a way to simulate healthcare processes and uses it to validate the approach.

Keywords: Process mining · Healthcare · HL7 FHIR

1 Introduction

We provide a process analytics approach to enable the mining of standardized audit trails of healthcare information systems by transforming them into eXtensible Event Stream (XES) logs via an automated mapping approach. We tested it by simulating a radiology practice workflow, and analyzed the results with a process mining tool.

With diverse use cases and different approaches, techniques, and algorithms, process mining became its own scientific discipline over the last 20 years [1]. With the goal of understanding and improving the real-world processes, process mining provides an evidence-based (i.e., data-driven) view on the processes recorded by information systems. An increasing number of case studies also show the applicability of process mining in the healthcare domain (cf. the reviews in [4,18]). Most of those case studies focus their analysis on single hospitals or even departments due to problems of data integration or data availability [4].

© Springer Nature Switzerland AG 2021
S. Leemans and H. Leopold (Eds.): ICPM 2020 Workshops, LNBIP 406, pp. 343–355, 2021.
https://doi.org/10.1007/978-3-030-72693-5_26

1.1   Problem Statement

Rebuge and Ferreira [17] conclude in their work that healthcare processes, both organizational and medical treatment, are *highly dynamic, highly complex, increasingly multi-disciplinary* and *generally ad-hoc*. All four characteristics make it hard to apply process mining techniques. In this work we focus on the aspect of high complexity, partly caused by the high number of participants, heterogeneous information systems, and the resulting lack of interoperability [4,17].

Rojas et al. [18] found in their review that three implementation strategies for process mining projects in healthcare exist: (1) The majority of case studies work with *direct implementations*, where data is gathered directly from hospital information systems (HIS) for building an event log. Data extraction and building the correct event log poses major challenges here. (2) The second, *semi-automated*, strategy involves the integration and extraction of data from different sources via custom-made developments. The disadvantage here is the ad-hoc, proprietary nature of these developments, as they only work for specific data sources and environments. Both strategies, direct implementation and semi-automated, share the need to understand process mining tools and algorithms for conducting process analytics. (3) The third strategy is the implementation of an *integrated suite*. Specific data sources are connected and integrated, and specific process mining algorithms are executed in order to perform defined analytics tasks. Once implemented, these solutions are easily applicable, but like the semi-automated strategy, fail to integrate other data sources and environments.

We conclude, that a major problem with starting a process mining project in healthcare is that one has to choose between either complex manual data extraction and integration, or locking oneself in on specific data sources and environments (i.e., vendor lock-in).

1.2   Related Work

To overcome the problems of process mining on heterogeneous data sources in healthcare, some studies tried to analyze standardized audit trails [3,7,16]. We will build on this work, using their concepts of audit events, mapping strategies, and multi-perspective process mining.

Cruz-Correia et al. [3] were the first to explicitly make the connection between standardized auditing in healthcare and process mining. They specifically looked at the Integrating the Healthcare Enterprise (IHE) integration profile Audit Trail and Node Authentication (ATNA). Being one of the core profiles dealing with IT infrastructure in healthcare, ATNA defines how to build up a secure domain that provides patient information confidentiality, data integrity, and user accountability. They analyzed ATNA audit trails from four different hospitals in Portugal and identified several data quality issues.

Later, Helm and Paster [7] investigated the suitability of event logs recorded by the means of IHE ATNA for process mining. They adopted a direct mapping approach, transforming IHE audit messages into XES event logs. They encountered issues regarding the determination of trace identifiers and semantics preserving mapping.

De Murillas et al. [16] took on the previous approach [7] and presented a method to overcome the problems of trace identification and incorrect mappings. By integrating the audit trail data into a generic meta model (OpenSLEX), they provided the means to query and analyze the data from different perspectives.

While these approaches try to solve the issue of heterogeneous data sources, they either lock the user in on a predefined mapping [7] or provide a non-standardized interface to the process data [16] – two shortcomings that can be avoided with our approach.

1.3   Proposed Solution

Supporting definition, instantiation, and execution of workflows is still a topic of vivid discussions in the respective standards development working groups. For the analysis part, first steps have been taken. Standardized Operational Log of Events (SOLE) is a recently developed IHE integration profile. It is a supplement for the radiology technical framework and currently in revision 1.2, published for trial implementation in mid 2018 [13]. SOLE describes the capture and retrieval of operational events in the radiology domain and utilizes transactions from the ATNA profile, including the new RESTful ATNA [12], based on the Health Level Seven (HL7) standard Fast Healthcare Interoperability Resources (FHIR). The profile authors' incentive for writing the SOLE integration profile was the strong desire of healthcare providers "to increase throughput and efficiency, both to improve the quality and timeliness of care and to control costs" [13]. They conclude, that workflow events must be captured in order to be able to apply *business intelligence tools* [13].

We propose an open standards-based process analytics approach for healthcare information systems to overcome the problems mentioned above. It enables the development of tools that combine the easy applicability of an integrated suite with the ability to integrate different data sources. This will make existing process mining tools the *business intelligence tools* the community wants.

To this end, this paper aims to show how existing concepts can be utilized and what changes in the standard are necessary to enable process mining based on HL7 FHIR. This paper also contributes to the field by presenting a novel approach to utilize a process simulation tool in a healthcare environment.

2   Background

This section provides a brief overview on the two major standards involved in building the open process analytics approach, HL7 FHIR and XES.

2.1   HL7 FHIR

FHIR[1] is the latest addition to the family of healthcare interoperability standards maintained and published by HL7 International [8]. FHIR provides a

---

[1] HL7, FHIR and the FHIR logo are the registered trademarks of Health Level Seven International and their use does not constitute endorsement by HL7.

comprehensive information model which is geared towards supporting semantic interoperability of clinical data. The fundamental building blocks for this information model are *resources*. A resource as described by Mandel et al. [15] is a coherent expression of clinical data and is based on a set of well-defined fields and data types. Every resource comprises the standard defined data content, a human-readable representation of respective content and has an identity. The FHIR specification defines resources for common clinical concepts, e.g., Patient, Medication, Observation, Condition. Besides that, FHIR leverages modern web technologies together with a strong foundation of web standards and offers support for RESTful architectures. Following the RESTful paradigm, FHIR allows to alter the state of a particular resource using a set of predefined actions for Create/Read/Update/Delete (CRUD). If required by a given use-case, it is also possible to apply a more Remote Procedure Call (RPC)-like interaction paradigm. This is achieved by defining operations that work on input and produce an output [9]. The operations can be executed on the server level, on the resource type level, or on the instance level of a specific resource and are typically invoked by a HTTP POST or can alternatively be invoked by a HTTP GET if no changes are caused on the server.

According to HL7 International [8], a central challenge for the FHIR specification is handling the wide variety and variability in diverse healthcare processes. This challenge is solved by offering a simple framework for extending the existing resources and describe use cases based on profiles. Profiling a resource allows to constrain and extend a resource specification for a given context [15]. By providing reference implementations for the specification, HL7 intends to reduce the entry barrier for developing FHIR conformant solutions. The development of the specification and the standard follows a developer first approach, which is reflected by the specification as a mixed standard comprising normative portions and parts still undergoing trial use [8].

2.2 XES

Log data is created from a variety of different systems with their own proprietary data models, formats, and semantics. Process mining techniques require their input data in a specific format. Some tools directly integrate data from (1) Enterprise Resource Planning (ERP) systems, (2) databases, or (3) Comma Separated Value (CSV) files, all three in a proprietary way. However, developed in 2010, XES became the IEEE standard for "achieving interoperability in event logs and event streams" [11]. Today, XES is supported by the majority of process mining tool vendors.

XES defines three basic objects: log, trace and event. Log (the process) contains a collection of traces (execution instances) and a trace contains a collection of events [20]. Each object can contain an arbitrary set of strongly typed attributes in the form of key-value pairs. Every attribute value has a data type, like string, boolean, or date. To add semantics to these data types, XES defines the concept of extensions. An extension defines a set of attributes, their types, and keys with a specific semantic meaning.

3   Materials and Methods

This section describes which standards and tools were used in building the analytics suite and how we utilized and extended them to enable process mining based on HL7 FHIR.

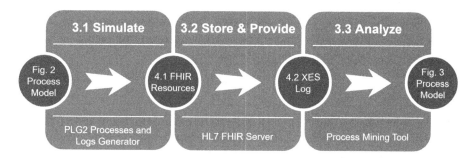

Fig. 1. The three steps of the interface test setting including the respective consumed and produced data. The numbers correspond to sections or figures in this paper.

Figure 1 depicts the three steps (1) simulate, (2) store&provide, and (3) analyze, that aim to show how the open standardized process analytics approach works. The circles represent data consumed and produced in those three steps.

To test the approach, a simple process was used. Figure 2 shows a simplified process model for an examination in a radiology practice using Business Process Model and Notation (BPMN). It shows the main steps from the appointment scheduling to the distribution of the diagnostic report. It is based on the work of Erickson et al. on business analytics in radiology [5] and on the process model used for evaluation in [7]. This is of course just an example and the approach is applicable to other healthcare domains as well.

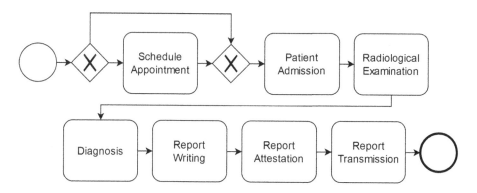

Fig. 2. BPMN process model of the radiology practice workflow based on [5,7].

In the first step, a patient that, e.g., received a referral for a radiological examination, calls the practice to *schedule an appointment*. In some cases of our simulation, this step can be skipped and the patient arrives without a scheduled appointment. On the day of the examination, the patient arrives at the reception and is placed on the waiting list (*patient admission*). When called, the patient enters the procedure room and the *radiological examination* takes place. Afterwards, the radiologist makes a *diagnosis* and dictates the report. The *report writing* is done by trained specialists. The resulting report is *attested* by the radiologist. Finally, the report is sent to a requesting physician or handed out directly to the patient (*report transmission*).

3.1 Simulate

In order to be able to automatically generate process data, some sort of process engine or simulator is required. Burattin [2] developed a tool specifically designed to simulate processes and generate event logs for process mining, the Processes and Logs Generator (PLG2). The tool allows to generate and simulate random BPMN models, and to add randomized noise (e.g., double activity execution, skipping activities, etc.). The tool also allows to load an existing model, in our case the model from Fig. 2, and simulate it.

To use PLG2 for the simulation, we needed to make REST calls to our HL7 FHIR server. PLG2 allows to specify the execution time of different activities using Python scripts [2]. We adapted those scripts to execute REST calls using Client for URLs (cURL). By default PLG2 provides a single parameter, that is, the case identifier (caseId), to these python functions. We used this parameter to make the process instances distinguishable by deriving resource identifiers from it (i.e., patientId and encounterId).

Each activity in the process from Fig. 2 was extended with REST calls, creating, reading, or updating resources and executing operations on the FHIR server (according to the mapping described in the next section). The process was then simulated 10 times without randomized noise, each run resulting in one process instance recorded on the server.

3.2 Store and Provide

We set up a FHIR server including the required extensions and operations to automatically record audit trails, and to transform and provide this information in the XES format for process mining.

FHIR Server. We implemented our FHIR Server based on the open-source project "HAPI-FHIR Starter"[2]. This project provides a fully working FHIR server, including a database connection, based on the HAPI FHIR JPA project. Adjustable configuration files and the interceptor framework [19] create high

---

[2] https://github.com/hapifhir/hapi-fhir-jpaserver-starter.

flexibility for custom changes and for adding extensions to the existing server implementation.

We utilized the Consent Interceptor, which amongst other functionalities has the ability to hook into the point of the server code, where a CRUD operation (e.g., creating an appointment or reading a patient record) has been finished. One of the Consent Interceptor's roles is to write audit trail records, creating an AuditEvent resource every time an operation has been finished successfully or with a failure.

In addition to the interceptor implementation, we provided the FHIR operation *$fhirToCDA* as part of our custom extensions to the server implementation. The operation can be executed on a specific instance of the DiagnosticReport resource and it returns an empty document to the client. An AuditEvent recording the execution of this operation in the context of a radiology workflow encounter will, for mapping purposes, be interpreted as a report transmission activity.

To query for an event log in the XES format, we extended our FHIR server by the *$xes* operation, which is defined to work on the AuditEvent resource type and is there to identify and transform all AuditEvents of the radiological workflow "rad-wf" into the XES format:

```
GET [fhirserver]/AuditEvent/$xes?plandefinition=PlanDefinition/rad-wf
```

Extending AuditEvent. We filled the AuditEvent resource with request details that are automatically provided for any standard CRUD operation. In order to be able to query for relevant AuditEvent resources, we needed to identify grouping elements. We decided to extend the AuditEvent resource by references to the Encounter and PlanDefinition resources (cf. Sect. 5.1). Geared to the other resources containing the Encounter resource reference as part of their standard FHIR resource definition, we named the extended AuditEvent element "encounter". An additional extension "basedon" is used to reference the PlanDefinition resource "rad-wf", that defines the radiological workflow. This element can later be used to filter AuditEvent resources related to the executions of the radiological workflow process, while Encounter references are used to distinguish the single process instances (i.e., the traces).

Mapping FHIR AuditEvent to XES. For the test setting, we base our mapping on the assumption that Encounter identifier can be utilized as trace identifiers and that recorded events refer to a common process description, i.e., a medical guideline or pathway defined as a PlanDefinition. Of course, this is just one perspective, and different perspectives can be taken on the data (cf. Sect. 5.3).

Let R be the set of all resources on the FHIR server. Let $A \subseteq R$ be the set of all AuditEvent resources, and $E \subseteq R$ be the set of all Encounter resources, and $P \subseteq R$ be the set of all PlanDefinition resources. All three subsets are disjoint,

Table 1. Mapping table of operations on specific FHIR resources to activities of the radiology practice workflow, ordered by occurence in the simulated model in Fig. 2.

Operation	FHIR Resource	$\mapsto$	Activity
Create	Appointment		Schedule appointment
Update	Appointment		Patient admission
Create	Procedure		Radiological examination
Create	Media		Diagnosis
Create	DiagnosticReport		Report writing
Update	DiagnosticReport		Report attestation
Execute	*$fhirToCDA		Report transmission

i.e., $A \cap P = \emptyset$, $A \cap E = \emptyset$, and $E \cap P = \emptyset$. Resources can refer to other resources via the predicate refersTo$(r, r') :\Leftrightarrow (r, r') \in R$, where r' is referenced by r, i.e., r contains the identifier of r'.

Let $p_w \in P$ be the PlanDefinition resource "rad-wf" defining the radiology workflow. Then, $A_w = \{a \in R | \forall_{a \in A} \text{refersTo}(a, p_w)\}$ is the set of all AuditEvent resources recorded during the execution of radiology workflows.

For our mapping, let A_w be a set of disjoint sets A_{wi}, where every A_{wi} represents a set of AuditEvents recorded during a specific radiology workflow encounter $\exists e \in E$ of one patient. Then, every A_{wi} will be mapped to a trace σ in an XES event log L.

For testing the approach, we only map to mandatory fields in L, e.g., concept: name of the event (providing the activity name) and time: timestamp of the event (for ordering). Table 1 describes which recorded combination of operation and resource is mapped to which activity name. The timestamp is mapped directly from the recording time AuditEvent.recorded.

3.3   Analyze

Querying the FHIR server for AuditEvent resources using the $xes operation returns an XES event log. Since the operation already utilizes XES standard extensions (i.e., Concept and Time), the semantics of the fields are clear for process mining tools. The next step is to analyze if the simulated process matches the one stored and provided by the HL7 FHIR server. Thus, we want to compare the input model with a model generated based on the retrieved XES event log. We use the process mining tool ProM 6.9 [20] with the Visual Inductive Miner plugin [14] to generate a model.

4   Results

This section shows three exemplary results of the implementation: (1) a FHIR resource generated by the simulator, (2) the corresponding event in the XES event log, and (3) the process model created based on the event log. All results and examples can also be found in our GitHub open-source project[3].

4.1   FHIR Resources

As described in the mapping in Table 1, the Report Writing activity is associated with creating a DiagnosticReport resource. The simulator thus executes the following cURL statement:

```
POST [fhirserver]/DiagnosticReport
{ "resourceType": "DiagnosticReport",
  "subject": { "reference": "Patient/[patientId]" },
  "encounter": { "reference": "Encounter/[encounterId]" },
  "status": "preliminary",
  "code": {
    "coding": [ {
        "system": "http://loinc.org",
        "code": "LP31534-8",
        "display": "Study report"
    } ]
  }
}
```

This triggers the creation of an AuditEvent resource. This one is shown in abbreviated form, focusing on the elements relevant for the mapping:

```
{ "resourceType": "AuditEvent",
  "extension": [
    { "url": "https://fhirserver.com/extensions/auditevent-encounter",
    "valueReference": { "reference": "Encounter/[encounterId]" }},
    { "url": "https://fhirserver.com/extensions/auditevent-basedon",
    "valueReference": { "reference": "PlanDefinition/rad-wf" }}
  ],
  "action": "C",
  "recorded": "2020-08-14T08:42:51.523+02:00",
  "entity": [ {
    "what": { "type": "DiagnosticReport" },
    "detail": [ {
      "type": "RequestedURL",
      "valueString": "[fhirserver]/DiagnosticReport/"
    } ]
  } ]
}
```

The created AuditEvent resource refers to the respective Encounter resource and to the PlanDefinition resource "rad-wf" that defines the radiology workflow. The *action* field indicates the type of operation (C = Create) and the *entity* element contains details about the manipulated resource, i.e., the DiagnosticReport.

---

[3] https://github.com/fhooeaist/ProcessMiningOnFHIR/.

4.2   XES Log

The query for AuditEvent resources with the $xes operation returns the following XES event log (only one trace with one event is shown, extensions left out):

```xml
<?xml version="1.0" encoding="UTF-8" ?>
<log xmlns="http://www.xes-standard.org/">
  <string key="concept:name" value="PlanDefinition/rad-wf"/>
  <trace>
    <string key="concept:name" value="Encounter/enccase55"/>
    <event>
      <string key="concept:name" value="Report Writing"/>
      <date key="time:timestamp" value="2020-08-14T08:42:51.523+02:00"/>
    </event>
  </trace>
</log>
```

This detail of the resulting XES log shows the concept: name attributes on log and trace level, derived from the referenced PlanDefinition and Encounter resources respectively. The event (*report writing*) was generated for the AuditEvent resource presented in the previous Sect. 4.1, that recorded the creation of a DiagnosticReport.

4.3   Process Model

Figure 3 shows the resulting model after importing the XES event log in ProM and analyzing it with the Inductive Visual Miner [14]. It is split up in two parts and highlights the similarity to the input model in Fig. 2. All traces were identified based on their Encounter reference and all AuditEvents were correctly mapped according to Table 1. All 10 recorded executions are visible, with 5 skipping the first (*schedule appointment*) activity.

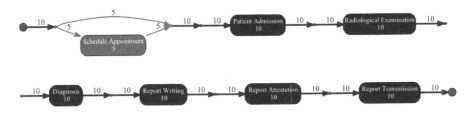

Fig. 3. Process model generated with the Inductive Visual Miner.

5   Discussion

The presented work is a proof of concept, making the case for a standards-based process analytics approach and making sure that the standard in development,

HL7 FHIR, is aware of the capabilities and requirements of process mining. We were able to show how only minor extensions, namely the addition of Encounter and PlanDefinition references, and a simple mapping, enabled the analysis of the radiology practice workflow with process mining tools.

5.1 Impact on Standardization

In the FHIR Workflow project, the authors made a case for checking the usability of FHIR resources for process mining. Together, the working group members proposed the addition of a trace identifier to the AuditEvent and Provenance resources[4]: "We want to be able to search on all events (creates, updates, deletes, etc.) that happened during a given encounter, that happened based on a particular protocol or as a result of a particular order". Based on the discussions in that group, we decided to use PlanDefinition and Encounter for the grouping and mapping approach. A proposal to extend AuditEvent to support this is currently under review for inclusion in the next FHIR release R5.

5.2 AuditEvent vs. Provenance

In this work we analyzed AuditEvent resources, building on existing approaches that aimed to analyze audit data [3,7,16]. However, HL7 FHIR also makes use of the concept of *provenance*, recording "information about entities, activities, and people involved in producing a piece of data or thing, which can be used to form assessments about its quality, reliability or trustworthiness" [6]. A Provenance resource is created by the client (i.e., the person or system conducting the work) as opposed to the AuditEvent resource, which is created automatically by a server. The client should explain for what purpose a resource was edited (created, updated, deleted). In addition, a client can add information about the process (or policy) behind the edit, and provide reasoning why something was done (i.e., which path of a process model was taken). However, Provenance is (1) not widely used (yet), and (2) not documenting non-changing access to a resource (i.e., read). To summarize, Provenance can provide more detailed information on a process, but relies on the clients to record it and might thus be not present at all. Further research on the utilization of the Provenance resource for process mining is needed.

5.3 Considering Different Perspectives

In our example, A_w, the set of all AuditEvent resources recorded during the execution of a radiology workflow (as defined by the referenced PlanDefinition "rad-wf"), was split to traces based on the referenced Encounter resources. However, in fact, A_w represents a multiset of traces, that can be split based on the perspective you take on the data. A more generic approach should thus indicate the grouping behaviour in the query, based on the concepts developed in [12].

---

[4] https://jira.hl7.org/projects/FHIR/issues/FHIR-28100.

Another viable perspective would be, for example, to look at the active participants of the workflow. AuditEvent.agent is described as "an actor taking an active role in the event or activity that is logged" [10]. Mapping name and role to the corresponding fields of the XES Organizational extension allows for additional analysis, e.g., social networks or handover of work for medical or care personnel.

References

1. van der Aalst, W.: Academic view: development of the process mining discipline. Process Mining in Action, pp. 181–196. Springer, Cham (2020). https://doi.org/10.1007/978-3-030-40172-6_21
2. Burattin, A.: PLG2: multi perspective process randomization with online and offline simulations. In: BPM (Demos), pp. 1–6 (2016)
3. Cruz-Correia, R., et al.: Analysis of the quality of hospital information systems audit trails. BMC Med. Inform. Decis. Making 13(1), 84 (2013)
4. Erdogan, T.G., Tarhan, A.: Systematic mapping of process mining studies in healthcare. IEEE Access 6, 24543–24567 (2018)
5. Erickson, B.J., Meenan, C., Langer, S.: Standards for business analytics and departmental workflow. J. Digit. Imaging 26(1), 53–57 (2013)
6. Groth, P., Moreau, L.: W3C Working Group Note "PROV-Overview" (2013). https://w3.org/TR/2013/NOTE-prov-overview-20130430/. Accessed 1 Aug 2020
7. Helm, E., Paster, F.: First steps towards process mining in distributed health information systems. Int. J. Electron. Telecommun. 61(2), 137–142 (2015)
8. HL7 International: FHIR Specification (v4.0.1: R4) Executive Summary (2019). https://hl7.org/fhir/summary.html. Accessed 1 Aug 2020
9. HL7 International: FHIR Specification (v4.0.1: R4) Operations (2019). https://hl7.org/fhir/operations.html. Accessed 1 Aug 2020
10. HL7 International: FHIR Specification (v4.0.1: R4) Resource AuditEvent - Detailed Desc. (2019), https://hl7.org/fhir/auditevent-definitions.html. Accessed 1 Aug 2020
11. IEEE Task Force on Process Mining: IEEE Standard for eXtensible Event Stream (XES) for Achieving Interoperability in Event Logs and Event Streams. IEEE Std. 1849–2016, 53–57 (2016). https://doi.org/10.1109/IEEESTD.2016.7740858
12. IHE IT Infrastructure Technical Committee: Add restful ATNA (query and feed). IHE ITI-TF Supplement, Rev. 3 (2019)
13. IHE Radiology Technical Committee: Standardized operational log of events (sole). IHE RAD-TF Supplement, Rev. 1.2 - Trial Implementation (2018)
14. Leemans, S.J., Fahland, D., van der Aalst, W.: Process and deviation exploration with inductive visual miner. In: BPM (Demos), vol. 1295, no. 8 (2014)
15. Mandel, J.C., Kreda, D.A., Mandl, K.D., Kohane, I.S., Ramoni, R.B.: SMART on FHIR: a standards-based, interoperable apps platform for electronic health records. J. Am. Med. Inf. Assoc. 23(5), 899–908 (2016)
16. González López de Murillas, E., Helm, E., Reijers, H.A., Küng, J.: Audit trails in OpenSLEX: paving the road for process mining in healthcare. In: Bursa, M., Holzinger, A., Renda, M.E., Khuri, S. (eds.) ITBAM 2017. LNCS, vol. 10443, pp. 82–91. Springer, Cham (2017). https://doi.org/10.1007/978-3-319-64265-9_7
17. Rebuge, Á., Ferreira, D.R.: Business process analysis in healthcare environments: a methodology based on process mining. Inf. Syst. 37(2), 99–116 (2012)

18. Rojas, E., Munoz-Gama, J., Sepúlveda, M., Capurro, D.: Process mining in health-care: a literature review. J. Biomed. Inform. **61**, 224–236 (2016)

19. Smile CDR Inc.: HAPI FHIR Interceptors: Overview (2020). https://hapifhir.io/hapi-fhir/docs/interceptors/interceptors.html. Accessed 25 Aug 2020

20. Verbeek, H.M.W., Buijs, J.C.A.M., van Dongen, B.F., van der Aalst, W.M.P.: XES, XESame, and ProM 6. In: Soffer, P., Proper, E. (eds.) CAiSE Forum 2010. LNBIP, vol. 72, pp. 60–75. Springer, Heidelberg (2011). https://doi.org/10.1007/978-3-642-17722-4_5

Deriving a Sophisticated Clinical Pathway Based on Patient Conditions from Electronic Health Record Data

Jungeun Lim[1], Kidong Kim[2], Minsu Cho[3], Hyunyoung Baek[2], Seok Kim[2], Hee Hwang[2], Sooyoung Yoo[2], and Minseok Song[1(✉)]

[1] Department of Industrial and Management Engineering, Pohang University of Science and Technology, Pohang, South Korea
mssong@postech.ac.kr
[2] Office of eHealth Research and Businesses, Seoul National University Bundang Hospital, Seongnam, South Korea
[3] Research Institute of Industry and SME Strategy, Korea Institute of Industrial Technology, Seoul, South Korea

Abstract. Clinical pathway (CP), a standardized treatment process based on a clinical guideline, is widely used to reduce costs while maintaining or improving patient care quality. However, there is a gap between the actual clinical process and the guideline, that causes CP application to be disturbed. A study on developing a data-driven automated clinical pathway to obtain insight into real clinical processes has been conducted. Still, patient characteristics and conditions, which could cause a variation, have not been fully considered. In this study, we aimed to develop a framework to derive a sophisticated clinical pathway from electronic health records (EHRs) data by exploring process variations according to the patient characteristics and conditions. To validate the applicability of the proposed framework, We conducted a case study using the Total Laparoscopic Hysterectomy (TLH) CP data, which was retrieved from an EHR system of a tertiary general hospital in South Korea between January 2012 and April 2016. We found that diabetic TLH patients show different medical performances with other TLH patients. We developed a tailored CP that adds eleven orders over the standard TLH CP, and experts evaluated it as meaningful.

Keywords: Clinical pathways · Electronic health records (EHR) · Statistical analysis · Evidence-based approach · Clinical features · Business process analysis

1 Introduction

A clinical pathway (CP) is a standardized care process in a specific setting such as a particular surgery [4,7]. The use of CPs is gaining interest to help decrease hospital costs and improve the quality of medical services by reducing undesired practice variability [12,13]. Additionally, CPs shorten the length of

© Springer Nature Switzerland AG 2021
S. Leemans and H. Leopold (Eds.): ICPM 2020 Workshops, LNBIP 406, pp. 356–367, 2021.
https://doi.org/10.1007/978-3-030-72693-5_27

hospital stays, lower costs, reduce complications and lower mortality [8,13]. As such, more than 80% of hospitals in the United States adopted CPs in the late 1990s [14], and currently, the implementation of CPs is widely contemplated by hospitals all over the world [21].

The traditional approach for developing a CP relied solely on the knowledge of clinical experts and clinical guidelines. Although the approach was a valuable method derived from solid theoretical backgrounds, it was limited by the time and effort required and the lack of generalization [17,19]. Due to the highly dynamic, highly complex and ad hoc features of the medical treatment process, there is also a gap between the actual clinical process and the CP. As such, an automated approach from data is needed, and researchers have tried to resolve these challenges using process mining and data mining.

Mans et al. [10] applied heuristic miner, and a further work [11] used fuzzy miner and trace clustering to obtain insights from CPs. Huang et al. [4] proposed a new approach for mining CP patterns with time information from chronicle mining. Rebuge et al. [16] suggested a framework to compare the discovered CP and its variants using sequence clustering. Xu et al. [18] developed a more straightforward CP using the Latent Dirichlet Allocation technique. Additionally, researchers have employed further data mining techniques to develop CPs, such as frequent itemset mining [15], sequential pattern mining [15], and a rule induction algorithm [5].

These studies have contributed to developing the automated and accurate CPs based on data, deriving a standardized CP for the majority of patients. Despite these efforts using the data-driven approaches, it is still challenging to apply and complete CP with little effort in practice. In general, most hospitals only implement a single universal CP for a specific surgery or procedure. But, given the various clinical features of diabetes, cardiovascular, age, and medical history, a single CP cannot cover all different patients even with the same surgery; thus, a CP needs to be subdivided according to the clinical features. Therefore, with the aim of the increase of practical use, it is required to implement an approach for CP segregation with clinical features.

This study aims to identify the distinctive clinical characteristics that affect to distinguish a new clinical pathway. To this end, this paper suggests a framework consisting of four phases: data preparation, feature engineering, statistical analysis, and CP development. We first define the outcome measures and explanatory variables from the data. The matching rate, which represents a similarity between clinical trace and reference CP, is adopted as one of the medical performances for process-oriented assessment. Then, statistical testing is conducted to identify the key features highly related to clinical performance measures. Based on decisive factors from the statistical results, we distinguish a new CP (i.e., CP development) after post-hoc analysis with trace alignment. To validate the proposed framework, we performed a case study with real data from a tertiary hospital in South Korea.

The remainder of this paper is organized as follows. Section 2 explains the proposed framework. Section 3 shows a case study, and Sect. 4 discusses the results. Finally, Sect. 5 concludes the paper with future work.

2 Proposed Framework

In this section, we propose a framework for CP segmentation by patient characteristics. As shown in Fig. 1, the framework consists of four phases: data preparation, feature engineering, statistical analysis, and post hoc analysis & CP development. Data preparation, the first phase, aims to identify the data that can be utilized for data analysis by wrangling the collected data. Then, dependent (i.e., outcomes) and explanatory variables (i.e., patient characteristics) are defined in the feature engineering step. The statistical analysis phase conducts experiment to identify the relationship between outcome and independent variables. Lastly, in the post hoc analysis and CP development phase, we distinguish the new CP based on the result of comparing the clinical orders by statistical analysis and trace alignment.

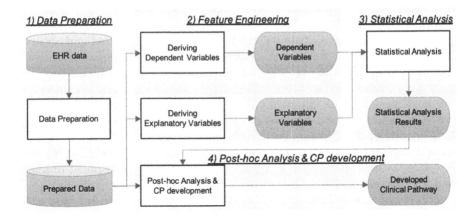

Fig. 1. The proposed framework in this paper.

2.1 Step 1: Data Preparation

The first phase of the framework aims to prepare data with a suitable format for statistical analysis by collecting and pre-processing records. Clinical data generally are complex and heterogeneous [3]. There are four kinds of quality issues: missing data, incorrect data, imprecise data, and irrelevant data [1]. Missing data indicates that data is missing from logs, while incorrect data signifies that information recorded is not correct. Imprecise data represents that the level of data is too coarse, whereas irrelevant data means that information is not related at all with the log. These four types of quality issues are explicitly connected with the healthcare environment, and it needs to be processed thoroughly. To resolve these issues, users can choose proper data repair and noise removal methods based on the data quality. In our case, the most of issues was relevant with missing data, and we tried to remove all problematic data. Details will be given in the Result section.

2.2 Step 2: Feature Engineering

One of the main parts in our framework is to identify the patient characteristics that are highly relevant to the outcomes. To this end, we perform feature engineering to build a research model before the data analysis. As such, the second phase aims to derive dependent and explanatory variables implied for statistical modelling. In more detail, dependent variables represent the outcomes, such as the length of stays or matching rate, i.e., an indicator that signifies the difference between the clinical pathway and relevant clinical log [20], while independent variables signify the patient characteristics. They are derived by selecting or refining records from the prepared data.

Dependent Variables (Outcome Measures). Dependent variables represent the materials to evaluate the outcomes, such as length of stays, hospital costs, the amount of antibiotics used, and matching rates with respect to efficiency and complication rates, re-hospitalization rates, and mortality with respect to quality of the clinical services. Among these variables, in this study, we only employed the length of stays and matching rates, i.e., the efficiency-focused, because of the insufficiency of data related to the quality perspective. More in detail, we were not able to collect the patients' records who re-visited the hospital with the same diagnosis within the 30 days (i.e., re-hospitalization) or were turned out to be dead (i.e., mortality).

The length of stays is one of the critical indicators in most hospitals because it lowers the risk of infection and medical costs for patients. In this study, we derived the length of stays by calculating the difference between the admission date and discharge date.

The matching rates signify how patient records collected from the logs coincide with the orders in the CPs. Thus, the rates can be used to evaluate the practical application of the CP in the quantitative approach. The matching rate is formalized as follows [20].

$$CP \ order \ matching \ rate = \frac{1}{2}(1 - \frac{M_{cp}}{N_{cp}}) + \frac{1}{2}(1 - \frac{R_{log}}{N_{log}}) \tag{1}$$

M_{CP} is the number of orders included in the CP but not shown in the log, N_{CP} is the number of orders included in the CP, R_{log} is the number of orders included in the log but not shown in the CP and N_{log} is the number of orders included in the log.

Explanatory Variables (Patient Characteristics). As introduced earlier, explanatory variables represent the materials that classify patients with their characteristics. Thus, regarding these characteristics, patients can be divided into groups. For example, patients are divided into age groups, such as infants, children, young adults, middle-aged adults, and older adults. Additionally, they may be classified by whether they have a specific history or not.

EHR system contains numerous patient characteristics, including age, sex, family history, past history, and they can be categorized into three types: background information, clinical events, and non-clinical events. The background patient information signifies historical records of patients before hospitalization. This group includes age, sex, allergy, operation history, medication history, family disease history, and chronic diseases (diabetes, hypertension, hyperlipidaemia, and cardiovascular and cerebrovascular diseases). The second group is the data derived from the clinical events during hospitalization, such as transfer of wards, transfer of departments, diagnosis from another department (not from obstetrics and gynaecology), and operation from another department. The last category is related to the administrative information during hospitalization, including severity, admission type, Diagnosis Related Group (DRG).

2.3   Step 3: Statistical Analysis

This step performs a statistical analysis to identify the distinctive patient characteristics considered for CP development. To this end, hypothesis testing is performed based on dependent and independent variables derived in Step 2. Regarding hypothesis testing, different types of methods are utilized considering the number of groups and shape of distributions. In this study, we applied two types of statistical analysis methods: Mann-Whitney U test and Jonckheere-Terpstra test.

Mann-Whitney U Test. The Mann-Whitney U test identifies whether two populations are equal or not [9]. As such, the test was applied when the patients were divided into two groups by a patient characteristic, such as sex and severity. Its null(H_0) and alternative(H_1) hypotheses are as follows.; H_0: Two populations are equal, H_1: Two populations are not equal.

Jonckheere-Terpstra Test. As a substitute for the Mann-Whitney U test, the Jonckheere-Terpstra test is applied when the number of groups is more than two (i.e., three or more) and they tend to increase or decrease [6]. For example, the changes of outcome variables can be identified by the increase in the number of operations. Letting di be the median for the population i, the null and alternative hypotheses are defined as follows; $H_0 : d_1 = d_2 = d_3 = \cdots = d_k, H_1 : d_1 \leq d_2 \leq d_3 \leq \cdots \leq d_k$ (where, k is the number of groups).

2.4   Step 4: Post Hoc Analysis and CP Development

The last step compares the selected patients' clinical orders based on their characteristics and derives a new CP. Here, the critical patient characteristics are employed from the statistically significant factors in Step 3. In this phase, patients are grouped by a specific feature, and the application rates of clinical orders are measured for each group. Then, the difference in the application rates of the orders between groups is identified. For example, if the order applies only to 90% of the severely ill group and 10% of non-severe patients, the order should be included in the CP of the severely ill group. Then, if a group of features differentiates multiple clinical orders, some traces from each group are sampled to visualize the differences and discuss with clinical experts. CP segmentation is performed when the clinical expert concludes that the functional group needs a new CP.

3   Case Study

3.1   Introduction

A general tertiary hospital in South Korea has developed and applied numerous electronic CPs based on clinical experience to provide appropriate medical services to patients. In this case study, we primarily analyzed the Total Laparoscopic Hysterectomy (TLH) CP, which has been in use since August 2009. From the hospital's EHR system, log data of patients determined as candidates to be applied to the TLH CP were extracted from January 2012 to April 2016, resulting in data collected from 1100 inpatients. EHR data of patients' demographics, hospitalization, applied CP, surgery, diagnosis, transfers, referrals, physician orders including medications and labs, and CP history was extracted.

3.2   Data Preparation

Based on the collected data from 1100 inpatients, we performed data preprocessing. Among the four types of data quality issues, e.g., missing data, incorrect data, imprecise data, and irrelevant data, our data included the first type as we lacked the medical history of patients, such as operations and medication history. Additionally, the second-hand data collected from surveys, such as drinking and smoking, had many blank spaces. As such, those characteristics were removed from the data to be analysed. Furthermore, part of the clinical orders had incorrect data, such as an unexpected hold (3.4%) and immediate removal by systems (2.5%). These were also excluded, and finally, the data was prepared.

3.3   Feature Engineering

Dependent Variables (Outcomes). As introduced earlier, we applied the length of stays and matching rates as dependent variables (i.e., outcomes). Regarding the length of stays, the average value was 4.57 days (median: 4 days and standard deviation (SD): 1.8 days). Regarding the matching rate, the average was 0.716 (median: 0.724 and SD: 0.053).

Explanatory Variables (Patient Characteristics). After preparing the data, we selected 11 explanatory variables based on a thorough discussion with clinical experts: diabetes, hypertension, hyperlipidaemia, cardiovascular, cerebrovascular, severity, operations, transfers of departments, transfers of wards, diagnosis from other departments (not from obstetrics and gynaecology), and referrals to other departments.

Only a small number of patients had chronic diseases, including diabetes, hypertension, hyperlipidaemia, cardiovascular, and cerebrovascular at 3.5%, 4.7%, 1.5%, 0.1%, and 0.6%, respectively. The number of patients with severity, however, was relatively high at 33.1%. Regarding the number of operations, most patients received only one operation while 0.9% of patients received two operations. Regarding transfers of departments, only four patients (0.4%) changed departments. Lastly, regarding the other characteristics (e.g., transfers of wards, diagnosis from other departments, and referrals to other departments), for each feature, more than 50% of the patients were not associated with the feature at all, but the remaining patients had more than one frequency.

3.4   Statistical Analysis

Among the 11 independent variables (i.e., patient characteristics), only six, e.g., diabetes, hypertension, severity, transfers of wards, diagnosis from other departments (not from obstetrics and gynaecology), and referrals to other departments, were considered for statistical testing because the sample size for testing should be sufficient (i.e., more than 30) [9], and the sample sizes for the other features are not sufficient.

We applied two different statistical testing methods: the Mann-Whitney U test and Jonckheere test. The Mann-Whitney U test was applied to diabetes, hypertension, and severity while the Jonckheere test was employed for the remaining variables. Table 1 presents the statistical testing results of the length of stays and matching rates on patient characteristics.

Table 1. Statistical testing results on patient characteristics.

Patient characteristics	p-value		Test type
	LOS	Matching rates	
Diabetes	<0.01	<0.01	Mann-Whitney U test
Hypertension	0.014	0.045	Mann-Whitney U test
Severity	<0.01	<0.01	Mann-Whitney U test
Transfers of wards	<0.01	0.035	Jonckheere test
Diagnosis from the other departments (not from obstetrics and gynaecology)	<0.01	0.149	Jonckheere test
Referrals to other departments	<0.01	<0.01	Jonckheere test

As a result of the statistical tests, diabetes, severity, transfers of wards, diagnosis from other departments and referrals to other departments significantly affected the length of stays while the matching rates were significantly affected by diabetes, severity, and referrals to other departments. Therefore, we concluded that only three features, e.g., severity, diabetes, and referrals to other departments, are key characteristics for CP segmentation.

Based on these results, we had a thorough discussion with clinical experts. First, regarding severity, we determined that the result was caused by incorrect application of the CP in cancer patients, not the CP target patients. In the hospital, clinicians sometimes applied the CP to cancer patients because there was no significant difference in clinical operation processes between the two. The cancer patients, however, required a longer stay and different routines from the CP patients. Thus, we determined that it was misleading that there was an impact on clinical outcomes. Additionally, regarding the referrals to other departments, the domain experts concluded that the feature needs to be managed by monitoring rather than CP development. For these reasons, we performed further post hoc analysis and CP development based on diabetes.

3.5 Post Hoc Analysis and CP Development

Considering diabetes, we analyzed the differences in clinical orders between diabetic and non-diabetic patients. The total number of diabetic and non-diabetic patients was 38 and 1062, respectively. We performed trace alignment to visualize how the order records of each group differ. For simplicity, in each group, 20 patients, who stay in the hospital for four days, are sampled, and the result of trace alignment is in Fig. 2.

Additionally, we employed the CP development methodology [2], which derives an optimal set of clinical orders that maximize the matching rates. Based on the exploited method, we received clinical orders for diabetic and non-diabetic patients. After, the developed CP for diabetes was compared with that for non-diabetes. We identified that 11 clinical orders, e.g., Pot chloride, Humalog, Palonosetron, Ephedrine, Electrolyte panel, Glucose, DM diet (for diabetes), BST, Infusion pump, Interceed, and Simple hysterectomy, were applied for most of the diabetic application rates. In contrast, two clinical orders (i.e., Granisetron and Other dermatological) were utilized only for non-diabetic patients.

Table 2 provides the clinical order application rates of diabetic and non-diabetic patients. Overall, we were able to identify a clear difference in each code's application rates by the group. Therefore, we concluded that the new CP for diabetes should be distinguished from the general one.

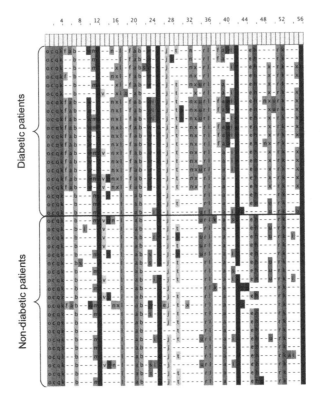

Fig. 2. Trace alignment result of diabetic patients and non-diabetic patients.

Table 2. Clinical order application rates of diabetic and non-diabetic patients.

Order information		Application rates (%)	
Type	Name	Diabetic	Non-diabetic
Medications	Pot chloride	84.2	2.7
	Humalog	84.2	2.0
	Palonosetron	57.9	47.1
	Ephedrine	50	36.3
	Granisetron	42.1	56.2
	Other dermatologicals	44.7	49.3
Lab test	Electrolyte panel	79.0	9.5
	Glucose	79.0	3.2
Diet	DM diet (for diabetes)	71.1	1.6
Treatment	BST	89.5	2.2
Procedures	Infusion pump	81.6	7.1
	Interceed	57.9	47.6
	Simple hysterectomy	50	39.9

4 Discussion

The results of the analysis showed that diabetes affects medical outcomes, such as the length of stays and matching rates. To this end, we identified that glucose control is the reason for the extended hospital stays and the lower matching rates. Patients with diabetes require a specific amount of time to control their blood sugar before surgery, which can lead to longer hospital stays. Additionally, the diabetic patients received surgery later than general patients.

Regarding the lower matching rate for diabetic patients, we found that controlling the patient's blood sugar affected the results through the post hoc analysis. We identified that diabetic patients received insulin (e.g., humalog) with Alberti regimen and dextrose fluid (e.g., pot chloride) containing potassium chloride to ensure adequate water, electrolyte, and feeding before operations. Additionally, diabetic patients received tests to check blood sugar and electrolytes for glucose control. Moreover, some materials (e.g., infusion pump) were also utilized for diabetic patients to inject the proper medicines. Therefore, we determined that these orders are required entirely for diabetic patients with both data and clinical perspectives.

This research has important contributions for both practice and research standpoints. As far as practical use is concerned, this research helps to develop the clinical decision support system by resolving the large demands from hospitals to continuously improve and manage CPs. Despite the facts that hospitals generally cannot develop and enhance CPs due to an insufficient workforce, time, and costs, however, it is required to implement a tool that gives accurate clinical pathways to clinicians, driving to provide high-qualified patient-centric services. In this standpoint, this paper is of value as it automatically recommends distinctive patient characteristics and develops a new CP with a data-driven approach.

Also, as far as the research standpoint is concerned, this paper is different from existing works that merely discover a one-off CP and provides a direction that enables the continuous development of improved CPs with a statistical approach. Furthermore, the patient characteristics and clinical outcome measures derived in this research are applicable to multiple clinical research disciplines, such as real-time monitoring and prescriptive analytics in hospitals.

Despite these contributions, this paper has some challenges. First, there has been a problem that the number of patients to be analyzed is reduced because latest data of short-term period data must be used to reflect the latest order information. Nonetheless, it is significant that we were able to segment the CP according to the patient condition of diabetes. The framework presented in this study considerably contributes in terms of managing the clinical pathway and practical use of the clinical pathway and will continue to demonstrate its usefulness through further data acquisition.

Also, this research did not address the inter-relationship between patient characteristics and thus only aimed at developing new CPs for each patient feature. However, it is possible to construct CPs that consider multiple patient characteristics at once (e.g., diabetic-female-TLH CP). Furthermore, we limited clinical outcome measures to length of stays and matching rates. Future studies should be expanded to more scalable methodologies, including patient costs and the use of antibiotics. Lastly, the analysis result presented in this paper was only based on a single hospital. As there are differences in CPs and data between hospitals, the study may lack generalizability. Thus, we need to perform more case studies using data from multiple hospitals. We believe that we can build a more robust framework for CP segmentation by resolving these issues.

5   Conclusion

In this paper, we proposed a framework for CP segmentation based on patient characteristics. In this process, we performed feature engineering to define the clinical outcome measures related to CPs (i.e., dependent variables) and patient characteristics (i.e., independent variables). We also conducted statistical testing using the Mann-Whitney U test and Jonckheere test, and finally a new CP was distinguished from the general CP.

This paper proposes guidelines to increase the applicability of CPs and suggests how to develop CP variants using patient characteristics and clinical outcomes. Additionally, the proposed framework has a distinctiveness that enables the continuous development of improved CPs different from existing works that merely discover a single CP. Therefore, we believe that our methodology is helpful for practical use.

In future studies, we will consider the inter-relationship between patient characteristics for CP segmentation. Additionally, other clinical outcomes, such as patient costs and the use of antibiotics, may be included. Furthermore, more case studies should be performed to validate our approach and make various use cases.

Acknowledgement. This work was approved (IRB No. B-1609/361-105) by the Institutional Review Board of the SNUBH, which waived patients' informed consent. This work was funded by the MSIT (Ministry of Science and ICT), Korea, under the ITRC (Information Technology Research Center) support program (IITP-2020-2018-0-01441) supervised by the IITP (Institute for Information & Communications Technology Promotion).

References

1. Bose, R.J.C., Mans, R.S., van der Aalst, W.M.: Wanna improve process mining results? In: 2013 IEEE Symposium on Computational Intelligence and Data Mining (CIDM), pp. 127–134. IEEE (2013)
2. Cho, M., et al.: Developing data-driven clinical pathways using electronic health records: the cases of total laparoscopic hysterectomy and rotator cuff tears. Int. J. Med. Inform. **133**, 104015 (2020)
3. Cios, K.J., Moore, G.W.: Uniqueness of medical data mining. Artif. Intell. Med. **26**(1–2), 1–24 (2002)
4. Huang, Z., Lu, X., Duan, H.: On mining clinical pathway patterns from medical behaviors. Artif. Intell. Med. **56**(1), 35–50 (2012)

5. Iwata, H., Hirano, S., Tsumoto, S.: Construction of clinical pathway based on similarity-based mining in hospital information system. Procedia Comput. Sci. **31**, 1107–1115 (2014)
6. Jonckheere, A.R.: A distribution-free k-sample test against ordered alternatives. Biometrika **41**(1/2), 133–145 (1954)
7. Lenz, R., et al.: It support for clinical pathways-lessons learned. Int. J. Med. Inform. **76**, S397–S402 (2007)
8. Macario, A., et al.: The effect of a perioperative clinical pathway for knee replacement surgery on hospital costs. Anesth. Analg. **86**(5), 978–984 (1998)
9. Mann, H.B., Whitney, D.R.: On a test of whether one of two random variables is stochastically larger than the other. Annals Math. Stat. 50–60 (1947)
10. Mans, R., et al.: Process mining techniques: an application to stroke care. MIE **136**, 573–578 (2008)
11. Mans, R.S., Schonenberg, M.H., Song, M., van der Aalst, W.M.P., Bakker, P.J.M.: Application of process mining in healthcare – a case study in a Dutch hospital. In: Fred, A., Filipe, J., Gamboa, H. (eds.) BIOSTEC 2008. CCIS, vol. 25, pp. 425–438. Springer, Heidelberg (2008). https://doi.org/10.1007/978-3-540-92219-3_32
12. Newman, B.: Enhancing patient care: case management and critical pathways. Aust. J. Adv. Nurs. Q. Publication R. Aust. Nurs. Fed. **13**(1), 16 (1995)
13. Panella, M., Marchisio, S., Di Stanislao, F.: Reducing clinical variations with clinical pathways: do pathways work? Int. J. Qual. Health Care **15**(6), 509–521 (2003)
14. Pearson, S.D., Goulart-Fisher, D., Lee, T.H.: Critical pathways as a strategy for improving care: problems and potential. Ann. Intern. Med. **123**(12), 941–948 (1995)
15. Perer, A., Wang, F., Hu, J.: Mining and exploring care pathways from electronic medical records with visual analytics. J. Biomed. Inform. **56**, 369–378 (2015)
16. Rebuge, Á., Ferreira, D.R.: Business process analysis in healthcare environments: a methodology based on process mining. Inf. Syst. **37**(2), 99–116 (2012)
17. Weiland, D.E.: Why use clinical pathways rather than practice guidelines? Am. J. Surg. **174**(6), 592–595 (1997)
18. Xu, X., Jin, T., Wei, Z., Wang, J.: Incorporating topic assignment constraint and topic correlation limitation into clinical goal discovering for clinical pathway mining. J. Healthcare Eng. (2017)
19. Yang, W., Su, Q.: Process mining for clinical pathway: Literature review and future directions. In: 2014 11th International Conference on Service Systems and Service Management (ICSSSM), pp. 1–5. IEEE (2014)
20. Yoo, S., et al.: Conformance analysis of clinical pathway using electronic health record data. Healthcare Inform. Res. **21**(3), 161–166 (2015)
21. Zhang, Y., Padman, R., Patel, N.: Paving the cowpath: learning and visualizing clinical pathways from electronic health record data. J. Biomed. Inform. **58**, 186–197 (2015)

Exploration with Process Mining on How Temperature Change Affects Hospital Emergency Departments

Juan José Lull[1]([⊠])(iD), Onur Dogan[2](iD), Angeles Celda[4], Jesus Mandingorra[4](iD),
Lenin Lemus[1](iD), Miguel Ángel Mateo Pla[1](iD), Javier F. Urchueguía[1](iD),
Gema Ibanez-Sanchez[1](iD), Vicente Traver[1](iD), and Carlos Fernandez-Llatas[1,3](iD)

[1] Instituto Universitario de Tecnologías de la Información y de las Comunicaciones
(ITACA), Universitat Politècnica de València, Camino de Vera S/N,
46022 Valencia, Spain
{jualulno,geibsan,vtraver,cfllatas}@itaca.upv.es
[2] Department of Industrial Engineering, Izmir Bakircay University,
35665 Izmir, Turkey
onur.dogan@bakircay.edu.tr
[3] Department of Clinical Science, Intervention and Technology (CLINTEC)
Karolinska Institutet, Stockholm, Sweden
carlos.fernandezllatas@ki.se
[4] Consorci Hospital General Universitari de València, Av. de les Tres Creus, 2,
46014 València, Spain

Abstract. The way patients are treated in Hospital Emergency Departments changes during the year, depending on many factors. One key component is weather temperature. Some seasonal maladies are tightly related to temperature, such as flu in cold weather or sunburn in hot weather. In this study, data from a hospital in Valencia was used to explore how harsh weather changes affect the emergency department, obtaining information about probable impacts of global warming effects in healthcare systems. Process mining techniques helped in the discovery of changes in the Emergency Departments. Some illnesses, such as heat stroke, are more prevalent during heatwaves, but more interestingly, the time to attend patients is also higher. Rapid changes in temperature are also analyzed through Process Mining techniques.

Keywords: Process mining · Emergency · Weather conditions · Healthcare system

1 Introduction

Emergency departments (EDs) work seven days, 24 hours a week. They are key departments that provide urgent care to the patients. Since many patients get further care after the ED's first response, they are regarded as the gateway to other hospital departments. The EDs aim to present urgent care to treat people

© Springer Nature Switzerland AG 2021
S. Leemans and H. Leopold (Eds.): ICPM 2020 Workshops, LNBIP 406, pp. 368–379, 2021.
https://doi.org/10.1007/978-3-030-72693-5_28

recover from their illnesses or at least alleviate the symptoms. Well-performed and standard processes can accomplish this aim in the ED, where healthcare professionals collaborate systematically. The increasing number of patients causes a crisis of agglomeration in the gateway of hospitals [20]. Although it is well-known among professionals and literature that most EDs are frequently crowded, many questions wait for their answers [2]. Among these questions, one is how global warming affects emergency departments.

The Intergovernmental Panel on Climate Change (IPCC) points out that weather conditions will probably become hotter or colder frequently and intensely during the next years [12, 18]. Large parts of the World, especially Asia, Europe and Australia have encountered an increased recurrence of heatwaves [14]. Besides, human mortality rates related to extreme hot weather have raised with global warming [14]. Several reasons may affect the correlation between disease and global warming, such as local demographics, economic welfare, underlying disease risk or weather variability in seasons [8]. Another reason is that steep changes in daily temperature may have an impact on ED processes. For this consideration, more reliable intellection of disease conditions during temperature changes is an essential tool for health practitioners and the investigation of ED processes is gaining more and more attention [6, 9].

Despite progress in the analysis of ED processes, novel strategies are required because of complexity, diversity and non-adaptability reasons [1, 15]. ED processes are not adjustable or adaptable from another process model because of their nature and complexity. This complexity makes it hard to provide a clear representation of the patient flow. Hence, most investigations focus on the observations to discover the process model, which is time-consuming and unreliable. Process models are the central part of crowded ED problems. Therefore, they should represent real and reusable patient flows to find acceptable solutions. Process mining (PM) automatically creates process models using real data stored in the IT system as event logs [19]. By applying PM methods, the actual ED processes followed by patients can be discovered to see the effects of environmental temperature.

The studies presented in the following sections show the relation between higher temperatures and extra attention time, and explore the connection between patient cases and harsh, sudden temperature changes. This shows the potentiality of using PM in the study of global warming and healthcare.

2 Materials and Methods

For the study data was collected from 483,229 visits to the Emergency Department at Hospital General of Valencia. These were records from the years 2015 till 2018. The records included: patient ID; date and time of arrival; date and time of the start of the triage and its end; waiting queue assigned to the patient in the triage; the specific service that attended the patient (e.g. surgery, dermatology) and timestamp, both at the beginning and end of the attention to the patient; patient destination (e.g. home, hospitalization, another medical service); patient's date of birth; patient's diagnostic.

Daily temperature information was also available, including per day: average temperature, minimum temperature and maximum temperature. Across the years, subjects usually go to the hospital more than once. Specifically, 192,884 patients generated the 483,229 visits, with an average of 2.5 visits to the hospital per patient. Process Mining [19] solutions facilitate a clear understanding of the care process and it can help build models that can be understood by humans who can modify those processes according to their expert understanding of the processes. It also lets them measure changes objectively. The doctor or technician can thus understand the models that show their patients' behaviour.

When the user is set at the center, specifically the user with expert knowledge about the processes but with little to no PM abilities, interactive and visual tools are needed in what is known as Interactive Process Mining (IPM). PMApp [5] is an application that facilitates IPM. With PMApp and the PM algorithm PALIA, the processes with the data are represented as Timed Parallel Automata (TPA) and can be outlined visually in workflows with color gradients, or with other representations. In this case, workflows representing the different events in an ordered way, their connections, the time spent at each activity and the flow that was followed, all were summarized in a specific workflow, as seen in Fig. 1. This powerful visualization is called an Interactive Process Indicator (IPI) and is explained in the following lines.

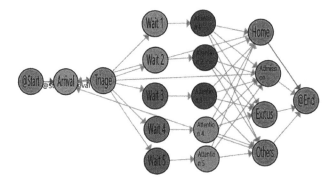

Fig. 1. IPI representing all the visits, including 393,963 traces after incorrect traces were discarded. Redder color in nodes, as opposed to green, represents higher time in that stage while redder color in transition means larger number of cases in that transition. (Color figure online)

In Fig. 1 a model of an ED is shown. The events (nodes) that exist, as seen in the IPI, are: 1. "Artificial" Start (@Start). Initial event, always present. 2. Arrival to the ED 3. Triage of the patient: Assessment of severity and urgency of the treatment 4. Wait 1–5, indicating the queue the patient was assigned to 5. Attention 1–5, indicating the attention associated to the corresponding queue 6. Final step, which can be Home, Admission into the Hospital, Exitus (decease) or Others (e.g. the patient ran away) 7. "Artificial" End (@End), always present.

In the IPI, it is observed that the patient can be assigned to different queues, different destinations, etc. A transition is needed so a visit can be traced e.g. a patient is assigned waiting queue 2 and then, after the attention, is sent home. This is represented by transitions (arrows) between the nodes.

There is key information that resides in the time spent at the nodes, and the different paths that the visits go through. Summarized information about e.g. waiting times, distribution of waiting queues, etc. can be seen through color gradients in the IPI for the nodes and transitions. In this specific study, there is a green-to-red gradient for the nodes that represents median duration in the node (e.g. time spent at waiting queue 1). As seen in Fig. 1, waiting time is lower the lower the wait queue number. Conversely, attention time is higher the lower the attention number.

The same gradient color coding exists for the transitions and it represents the number of visits that went through any transition. Thus, all the 393,963 visits went from arrival to triage. Most visits were assigned to wait queues 3 and 4, as seen color-coded in the transition.

The Waiting 1 to 5 and Attention 1 to 5 nodes represent the queues (and level of emergency, top to bottom) that each patient is assigned after the arrival, at the triage step. After attention the patient usually returns home, though he or she could also be admitted into the hospital, or finish in *exitus*, among other possibilities. The Emergency Department modelling has been described elsewhere [7].

The IPI shown at Fig. 1 comprises the whole dataset that generated the model.

2.1   Assigning Temperature to Cases and Discretization

With the help of PMApp, daily temperature information about Valencia city, where the hospital is located, was fused with the ED data, by assigning temperature to the date of each case. Average temperature information was then discretized, generating sub-groups of cases: 15–20 °C, 20–25 °C, 25–30 °C, 30–35 °C. Also, taking into account the average, minimum and maximum temperature, day to day steep temperature changes were selected and then divided into sudden temperature increases and sudden temperature decreases.

Inaccurate data (i.e. blank information, wrong dates) were removed, leaving 393,963 correct traces corresponding to visits to the ED. The other 89,266 traces either had empty values for any needed date or had an incorrect process flow (e.g. date of attention was prior to the date of arrival). PMApp let the user create groups dividing the TPA into groups by different patterns e.g. average temperature group, diagnosis, etc. and a combination of those.

Different models were explored relating temperature to diagnosis. Some of them did not show any interesting information. Others did and are exposed in the Results section. Firstly, high temperature and heat strokes was analyzed. An IPI for each temperature group was generated and inspected. In an exploratory way, correlations between diagnostics and temperature increases were also looked for and the otitis cases related to temperature are also shown. Finally, in order to study how sudden changes in the weather (a phenomenon related to global warming) affect the ED, harsh changes in day-to-day temperature were detected and those with a higher sudden change were selected and their processes were visualized and studied.

In order to select the days with higher changes, they were compared: Each day's minimum temperature was subtracted to the minimum temperature from the previous day. The same calculation was performed for the maximum and the average temperatures. Finally, those values were multiplied and a threshold of 100 was introduced, accounting for 14 days with increased temperature and 17 days with reduced temperature. Two groups were created according to temperature increase and decrease and an IPI was generated for each along with another one for stable temperature.

For the three mentioned IPIs, in order to assess the significance in the differences for specific nodes, the normality of each population of durations was assessed by the Kolmogorov-Smirnov test. In case both populations are normal, a Student T test is applied. Otherwise, a Mann-Whitney-Wilcoxon is applied. In any case, a p value is applied as a threshold to determine statistical significance between the populations. In this study a p value of 0.05 was set as the threshold for statistical significance. Yellow circles around nodes indicate a significant statistically difference between the duration in the population of days with a steep change compared to days without important changes in temperature. These are also shown in the results section.

3   Results

3.1   Temperature and Heat Strokes

The first study is related to daily weather temperature and heat strokes. As the WHO relates [21], heat is one important factor that affects mainly the elder population, causing cardiovascular and respiratory diseases. According to the same report, in 2003 an excess 70,000 elderly people died due to a heat wave.

IPIs were generated for each temperature range and it was visually observed that the attention of patients with heat stroke took longer the higher the temperature. The three IPIs corresponding to the ranges 20–25 °C 25–30 °C are shown in Fig. 2. The color gradients are common for the three IPIs so it can be seen that attention time for the 25–30 °C was highest, especially in Attention 2 (the following highest duration was Attention 3 from the range 25–30 °C, then came Attention 4 from the range 25–30 °C). The color for number of executions (coded in the transitions) was also common to all IPIs, so it can be observed that most of the visits corresponded to temperatures between 25 °C and 30 °C (mostly in Wait queue 3, then queue 4). There were very few cases in the 30–35 °C range, so they were not included in the Figure. Generally, waiting times were low compared to the time spent at Attention.

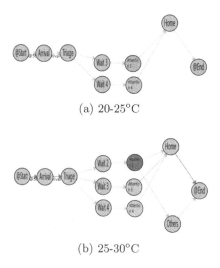

Fig. 2. Interactive Process Indicators (IPIs) with groups of average temperature per day, in heat stroke patients.

3.2   Otitis Cases Related to Temperature

A high number of otitis (inflammation of the ear due to infection) cases were detected that related to high temperature. Their IPI is shown in Fig. 3.

Fig. 3. IPI for otitis patients. Gradient colors represent the same durations as in Fig. 1.

As can be seen in Fig. 3, there were no otitis cases among the 568 diagnosed ones that were considered as serious, since none were triaged in the most urgent queues, 1 and 2. It was observed that the number of otitis patients related to the number of general patients increased with temperature, as seen in Fig. 4. Although the attention of those patients took little time, their wait time was very high (as seen in Fig. 3). This indicates a higher load of waiting rooms.

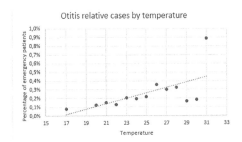

Fig. 4. Relative number of otitis cases (percentage of cases by general emergency patients), and its increase with temperature.

3.3   Harsh Changes in Temperature and ED

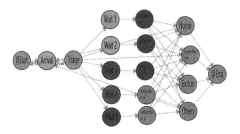

Fig. 5. IPI with information on days without harsh temperature changes. The redder the color, the longer the time at a node. Node colors (i.e. durations), are directly comparable between this Figure and Figs. 6 and 7. (Color figure online)

In the IPIs that compare to the baseline, greener means higher times while redder means lower (negative) subtracted times (Figs. 6(b) and 7(b)).

The IPI that represents days without steep day-to-day changes in temperature is shown in Fig. 5.

The comparison between days with a high change in temperature compared to the previous one, generally showed higher attention times. Specifically, for increases in temperature, attention in queues 2 and 3 took longer than days with no significant change in temperature (see Fig. 6). The same situation happened for steep temperature decreases, with a higher attention time and wait time for patients classified in queue number 3, and wait time for queue 5 was also higher, along with the triage time (see Fig. 7). These findings were statistically significant. There were no statistically significant reductions in time.

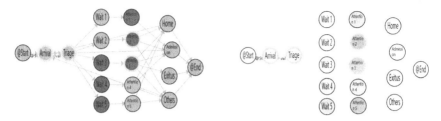

(a) IPI for harsh increase in temperature   (b) Same IPI, compared to baseline

Fig. 6. IPIs for ED patients in days with steep increases in temperature. (Color figure online)

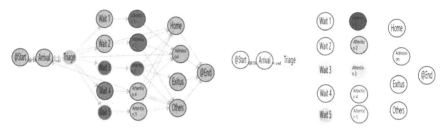

(a) IPI for harsh decrease in temperature   (b) Same IPI, compared to baseline

Fig. 7. IPIs for ED patients in days with steep decreases in temperature. (Color figure online)

4   Conclusion and Discussion

This study considered data collected from an emergency department (ED) at a Hospital in Valencia city, from 2015 to 2018. 483,229 visits created by 192,884 patients were investigated for four years. The main goal of the study was to analyze a large dataset that had ED data along with weather temperature information through PM, specifically through a tool that allowed for visual inspection and surveillance of the processes discovered by PM, PALIA.

The effects of temperature on heat stroke and otitis cases are investigated in this study by analyzing process flows, along with a general investigation about the effects of steep changes in temperature on the Emergency Departments. Valencia is a warm Mediterranean coastal city, so the weather is generally mild. However, we could detect changes in the processes inside the Emergency Department that depended on weather temperature. This is especially interesting for our study, since extreme weather conditions such as heatwaves are more prevalent with global warming [4]. According to the Weather Meteorological Organization, heatwaves are the meteorological hazards that have created the maximum number of deaths in the recent years [22]. The effects of heatwaves and other steep temperature changes on the EDs can be observed and may be interpolated to places with more extreme weather conditions and to how attention may change in the future.

In the study, temperature data was categorized and linked to process cases to explore possible effects of heatwaves. Then PALIA algorithm created process flows of patients under categorized (discretized) temperature data. It was observed that Heat Stroke processes in EDs took longer the higher the temperature. There were also many more cases in the range of 25 to 30 °C. The number of cases confirms the intuition that sunburns are more prevalent the higher the temperature (it should be considered that roughly 20% of days had an average temperature at or above 25 °C, across the years of the study). The higher treatment time span per case could be thought of as intuitive too, but in this case the potential effect of global warming on the EDs is clear: with the increase of heatwaves, sunburns are expected to grow in number and EDs will have more cases that will need extra attention time. This is also compatible with the study by [10] where they analyzed the effect in a hospital during the most extreme heatwave in Melbourne. It was unexpected that wait time for these cases was very low compared to the usual wait times. In fact, heat stroke patients did not wait more in lower urgency waiting queues. This suggests that the ED treats those patients as soon as possible independently of the waiting queue.

In the case of otitis, this was an unexpected finding. It should be reviewed how much confounding factors played a role in the cases, such as infections due to longer times spent in swimming pools, as is usually the case during the summer vacations. This could nevertheless be attributed to overcrowded waiting rooms in humid areas with high temperatures.

The study also presented the effects of sudden changes in weather conditions to the ED. Generally, time spent at the waiting rooms and while being attended were longer for both sudden temperature increases and decreases. This exploration points in the direction that the more the sudden changes in temperature, the more collapsed EDs will be. And sudden changes in temperature are more and more frequent due to climatic change.

In future studies a correction for multiple condition tests should be added, such as a Bonferroni correction. The authors of PMApp have this in mind for the tool so these adjustments could be applied directly from the PM tool in future developments [7].

A formal definition of heatwave could have been used and the temperature ranges could have been defined accordingly. There are proposals for such a definition as in [17,23]. However, the purpose of this study was mainly focused on the exploration, through PM, of the effects of temperature changes. Since the importance of heatwaves has been found, this require further refinement. The use of IPIs for the detection of changes in processes has been proposed in other realms such as Operating Rooms processes [11], Type 2 Diabetes Primary Care [3], etc. In fact, as stated in [16], it is important that there exist visual tools that let the care professionals analyze complex processes in a simple visual way, which are currently lacking. This would help and sometimes enable the user to interpret the outcomes of the PM techniques applied to their data.

Visual inspection, through IPIs, as shown through the figures, let us see differences in processes at a glance. In this study, duration in each node (e.g. a waiting queue) was accounted for by a color gradient where green meant 0 time and red the maximum duration in the specific graph. Another visual cue was the coloring of the transitions. The arrows were redder the higher the number of cases and greener the lower the number of cases. This turns the graphs into powerful visual tools that the doctor (or any other user) can manage easily. Furthermore, as shown in some figures, difference maps can also be created where colors represent the difference in duration for two process groups in the case of nodes, and the difference between the number of occurrences in the case of transitions. The information in the flows could be enhanced by adding a legend to the graphics, that would allow the user to see the exact amount of traces or the amount of time spent at a specific activity. These are things that will be incorporated to PALIA in future releases.

With the presented results, the study puts forwards that global warming may have a significant impact on Emergency Department processes. This is an exploratory study that shows how PM enables ED workers, mainly medical doctors and technicians, to explore ED event data related to weather. The incidence of global warming on the treatments has been analyzed by watching the cases related to high temperature and also through one key hazard that is caused by global warming: steep temperature changes (mainly heatwaves). Direct observation of year-to-year increase in temperature and its effect on EDs is not feasible, since the average increase of $0.18\,^{\circ}$C per decade since 1981 (according to the American National Oceanic and Atmospheric Administration) [13] per decade would take several decades of data to extract valuable information. However, although the year by year increase is out of the scope of this paper, the effects that global warming in the next decades can be analyzed showing how temperature affects the citizens' health in a critical service as ED. This information is central to make the scientific community aware of the effects of global warming on our health.

The study also highlights the power of visual tools to understand the dynamics of the processes in the EDs and how these tools can help the people working in the healthcare domain to inspect large amounts of event data in a friendly but comprehensive way.

Acknowledgment. This activity has received funding from EIT Health (www.eithealth.eu) ID 20328, the innovation community on Health of the European Institute of Innovation and Technology (EIT, eit.europa.eu), a body of the European Union, under Horizon 2020, the EU Framework Programme for Research and Innovation.EIT Health—Promoting innovation in health https://www.eithealth.eu.

References

1. Abo-Hamad, W.: Patient pathways discovery and analysis using process mining techniques: an emergency department case study. In: Cappanera, P., Li, J., Matta, A., Sahin, E., Vandaele, N.J., Visintin, F. (eds.) ICHCSE 2017. SPMS, vol. 210, pp. 209–219. Springer, Cham (2017). https://doi.org/10.1007/978-3-319-66146-9_19

2. Bergs, J., et al.: Emergency department crowding: time to shift the paradigm from predicting and controlling to analysing and managing. Int. Emerg. Nurs. **24**, 74–77 (2016)

3. Conca, T., et al.: Multidisciplinary collaboration in the treatment of patients with type 2 diabetes in primary care: analysis using process mining. J. Medi. Internet Res. **20**(4), e127 (2018). https://doi.org/10.2196/jmir.8884, http://www.pubmedcentral.nih.gov/articlerender.fcgi?artid=PMC5915667, http://www.jmir.org/2018/4/e127/

4. Coumou, D., Rahmstorf, S.: A decade of weather extremes, July 2012. https://doi.org/10.1038/nclimate1452, www.nature.com/natureclimatechange

5. Fernández-Llatas, C., Benedi, J.M., García-Gómez, J., Traver, V.: Process mining for individualized behavior modeling using wireless tracking in nursing homes. Sensors **13**(11), 15434–15451 (2013)

6. Frumkin, H., Hess, J., Luber, G., Malilay, J., McGeehin, M.: Climate change: the public health response. Am. J. Public Health **98**(3), 435–445 (2008)

7. Ibanez-Sanchez, G., et al.: Toward value-based healthcare through interactive process mining in emergency rooms: the stroke case. Int. J. Environ. Res. Public Health **16**(10), 1783 (2019)

8. Kinney, P.L., O'Neill, M.S., Bell, M.L., Schwartz, J.: Approaches for estimating effects of climate change on heat-related deaths: challenges and opportunities. Environ. Sci. Policy **11**(1), 87–96 (2008)

9. Knowlton, K., et al.: The 2006 California heat wave: impacts on hospitalizations and emergency department visits. Environ. Health Perspect. **117**(1), 61–67 (2009)

10. Lindstrom, S.J., Nagalingam, V., Newnham, H.H.: Impact of the 2009 Melbourne heatwave on a major public hospital. Intern. Med. J. **43**(11), 1246–1250 (2013). https://doi.org/10.1111/imj.12275, http://doi.wiley.com/10.1111/imj.12275

11. Martinez-Millana, A., Lizondo, A., Gatta, R., Vera, S., Salcedo, V., Fernandez-Llatas, C.: Process mining dashboard in operating rooms: analysis of staff expectations with analytic hierarchy process. Int. J. Environ. Res. Public Health **16**(2), 199 (2019). https://doi.org/10.3390/ijerph16020199, http://www.mdpi.com/1660-4601/16/2/199

12. Meehl, G.A., Tebaldi, C.: More intense, more frequent, and longer lasting heat waves in the 21st century. Science **305**(5686), 994–997 (2004)

13. National Oceanic And Atmospheric Administration: Global Climate Report - Annual 2019. https://www.ncdc.noaa.gov/sotc/global/201913 (2019). Accessed 16 Nov 2020

14. Pachauri, R.K., et al.: Climate change 2014: synthesis report. Contribution of Working Groups I, II and III to the fifth assessment report of the Intergovernmental Panel on Climate Change. IPCC (2014)

15. Rebuge, Á., Ferreira, D.R.: Business process analysis in healthcare environments: a methodology based on process mining. Inf. Syst. **37**(2), 99–116 (2012)

16. Rojas, E., Munoz-Gama, J., Sepúlveda, M., Capurro, D.: Process mining in healthcare: a literature review. J. Biomed. Inf. **61**, 224–236 (2016)

17. Scalley, B.D., et al.: Responding to heatwave intensity: excess heat factor is a superior predictor of health service utilisation and a trigger for heatwave plans. Aust. N. Z. J. Public Health **39**(6), 582–587 (2015). https://doi.org/10.1111/1753-6405.12421, http://doi.wiley.com/10.1111/1753-6405.12421

18. Stocker, T.F., et al.: The physical science basis. Contribution of working group I to the fifth assessment report of the intergovernmental panel on climate change. Comput. Geom. **18**, 95–123 (2013)

19. Aalst, W.: Data science in action. Process Mining, pp. 3–23. Springer, Heidelberg (2016). https://doi.org/10.1007/978-3-662-49851-4_1

20. Verelst, S., Pierloot, S., Desruelles, D., Gillet, J.B., Bergs, J.: Short-term unscheduled return visits of adult patients to the emergency department. J. Emerg. Med. **47**(2), 131–139 (2014)

21. World Health Organization: WHO temperature report. https://www.who.int/news-room/fact-sheets/detail/climate-change-and-health (2020). Accessed 30 Aug 2020

22. World Meteorological Organization: Global Climate in 2015–2019: Climate change accelerates. Press release by the World Meteorological Organization. https://www.who.int/news-room/fact-sheets/detail/climate-change-and-health (2020). Accessed 15 Sep 2020

23. Xu, Z., Cheng, J., Hu, W., Tong, S.: Heatwave and health events: a systematic evaluation of different temperature indicators, heatwave intensities and durations. Sci. Total Environ. **630**, 679–689 (2018). https://doi.org/10.1016/j.scitotenv.2018.02.268

1st International Workshop on Trust and Privacy in Process Analytics (TPPA)

1st Workshop on Trust and Privacy in Process Analytics (TPPA)

Process mining has been successfully applied in analysing and improving processes based on event logs in all kinds of environments. However, the impact of trust and privacy on the technical design as well as the organizational application of process mining has been largely neglected. Both topics are closely related to the responsible application of data science, a topic that has received more attention in recent years as data-driven methods have started to permeate our society.

Privacy relates to the concern that event logs may contain personal data of both customers and employees and the challenge of protecting the information about individuals while still being useful for process mining (e.g., differential privacy, k-anonymity, homomorphic encryption, secure multi-party computing). Often, security aspects (e.g., encryption) are closely connected when processing personal data. On the other hand, the workshop is about the concept of trust, which is required both from the perspective of trust in organizational and technological measures that ensure event logs are not misused (e.g., for worker surveillance) as well as from the perspective of trust that the results of a process mining analysis faithfully reflect reality (e.g., data quality, traceability, auditability).

The main objective of the TPPA workshop was to give a forum for the trust and privacy aspects and responsible application of process mining. Finally, one paper was accepted for publication and a second paper was accepted for presentation only. The paper "Towards Quantifying Privacy in Process Mining" authored by M. Rafiei and W.M.P. van der Aalst proposes an approach to quantify the effectiveness of privacy preservation techniques. For this purpose, two measures for quantifying disclosure risks are introduced. The second paper (accepted only for presentation) entitled "Differentially-Private Process Mining (DPPM): Using A Real-World Sepsis Dataset in the Context of Privacy Preserving Process Mining" authored by S. Amna Sohail, F. Allah Bukhsh, and M. van Keulen summarizes the evaluation of the data utility of noise-generating plugins for privacy-preserving process mining.

In addition to these two papers, the program of the workshop included a panel discussion on "Trust and Privacy in Process Analytics". A summary of the panel discussion has been published in the special issue "Towards Privacy Preservation and Data Protection in Information System Design" of the EMISA Journal[1]. Around 50 attendees were present during the workshop presentations and panel discussion.

---

[1] The summary can be accessed at https://doi.org/10.18417/emisa.15.8.

Organization

Organizing Committee

Felix Mannhardt	Eindhoven University of Technology
Agnes Koschmider	Kiel University
Nathalie Baracaldo	IBM Almaden Research Center

Program Committee

Luciano García-Bañuelos	Tecnologico de Monterrey
Olivia Choudhury	Amazon Inc.
Stephan Fahrenkrog-Petersen	Humboldt University of Berlin
Marwan Hassani	Eindhoven University of Technology
Judith Michael	RWTH Aachen
Florian Tschorsch	Technische Universität Berlin
Melanie Volkamer	Karlsruhe Institute of Technology
Moe Wynn	Queensland University of Technology
Sebastiaan van Zelst	RWTH Aachen/FIT

Towards Quantifying Privacy in Process Mining

Majid Rafiei$^{(\boxtimes)}$ and Wil M. P. van der Aalst

Chair of Process and Data Science, RWTH Aachen University, Aachen, Germany
majid.rafiei@pads.rwth-aachen.de

Abstract. Process mining employs event logs to provide insights into the actual processes. Event logs are recorded by information systems and contain valuable information helping organizations to improve their processes. However, these data also include highly sensitive private information which is a major concern when applying process mining. Therefore, privacy preservation in process mining is growing in importance, and new techniques are being introduced. The effectiveness of the proposed privacy preservation techniques needs to be evaluated. It is important to measure both sensitive data protection and data utility preservation. In this paper, we propose an approach to quantify the effectiveness of privacy preservation techniques. We introduce two measures for quantifying disclosure risks to evaluate the sensitive data protection aspect. Moreover, a measure is proposed to quantify data utility preservation for the main process mining activities. The proposed measures have been tested using various real-life event logs.

Keywords: Responsible process mining · Privacy preservation · Privacy quantification · Data utility · Event logs

1 Introduction

Process mining bridges the gap between traditional model-based process analysis (e.g., simulation), and data-centric analysis (e.g., data mining) [1]. The three basic types of process mining are *process discovery*, where the aim is to discover a process model capturing the behavior seen in an event log, *conformance checking*, where the aim is to find commonalities and discrepancies between a process model and an event log, and *process re-engineering* (*enhancement*), where the idea is to extend or improve a process model using event logs.

An event log is a collection of events. Each event has the following mandatory attributes: a *case identifier*, an *activity name*, a *timestamp*, and optional attributes such as *resources* or *costs*. In the human-centered processes, case identifiers refer to individuals. For example, in a patient treatment process, the case identifiers refer to the patients whose data are recorded. Moreover, other attributes may also refer to individuals, e.g., *resources* often refer to persons performing activities. When event logs explicitly or implicitly include personal data,

© Springer Nature Switzerland AG 2021
S. Leemans and H. Leopold (Eds.): ICPM 2020 Workshops, LNBIP 406, pp. 385–397, 2021.
https://doi.org/10.1007/978-3-030-72693-5_29

privacy concerns arise which should be taken into account w.r.t. regulations such as the European General Data Protection Regulation (GDPR).

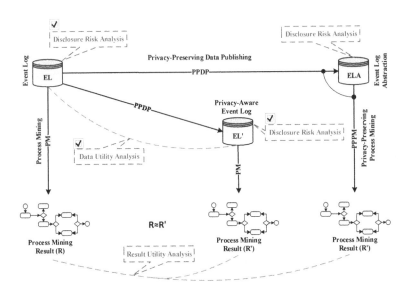

Fig. 1. Overview of privacy-related activities in process mining. Privacy preservation techniques are applied to event logs to provide desired privacy requirements. The aim is to protect sensitive personal data, yet, at the same time, preserve data utility, and generate as similar as possible results to the original ones. The parts indicated by dashed callouts show the analyses that need to be performed to evaluate the effectiveness of privacy preservation techniques.

The *privacy* and *confidentiality* issues in process mining are recently receiving more attention and various techniques have been proposed to protect sensitive data. Privacy preservation techniques often apply anonymization operations to modify the data in order to fulfill desired privacy requirements, yet, at the same time, they are supposed to preserve data utility. To evaluate the effectiveness of these techniques, their effects on *sensitive data protection* and *data utility preservation* need to be measured. In principle, privacy preservation techniques always deal with a trade-off between data utility and data protection, and they are supposed to balance these aims.

Figure 1 shows the general view of privacy in process mining including two main activities: *Privacy-Preserving Data Publishing* (PPDP) and *Privacy-Preserving Process Mining* (PPPM). PPDP aims to hide the identity and the sensitive data of record owners in event logs to protect their privacy. PPPM aims to extend traditional process mining algorithms to work with the non-standard event data so-called *Event Log Abstraction* (ELA) [16] that might result from PPDP techniques. *Abstractions* are intermediate results, e.g., a directly follows graph could be an intermediate result of a process discovery algorithm. Note that PPPM algorithms are tightly coupled with the corresponding PPDP techniques.

In this paper, our main focus is on the analyses indicated by the checkboxes in Fig. 1. Note that *disclosure risk analysis* is done for a single event log, while for *data/result utility analysis*, the original event log/result need to be compared with the privacy-aware event log/result. We consider simple event logs containing basic information for performing two main process mining activities: *process discovery* and *conformance checking*. We introduce two measures for quantifying disclosure risks in a simple event log: *identity (case) disclosure* and *attribute (trace) disclosure*. Using these measures, we show that even simple event logs could disclose sensitive information. We also propose a measure for quantifying *data utility* which is based on the *earth mover's distance*. So far, the proposed privacy preservation techniques in process mining use the *result utility* approach to demonstrate the utility preservation aspect which is not as precise and general as the *data utility* approach, since it is highly dependent on the underlying algorithms. We advocate the proposed measures by assessing their functionality for quantifying the disclosure risks and data utility on real-life event logs before and after applying a privacy preservation technique with different parameters.

The remainder of the paper is organized as follows. Section 2 outlines related work. In Sect. 3, formal models for event logs are presented. We explain the measures in Sect. 4. The experiments are described in Sect. 5, and Sect. 6 concludes the paper.

2   Related Work

In process mining, the research field of confidentiality and privacy is growing in importance. In [2], *Responsible Process Mining* (RPM) is introduced as the sub-discipline focusing on possible negative side-effects of applying process mining. In [12], the authors propose a privacy-preserving system design for process mining, where a user-centered view is considered to track personal data. In [18], a framework is introduced providing a generic scheme for confidentiality in process mining. In [14], the authors introduce a privacy-preserving method for discovering roles from event data. In [6], the authors apply k-anonymity and t-closeness on event data to preserve the privacy of *resources*. In [11], the notion of *differential privacy* is employed to preserve the privacy of *cases*. In [17], the TLKC-privacy model is introduced to deal with high variability issues in event logs for applying group-based anonymization techniques. In [5], a secure multi-party computation solution is proposed for preserving privacy in an inter-organizational setting. In [13], the authors analyze data privacy and utility requirements for healthcare event data, and the suitability of privacy-preserving techniques is assessed. In [16], privacy metadata in process mining are discussed and a privacy extension for the XES standard (https://xes-standard.org/) is proposed.

Most related to our work is [22], where a uniqueness-based measure is proposed to evaluate the re-identification risk of event logs. Privacy quantification in data mining is a well-developed field where the effectiveness of privacy preservation techniques is evaluated from different aspects such as *dissimilarity* [3], *information loss* [7], *discernibility* [8], and etc. We utilize the experiences achieved in this field and propose a trade-off approach as suggested in [4].

3   Preliminaries

In this section, we provide formal definitions for event logs used in the remainder. An event log is a collection of events, composed of different attributes, such that they are uniquely identifiable. In this paper, we consider only the mandatory attributes of events including *case identifier*, *activity name*, and *timestamp*. Accordingly, we define a simple event, trace, and event log. In the following, we introduce some basic concepts and notations.

Let A be a set. A^* is the set of all finite sequences over A, and $\mathcal{B}(A)$ is the set of all multisets over the set A. For $A_1, A_2 \in \mathcal{B}(A)$, $A_1 \subseteq A_2$ if for all $a \in A$, $A_1(a) \leq A_2(a)$. A finite sequence over A of length n is a mapping $\sigma \in \{1, ..., n\} \rightarrow A$, represented as $\sigma = \langle a_1, a_2, ..., a_n \rangle$ where $\sigma_i = a_i = \sigma(i)$ for any $1 \leq i \leq n$, and $|\sigma| = n$. $a \in \sigma \Leftrightarrow a = a_i$ for $1 \leq i \leq n$. For $\sigma_1, \sigma_2 \in A^*$, $\sigma_1 \sqsubseteq \sigma_2$ if σ_1 is a subsequence of σ_2, e.g., $\langle a, b, c, x \rangle \sqsubseteq \langle z, x, a, b, b, c, a, b, c, x \rangle$. For $\sigma \in A^*$, $\{a \in \sigma\}$ is the set of elements in σ, and $[a \in \sigma]$ is the multiset of elements in σ, e.g., $[a \in \langle x, y, z, x, y \rangle] = [x^2, y^2, z]$.

Definition 1 (Simple Event). *A simple event is a tuple $e = (c, a, t)$, where $c \in \mathcal{C}$ is the case identifier, $a \in \mathcal{A}$ is the activity associated to event e, and $t \in \mathcal{T}$ is the timestamp of event e. $\pi_X(e)$ is the projection of event e on the attribute from domain X, e.g., $\pi_\mathcal{A}(e) = a$. We call $\xi = \mathcal{C} \times \mathcal{A} \times \mathcal{T}$ the event universe.*

Definition 2 (Simple Trace). *Let ξ be the universe of events. A trace $\sigma = \langle e_1, e_2, ..., e_n \rangle$ in an event log is a sequence of events, i.e., $\sigma \in \xi^*$, s.t., for each $e_i, e_j \in \sigma$: $\pi_\mathcal{C}(e_i) = \pi_\mathcal{C}(e_j)$, and $\pi_\mathcal{T}(e_i) \leq \pi_\mathcal{T}(e_j)$ if $i < j$. A simple trace is a trace where all the events are projected on the activity attribute, i.e., $\sigma \in \mathcal{A}^*$.*

Definition 3 (Simple Event Log). *A simple event log is a multiset of simple traces, i.e., $L \in \mathcal{B}(\mathcal{A}^*)$. We assume each trace in an event log belongs to an individual and $\sigma \neq \langle \rangle$ if $\sigma \in L$. $A_L = \{a \in \mathcal{A} \mid \exists_{\sigma \in L} a \in \sigma\}$ is the set of activities in the event log L. $\tilde{L} = \{\sigma \in L\}$ is the set of unique traces (variants) in the event log L. We denote $\mathcal{U}_L$ as the universe of event logs.*

Definition 4 (Trace Frequency). *Let L be an event log, $f_L \in \tilde{L} \rightarrow [0, 1]$ is a function which retrieves the relative frequency of a trace in the event log L, i.e., $f_L(\sigma) = L(\sigma)/|L|$ and $\sum_{\sigma \in \tilde{L}} f_L(\sigma) = 1$.*

Definition 5 (Event Log Entropy). *$ent \in \mathcal{U}_L \rightarrow \mathbb{R}_{\geq 0}$ is a function which retrieves the entropy of traces in an event log, s.t., for $L \in \mathcal{U}_L$, $ent(L) = -\sum_{\sigma \in \tilde{L}} f_L(\sigma) log_2 f_L(\sigma)$. We denote $max\_ent(L)$ as the maximal entropy achieved when all the traces in the event log are unique, i.e., $|\tilde{L}| = |L|$.*

4   Privacy Quantification

We employ a *risk-utility* model for quantifying privacy in process mining where *disclosure risk* and *utility loss* are measured to assess the effectiveness of privacy preservation techniques before and after applying the techniques.

4.1   Disclosure Risk

In this subsection, we introduce *identity/case disclosure* and *attribute/trace disclosure* for quantifying disclosure risk of event logs. Identity disclosure quantifies how uniquely the trace owners, i.e., cases, can be re-identified. Attribute disclosure quantifies how confidently the sensitive attributes of cases (as individuals) can be specified. As discussed in [17], traces play the role of both quasi-identifiers and sensitive attributes. That is, a complete sequence of activities, which belongs to a case, is sensitive person-specific information. At the same time, knowing a part of this sequence, as background knowledge, can be exploited to re-identify the trace owner. In a simple event log, traces, i.e., sequence of activities, are the only available information. Therefore, *attribute disclosure* can be seen as *trace disclosure*.

In the following, we define *set, multiset,* and *sequence* as three types of background knowledge based on traces in simple event logs that can be exploited for uniquely re-identifying the trace owners or certainly specifying their complete sequence of activities. Moreover, we consider a size for different types of background knowledge as their power, e.g., the *set* background knowledge of size 3 is more powerful than the same type of background knowledge of size 2. Note that the assumed types of background knowledge are the most general ones, and more types can be explored. However, the general approach will be the same.

Definition 6 (Background Knowledge 1 - Set). *In this scenario, we assume that an adversary knows a subset of activities performed for the case, and this information can lead to the identity or attribute disclosure. Let L be an event log, and A_L be the set of activities in the event log L. We formalize this background knowledge by a function $proj_{set}^L \in 2^{A_L} \to 2^L$. For $A \subseteq A_L$, $proj_{set}^L(A) = [\sigma \in L \mid A \subseteq \{a \in \sigma\}]$. We denote $cand_{set}^l(L) = \{A \subseteq A_L \mid |A| = l \wedge proj_{set}^L(A) \neq []\}$ as the set of all subsets over the set A_L of size l for which there exists matching traces in the event log.*

Definition 7 (Background Knowledge 2 - Multiset). *In this scenario, we assume that an adversary knows a sub-multiset of activities performed for the case, and this information can lead to the identity or attribute disclosure. Let L be an event log, and A_L be the set of activities in the event log L. We formalize this background knowledge by a function $proj_{mult}^L \in \mathcal{B}(A_L) \to 2^L$. For $A \in \mathcal{B}(A_L)$, $proj_{mult}^L(A) = [\sigma \in L \mid A \subseteq [a \in \sigma]]$. We denote $cand_{mult}^l(L) = \{A \in \mathcal{B}(A_L) \mid |A| = l \wedge proj_{mult}^L(A) \neq []\}$ as the set of all sub-multisets over the set A_L of size l for which there exists matching traces in the event log.*

Definition 8 (Background Knowledge 3 - Sequence). *In this scenario, we assume that an adversary knows a subsequence of activities performed for the case, and this information can lead to the identity or attribute disclosure. Let L be an event log, and A_L be the set of activities in the event log L. We formalize this background knowledge by a function $proj_{seq}^L \in A_L^* \to 2^L$. For $\sigma \in A_L^*$, $proj_{seq}^L(\sigma) = [\sigma' \in L \mid \sigma \sqsubseteq \sigma']$. We denote $cand_{seq}^l(L) = \{\sigma \in A_L^* \mid |\sigma| = l \land proj_{seq}^L(\sigma) \neq []\}$ as the set of all subsequences of size (length) l, based on the activities in A_L, for which there exists matching traces in the event log.*

Example 1 (background knowledge). *Let $L = [\langle a,b,c,d \rangle^{10}, \langle a,c,b,d \rangle^{20}, \langle a,d,b,d \rangle^5, \langle a,b,d,d \rangle^{15}]$ be an event log. $A_L = \{a,b,c,d\}$ is the set of unique activities, and $cand_{set}^2(L) = \{\{a,b\},\{a,c\},\{a,d\},\{b,c\},\{b,d\},\{d,c\}\}$ is the set of candidates of the set background knowledge of size 2. For $A = \{b,d\} \in cand_{set}^2(L)$ as a candidate of the set background knowledge of size 2, $proj_{set}^L(A) = [\langle a,b,c,d \rangle^{10}, \langle a,c,b,d \rangle^{20}, \langle a,d,b,d \rangle^5, \langle a,b,d,d \rangle^{15}]$. For $A = [b,d^2]$ as a candidate of the multiset background knowledge, $proj_{mult}^L(A) = [\langle a,d,b,d \rangle^5, \langle a,b,d,d \rangle^{15}]$. Also, for $\sigma = \langle b,d,d \rangle$ as a candidate of the sequence background knowledge, $proj_{seq}^L(\sigma) = [\langle a,b,d,d \rangle^{15}]$.*

As Example 1 shows, the strength of background knowledge from the weakest to the strongest w.r.t. the type is as follows: *set*, *multiset*, and *sequence*, i.e., given the event log L, $proj_{seq}^L(\langle b,d,d \rangle) \subseteq proj_{mult}^L([b,d^2]) \subseteq proj_{set}^L(\{b,d\})$.

Identity (Case) Disclosure. We use the uniqueness of traces w.r.t. the background knowledge of size l to measure the corresponding case disclosure risk in an event log. Let L be an event log and $type \in \{set, mult, seq\}$ be the type of background knowledge. The case disclosure based on the background knowledge $type$ of size l is calculated as follows:

$$cd_{type}^l(L) = \sum_{x \in cand_{type}^l(L)} \frac{1/|proj_{type}^L(x)|}{|cand_{type}^l(L)|} \tag{1}$$

Equation (1) calculates the average uniqueness based on the candidates of background knowledge, i.e., $x \in cand_{type}^l(L)$. Note that we consider equal weights for the candidates of background knowledge. However, they can be weighted based on the various criteria, e.g., the sensitivity of the activities included. One can also consider the worst case, i.e., the maximal uniqueness, rather than the average value.

Example 2 (insufficiency of case disclosure analysis). *Consider $L_1 = [\langle a,b,c,d \rangle, \langle a,c,b,d \rangle, \langle a,b,c,c,d \rangle, \langle a,b,b,c,d \rangle]$ and $L_2 = [\langle a,b,c,d \rangle^4, \langle e,f \rangle^4, \langle g,h \rangle^4]$ as two event event logs. $A_{L_1} = \{a,b,c,d\}$ and $A_{L_2} = \{a,b,c,d,e,f,g,h\}$ are the set of unique activities in L_1 and L_2, respectively. $cand_{set}^1(L_1) = \{\{a\},\{b\},\{c\},\{d\}\}$ and $cand_{set}^1(L_2) = \{\{a\},\{b\},\{c\},\{d\},\{e\},\{f\},\{g\},\{h\}\}$ are the set of candidates of the set background knowledge of size 1. Both event logs have the same value as the case disclosure for the set background knowledge of* *size* 1 $(cd^1_{set}(L_1) = cd^1_{set}(L_2) = 1/4)$. *However, in* L_2, *the complete sequence of activities performed for a victim case is disclosed by knowing only one activity without uniquely identifying the corresponding trace.*

Example 2 clearly shows that measuring the uniqueness alone is insufficient to demonstrate disclosure risks in event logs and the uncertainty in the set of sensitive attributes matching with the assumed background knowledge need to be measured, as well. In the following, we define a measure to quantify the uncertainty in the set of matching traces. Note that, the same approach can be exploited to quantify the disclosure risk of any other sensitive attribute matching with some background knowledge.

Attribute (Trace) Disclosure. We use the entropy of matching traces w.r.t. background knowledge of size l to measure the corresponding trace disclosure risk in an event log. Let L be an event log and $type \in \{set, mult, seq\}$ be the type of background knowledge. The trace disclosure based on the background knowledge $type$ of size l is calculated as follows:

$$td^l_{type}(L) = 1 - \sum_{x \in cand^l_{type}(L)} \frac{ent(proj^L_{type}(x))/max\_ent(proj^L_{type}(x))}{|cand^l_{type}(L)|} \tag{2}$$

In (2), $max\_ent(proj^L_{type}(x))$ is the maximal entropy for the matching traces based on the type and size of background knowledge, i.e., uniform distribution of the matching traces. As discussed for (1), in (2), we also assume equal weights for the candidates of background knowledge. However, one can consider different weights for the candidates. Also, the worst case, i.e., the minimal entropy, rather than the average entropy can be considered.

The trace disclosure of the event logs in Example 2 is as follows: $td^1_{set}(L_1) = 0$ (the multiset of matching traces has the maximal entropy) and $td^1_{set}(L_2) = 1$ (the entropy of matching traces is 0). These results distinguish the disclosure risk of the event logs.

4.2 Utility Loss

In this subsection, we introduce a measure based on the *earth mover's distance* [19] for quantifying the utility loss after applying a privacy preservation technique to an event log. The *earth mover's distance* describes the distance between two distributions. In an analogy, given two piles of earth, it expresses the effort required to transform one pile into the other. First, we introduce the concept of reallocation indicating how an event log is transformed into another event log. Then, we define a trace distance function expressing the cost of transforming one trace into another one. Finally, we introduce the utility loss measure that indicates the entire cost of transforming an event log to another one using the introduced reallocation and distance functions.

Reallocation. Let L be the original event log and L' be an anonymized event log derived from the original event log. We introduce $r \in \tilde{L} \times \tilde{L}' \rightarrow [0,1]$ as a function that indicates the movement of frequency between two event logs. $r(\sigma, \sigma')$ describes the relative frequency of $\sigma \in \tilde{L}$ that should be transformed to $\sigma' \in \tilde{L}'$. To make sure that a reallocation function properly transforms L into L', the frequency of each $\sigma \in \tilde{L}$ should be considered, i.e., for all $\sigma \in \tilde{L}$, $f_L(\sigma) = \sum_{\sigma' \in \tilde{L}'} r(\sigma, \sigma')$. Similarly, the probability mass of traces $\sigma' \in \tilde{L}'$ should be preserved, i.e., for all $\sigma' \in \tilde{L}'$, $f_{L'}(\sigma') = \sum_{\sigma \in \tilde{L}} r(\sigma, \sigma')$. We denote $\mathcal{R}$ as the set of all reallocation functions which depends on L and L'.

Table 1. The dissimilarity between two event logs based on the earth mover's distance assuming r_s as a reallocation function and d_s as the normalized Levenshtein distance.

$r_s \cdot d_s$	$\langle a, b, c, d \rangle$	$\langle a, c, b, d \rangle$	$\langle a, e, c, d \rangle^{49}$	$\langle a, e, b, d \rangle^{49}$
$\langle a, b, c, d \rangle^{50}$	$0.01 \cdot 0$	$0 \cdot 0.5$	$0.49 \cdot 0.25$	$0 \cdot 0.5$
$\langle a, c, b, d \rangle^{50}$	$0 \cdot 0.5$	$0.01 \cdot 0$	$0 \cdot 0.5$	$0.49 \cdot 0.25$

Trace Distance. A trace distance function $d \in \mathcal{A}^* \times \mathcal{A}^* \rightarrow [0,1]$ expresses the distance between traces. This function is 0 if and only if two traces are equal, i.e., $d(\sigma, \sigma') = 0 \Longleftrightarrow \sigma = \sigma'$. This function should also be symmetrical, i.e., $d(\sigma, \sigma') = d(\sigma', \sigma)$. Different distance functions can be considered satisfying these conditions. We use the *normalized string edit distance* (Levenshtein) [9].

Utility Loss. Let L be an original event log, and L' be an anonymized event log derived from the original event log. Several reallocation functions might exist. However, the *earth mover's distance* problem aims to express the shortest distance between the two event logs, i.e., the least mass movement over the least distance between traces. Therefore, the difference between L and L' using a reallocation function r is the inner product of reallocation and distance. The data utility preservation is considered as $du(L, L') = 1 - \min_{r \in \mathcal{R}} ul(r, L, L')$.

$$ul(r, L, L') = r \cdot d = \sum_{\sigma \in \tilde{L}} \sum_{\sigma' \in \tilde{L}'} r(\sigma, \sigma') d(\sigma, \sigma') \qquad (3)$$

Example 3 (using earth mover's distance to calculate dissimilarity between event logs). Let $L = [\langle a, b, c, d \rangle, \langle a, c, b, d \rangle, \langle a, e, c, d \rangle^{49}, \langle a, e, b, d \rangle^{49}]$ and $L' = [\langle a, b, c, d \rangle^{50}, \langle a, c, b, d \rangle^{50}]$ be the original and aninymized event logs, respectively. Table 1 shows the calculations assuming r_s as a reallocation function and d_s as the normalized Levenshtein distance, e.g., $r_s(\langle a, b, c, d \rangle, \langle a, e, c, d \rangle) = 0.49$ and $d_s(\langle a, b, c, d \rangle, \langle a, e, c, d \rangle) = 0.25$. $ul(r_s, L, L') = 0.24$ and $du(L, L') = 0.76$.

5   Experiments

In this section, we demonstrate the experiments on real-life event logs to advocate the proposed measures. We employ two human-centered event logs, where the *case identifiers* refer to individuals. Sepsis-Cases [10] is a real-life event log containing events of sepsis cases from a hospital. BPIC-2017-APP [21] is also a real-life event log pertaining to a loan application process of a Dutch financial institute. We choose these event logs because they are totally different w.r.t. the uniqueness of traces. Table 2 shows the general statistics of these event logs. Note that *variants* are the unique traces, and $trace\_uniquness = {}^{\#variants}/_{\#traces}$. The implementation as a Python program is available on GitHub.[1]

Table 2. The general statistics of the event logs used in the experiments.

Event log	#traces	#variants	#events	#unique_activities	trace_uniqueness
Sepsis-Cases [10]	1050	845	15214	16	80%
BPIC-2017-APP [21]	31509	102	239595	10	0.3%

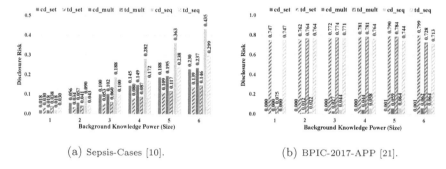

(a) Sepsis-Cases [10].   (b) BPIC-2017-APP [21].

Fig. 2. Analyses of the case disclosure (*cd*) and the trace disclosure (*td*) based on the three types of background knowledge (i.e., *set*, *mult*, and *seq*) when we vary the background knowledge power (size) from 1 to 6. For example, in the Sepsis-Cases event log, the case disclosure risk of the background knowledge *seq* (*cd_seq*) of size 3 is 0.188.

5.1   Disclosure Risk Analysis

In this subsection, we show the functionality of the proposed measures for disclosure risk analysis. To this end, we consider three types of background knowledge (*set*, *multiset*, and *sequence*) and vary the background knowledge power (size) from 1 to 6. Figure 2a shows the results for the Sepsis-Cases event log where the uniqueness of traces is high. As shown, the disclosure risks are higher for the more powerful background knowledge w.r.t. the *type* and *size*.

---

[1] https://github.com/m4jidRafiei/privacy_quantification.

Figure 2b demonstrates the results for the BPIC-2017-APP event log, where the uniqueness of traces is low. As shown, the case disclosure risk is low, which is expected regarding the low uniqueness of traces. However, the trace disclosure risk is high which indicates low entropy (uncertainty) of the traces. Moreover, for the stronger background knowledge w.r.t. the size, one can assume a higher case disclosure risk. However, the trace disclosure risk is correlated with the entropy of the sensitive attribute values and can be a high value even for weak background knowledge. The above-mentioned analyses clearly show that uniqueness alone cannot reflect the actual disclosure risk in an event log.

5.2 Utility Loss Analysis

In this subsection, we demonstrate the functionality of the proposed measure in Sect. 4.2 for quantifying data utility preservation after applying a privacy preservation technique. We use PPDP-PM [15] as a privacy preservation tool for process mining to apply the TLKC-privacy model [17] to a given event log. The TLKC-privacy model is a group-based privacy preservation technique which provides a good level of flexibility through various parameters such as the type and size (power) of background knowledge. The T in this model refers to the accuracy of timestamps in the privacy-aware event log, L refers to the power of background knowledge[2], K refers to the k in the k-anonymity definition [20], and C refers to the bound of confidence regarding the sensitive attribute values in an equivalence class.

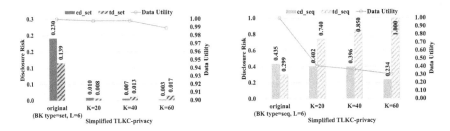

(a) Using *set* as background knowledge.   (b) Using *sequence* as background knowledge.

Fig. 3. The utility loss and disclosure risk analyses for the Sepsis-Cases event log where the background knowledge is *set* or *sequence*, and the power (size) of background knowledge is 6.

Assuming *set* (Definition 6) and *sequence* (Definition 8) as the types of background knowledge, we apply the TLKC-privacy model to the Sepsis-Cases event log with the following parameters: L = 6 (as the maximum background knowledge power in our experiments), K = {20, 40, 60}, C = 1 (there is no additional

---

[2] Note that this L is identical to the l introduced as the power (size) of background knowledge and should not be confused with L as the event log notation.

Table 3. The general statistics before and after applying the TLKC-privacy model.

Event log			#traces	#variants	#events	#unique_activities
Original Sepsis-Cases			1050	845	15214	16
Anonymized Sepsis-Cases	BK type = set BK size (L) = 6	K = 20	1050	842	15103	12
		K = 40	1050	842	14986	11
		K = 60	1050	818	14809	11
	BK type = seq BK size (L) = 6	K = 20	1050	34	3997	6
		K = 40	1050	33	4460	5
		K = 60	1050	18	3448	4

sensitive attribute in a simple event log), and T is set to the maximal precision (T has no effect on a simple event log). That is, the TLKC-privacy model is simplified to k-anonymity where the *quasi-identifier* (background knowledge) is the *set* or *sequence* of activities. Table 3 demonstrates the general statistics of the event logs before and after applying the privacy preservation technique.

Figure 3a shows disclosure risk and data utility analyses for the background knowledge *set*, and Fig. 3b shows the same analyses for the background knowledge *sequence*. In both types of background knowledge, the data utility value decreases. For the stronger background knowledge, i.e., *sequence*, the utility loss is much higher which is expected w.r.t. the general statistics in Table 3. However, the data utility for the weaker background knowledge remains high which again complies with the general statistics. Note that since we apply k-anonymity (simplified TLKC-model) only *case disclosure*, which is based on the uniqueness of traces, decreases. Moreover, for the *sequence* background knowledge, higher values for K result in more similar traces. Therefore, the *trace disclosure* risk, in the anonymized event logs, drastically increases. These analyses demonstrate that privacy preservation techniques should consider different aspects of disclosure risk while balancing data utility preservation and sensitive data protection.

6 Conclusion

Event logs often contain highly sensitive information, and regarding the rules imposed by regulations, these sensitive data should be analyzed responsibly. Therefore, privacy preservation in process mining is recently receiving more attention. Consequently, new measures need to be defined to evaluate the effectiveness of the privacy preservation techniques both from the sensitive data protection and data utility preservation point of views. In this paper, using a trade-off approach, we introduced two measures for quantifying disclosure risks: *identity/case disclosure* and *attribute/trace disclosure*, and one measure for quantifying *utility loss*. The introduced measures were applied to two real-life event logs. We showed that even simple event logs could reveal sensitive information. Moreover, for the first time, the effect of applying a privacy preservation technique on *data utility* rather than *result utility* was explored. The *data utility* measure is based on the *earth mover's distance* and can be extended to evaluate the utility w.r.t. the different perspectives of process mining, e.g., *time, resource*, etc.

Acknowledgment. Funded under the Excellence Strategy of the Federal Government and the Länder. We also thank the Alexander von Humboldt (AvH) Stiftung for supporting our research.

References

1. van der Aalst, W.M.P.: Process Mining - Data Science in Action. 2nd edn. Springer, Heidelberg (2016). https://doi.org/10.1007/978-3-662-49851-4
2. Aalst, W.M.P.: Responsible data science: using event data in a "people friendly" manner. In: Hammoudi, S., Maciaszek, L.A., Missikoff, M.M., Camp, O., Cordeiro, J. (eds.) ICEIS 2016. LNBIP, vol. 291, pp. 3–28. Springer, Cham (2017). https://doi.org/10.1007/978-3-319-62386-3_1
3. Bertino, E., Fovino, I.N., Provenza, L.P.: A framework for evaluating privacy preserving data mining algorithms. Data Min. Knowl. Discov. **11**(2), 121–154 (2005)
4. Bertino, E., Lin, D., Jiang, W.: A survey of quantification of privacy preserving data mining algorithms. In: Aggarwal, C.C., Yu, P.S. (eds.) Privacy-Preserving Data Mining - Models and Algorithms, Advances in Database Systems, vol. 34, pp. 183–205. Springer, Boston (2008). https://doi.org/10.1007/978-0-387-70992-5_8
5. Elkoumy, G., Fahrenkrog-Petersen, S.A., Dumas, M., Laud, P., Pankova, A., Weidlich, M.: Secure multi-party computation for inter-organizational process mining. In: Nurcan, S., Reinhartz-Berger, I., Soffer, P., Zdravkovic, J. (eds.) BPMDS/EMMSAD -2020. LNBIP, vol. 387, pp. 166–181. Springer, Cham (2020). https://doi.org/10.1007/978-3-030-49418-6_11
6. Fahrenkrog-Petersen, S.A., van der Aa, H., Weidlich, M.: PRETSA: event log sanitization for privacy-aware process discovery. In: International Conference on Process Mining, ICPM 2019, Aachen, Germany (2019)
7. Iyengar, V.S.: Transforming data to satisfy privacy constraints. In: Proceedings of the Eighth ACM SIGKDD International Conference on Knowledge Discovery and Data Mining, pp. 279–288. ACM (2002)
8. Jr., R.J.B., Agrawal, R.: Data privacy through optimal k-anonymization. In: Proceedings of the 21st International Conference on Data Engineering, ICDE (2005)
9. Levenshtein, V.I.: Binary codes capable of correcting deletions, insertions, and reversals. In: Soviet physics doklady, vol. 10, pp. 707–710 (1966)
10. Mannhardt, F.: Sepsis cases-event log. Eindhoven University of Technology (2016)
11. Mannhardt, F., Koschmider, A., Baracaldo, N., Weidlich, M., Michael, J.: Privacy-preserving process mining - differential privacy for event logs. Bus. Inf. Syst. Eng. **61**(5), 595–614 (2019)
12. Michael, J., Koschmider, A., Mannhardt, F., Baracaldo, N., Rumpe, B.: User-centered and privacy-driven process mining system design for IoT. In: Information Systems Engineering in Responsible Information Systems, pp. 194–206 (2019)
13. Pika, A., Wynn, M.T., Budiono, S., ter Hofstede, A.H., van der Aalst, W.M.P., Reijers, H.A.: Privacy-preserving process mining in healthcare. Int. J. Environ. Res. Public Health **17**(5), 1612 (2020)
14. Rafiei, M., van der Aalst, W.M.P.: Mining roles from event logs while preserving privacy. In: Business Process Management Workshops - BPM 2019 International Workshops, Vienna, Austria, pp. 676–689 (2019)

15. Rafiei, M., van der Aalst, W.M.P.: Practical aspect of privacy-preserving data publishing in process mining. In: Proceedings of the Best Dissertation Award, Doctoral Consortium, and Demonstration & Resources Track at BPM 2020 Co-Located with the 18th International Conference on Business Process Management (BPM 2020) (2020). CEUR-WS.or

16. Rafiei, M., van der Aalst, W.M.P.: Privacy-preserving data publishing in process mining. In: Fahland, D., Ghidini, C., Becker, J., Dumas, M. (eds.) BPM 2020. LNBIP, vol. 392, pp. 122–138. Springer, Cham (2020). https://doi.org/10.1007/978-3-030-58638-6_8

17. Rafiei, M., Wagner, M., van der Aalst, W.M.P.: $TLKC$-privacy model for process mining. In: Dalpiaz, F., Zdravkovic, J., Loucopoulos, P. (eds.) RCIS 2020. LNBIP, vol. 385, pp. 398–416. Springer, Cham (2020). https://doi.org/10.1007/978-3-030-50316-1_24

18. Rafiei, M., von Waldthausen, L., van der Aalst, W.M.P.: Supporting condentiality in process mining using abstraction and encryption. In: Data-Driven Process Discovery and Analysis - 8th IFIP WG 2.6 International Symposium, SIMPDA 2018, and 9th International Symposium, SIMPDA 2019, Revised Selected Papers (2019)

19. Rüschendorf, L.: The Wasserstein distance and approximation theorems. Probab. Theory Related Fields **70**(1), 117–129 (1985). https://doi.org/10.1007/BF00532240

20. Sweeney, L.: k-anonymity: a model for protecting privacy. Int. J. Uncertainty Fuzziness Knowl. Based Syst. **10**(05), 557–570 (2002)

21. Van Dongen, B.F.: BPIC 2017. Eindhoven University of Technology (2017)

22. von Voigt, S.N., et al.: Quantifying the re-identification risk of event logs for process mining - empiricial evaluation paper. In: Advanced Information Systems Engineering, CAiSE (2020)

Author Index

Printed in the United States
by Baker & Taylor Publisher Services